SpringerWienNewYork

Maria Stepanova • Steven Dew
Editors

Nanofabrication

Techniques and Principles

SpringerWienNewYork

Editors
Dr. Maria Stepanova
National Institute for Nanotechnology
National Research Council of Canada
Saskatchewan Drive 1142
T6G 2M9 Edmonton Alberta
Canada
maria.stepanova@nrc-cnrc.gc.ca

Dr. Steven Dew
University of Alberta
Electrical and Computer Engineering
2nd Floor ECERF
T6G 2V4 Edmonton Alberta
Canada
steven.dew@ualberta.ca

SpringerWienNewYork is a part of Springer Science+Business Media
springer.at

Typesetting: SPi, Pondicherry, India

Printed on acid-free and chlorine-free bleached paper
SPIN: 80025917

With 250 Figures

Library of Congress Control Number: 2011941772

ISBN 978-3-7091-0423-1 e-ISBN 978-3-7091-0424-8
DOI 10.1007/978-3-7091-0424-8
SpringerWienNewYork

Preface

This book came about from a realization that a gap had emerged in our collective library of techniques for structuring matter. The well-established but nonetheless still rapidly evolving field of microelectronics fabrication has given us a rich set of tools for depositing, patterning and etching thin layers of semiconductors, insulators and metals. These tools are well documented and continue to be invaluable to the emerging nanotechnology industry. However, in addition to these old standby methods, a new set of tools has emerged specifically around the structuring of matter with near nanometer precision. Some of these, such as atomic layer deposition, are already appearing in production microelectronics facilities. Others, such as helium ion beam nanomachining, are unlikely to hit high volume production in the foreseeable future, but nonetheless are emerging as powerful tools for producing nanoscale devices the likes of which we have not yet imagined. In between lies a set of techniques that have heretofore not been part of a standard microfabrication handbook, and hence we felt a need to produce a reference dealing with these techniques whose common thread is the structuring of matter with near atomic scale control. The intent is to complement more traditional works dealing with the state-of-the-art processing of CMOS integrated circuits. The exclusion of CMOS processing is not because it is not 'true nanofabrication', but because it is more evolutionary and more familiar and covered in many other works.

The book provides a balance of the fundamental principles and example applications of nanofabrication techniques. The intended audience is the process engineer in industry, the academic researcher, or the graduate or senior undergraduate student who is finding himself or herself drawn into learning about these new methods for nanostructuring matter. While not necessarily so intended, it could well serve as a text for a one term graduate course in nanofabrication.

In selecting authors for the individual sections, we have drawn on both local and international experts in their respective areas. In all, eight countries are represented across the 13 chapters. Each set of authors drew on their own respective communities to bring different perspectives and a highly authoritative treatment of the topic. As editors, we were exceptionally fortunate to have such dedicated professionals to work with. We are extremely grateful for their hard work and outstanding contribution.

We are also very grateful to our local community for its support. We are happy to acknowledge the staff of the University of Alberta NanoFab, the Integrated Nanosystems Research Facility, the Alberta Centre for Surface Engineering and Science, and the National Institute for Nanotechnology. Equally important are our respective teams of graduate students and researchers and colleagues who have provided the intellectual environment (and the hard work) to build up our own expertises in the field. In addition, we would like to thank our publisher, Springer Verlag, for believing and enabling this project. Finally, we both sincerely acknowledge the support of our respective families who have put up with us as we have travelled down the path that has made this book possible.

Edmonton, Canada

Maria Stepanova
(Maria.Stepanova@nrc-cnrc.gc.ca)
Steven Dew
(steven.dew@ualberta.ca)

Contents

Part I

Introduction

Directions in Nanofabrication 1

Steven K. Dew and Maria Stepanova

Abstract

Nanotechnology is dependent on nanofabrication techniques to structure matter at the 1–100 nm lengthscale. In this book, a variety of nanofabrication techniques are described, including those for lithography using electron and helium ion beams or nanoimprint lithography; for deposition using atomic layer deposition, surface monolayer formation, block copolymers, or electroepitaxy; and for etching using chemical mechanical polishing, ion beam induced reactions, laser nanopatterning, or anodic oxide templating. In each case, principles, methodology, equipment and process parameters are described, along with capabilities and example applications. Taken together, these represent a powerful toolbox of the rapidly evolving pool of nanofabrication techniques.

1.1 Nanofabrication

Nanotechnology is about the controlled assembly and structuring of matter at a lengthscale of approximately 1–100 nm in order to exploit unique properties that manifest at these dimensions. It is well recognized that this field will enable a vast range of new capabilities for healthcare, new materials for transportation, new categories and capabilities of consumer products, new processes for industry, new

S.K. Dew (✉)
Department of Electrical and Computer Engineering, University of Alberta, Edmonton, Alberta, Canada
e-mail: steven.dew@ualberta.ca

M. Stepanova
National Institute for Nanotechnology, National Research Council, and Department of Electrical and Computer Engineering, University of Alberta, Edmonton, AB, Canada
e-mail: Maria.Stepanova@nrc-cnrc.gc.ca

M. Stepanova and S. Dew (eds.), *Nanofabrication*,
DOI 10.1007/978-3-7091-0424-8_1, © Springer-Verlag/Wien 2012

devices and sensors for information technology, and more [1]. An estimate of the expected annual global economic impact of nanotechnology by 2020 is $3 trillion US dollars [2]. Nanofabrication is about the tool box that makes this exquisite control of matter at the nanoscale possible. It is the set of techniques to pattern, grow, form and remove material with near nanometer control, repeatability and precision.

This book is about the techniques for nanofabrication. It is intended for the practitioner, the researcher or the student who wishes to learn more about the major techniques that are driving the nanotechnology revolution. The toolbox is both diverse and growing rapidly, so we have been forced to choose just the most important set of tools to present here. There are other excellent books and sources that have taken different approaches or that focus solely on a particular topic at a high level of detail, and we recommend the reader consider them as well [3–9]. Notwithstanding these existing works, however, we felt there was still a strong need for a book that explored both the capabilities and underlying principles of the most important set of nanofabrication techniques. We also felt that the area of microelectronics, while itself incredibly important to the field of nanofabrication, drives a set of requirements that are highly specialized and exclusive of a large range of non-CMOS related applications. Hence, we felt there was a need for a resource that explored a broader range of nanofabrication techniques, to the exclusion of conventional microelectronics ones, if necessary. This book is our attempt to respond to that need.

Consisting of topical chapters written by international experts from eight countries, this book is divided into three parts. Crudely speaking, these are nanolithography, nanoscale deposition, and nanoscale etching and pattern transfer. The divisions are somewhat arbitrary as a number of the topics covered do require capabilities from two or more of these categories or can be used in more than one way. For example, ion beams can be used for lithography, for localized etching, or for deposition, thereby spanning all three areas. Nonetheless, the paradigm of deposition, lithography and etch is well-ingrained in the field of micro- and nanofabrication, and it is hard to break out of the mold.

1.2 Nanolithography

Lithography is a process whereby an arbitrary (usually 2D) pattern can accurately and reproducibly be generated in a specialized layer of material called the resist. Typically, that pattern is then transferred to another functional layer through a traditional etching or lift-off process. Conventionally, this is done with ultraviolet (UV) radiation modulated by opaque patterns on an otherwise transparent mask, converting the solubility of a polymer resist material. Through adaptations such as high numerical aperture optics and immersion lenses, deep UV excimer laser sources, phase shifting masks, optical proximity correction and double patterning approaches, conventional optical lithography has pushed well into the nanoscale. However, extension to the deep subnanometer lengthscales, especially for non-microelectronic applications that cannot necessarily tolerate the high costs, high volume and high planarity requirements, other lithography techniques are required. Extreme UV and x-ray lithography can extend to lower dimensions, but their very high costs will preclude their use for most non-microelectronic applications, at least for a while. That leaves three lithographic choices for 'low-cost', highly flexible, deep nanoscale lithography: electron beam, ion beam and nanoimprint lithography. These techniques are covered over four chapters in the first part of the book.

In Chap. 2, we cover the principles and major mechanisms of electron beam lithography (EBL). This technique has been the mainstay of nanolithography from the outset and will continue to be used to produce masks and templates even if volume production ultimately relies on other techniques (EUV or NIL). This chapter covers the effects of the various parameters (energy, dose, development temperature and time, etc.) and their impacts on resolution, sensitivity, line-edge roughness, and process windows.

Simulation and modeling is an essential component of understanding and optimizing processes at the nanoscale, and Chap. 3 addresses the modeling needs for electron beam lithography. Here, Ioannis Raptis and George Patsis take us through the essential steps of EBL simulation, including electron transport and energy deposition in the resist accounting for forward and backscattering, proximity correction, thermal processing and development, line edge roughness and stochastic effects. Different simulation approaches are presented, such as Monte Carlo and analytic treatments, continuous and discretized resist representations, and macroscopic and mesoscopic models. Existing commercial software available for EBL simulation is summarized.

Despite its flexibility and widespread acceptance, EBL does face important limitations on resolution, especially related to its proximity effect. With the recent development of sub-nanometer spot sizes, helium ion beam lithography (HIBL) has a potential advantage on this score. Chapter 4 describes the principle and capabilities of HIBL. Emile van der Drift and Diederik Maas compare EBL and HIBL explaining the higher sensitivity, reduced scattering, and improved proximity and resolution that the latter technique can achieve. Principles of He^+ beam generation and control are discussed along with ion-matter interactions and the implications for substrate damage. Various resists are explored including conventional HSQ and PMMA as well as inorganic Al_2O_3.

Perhaps the most promising approach in terms of 'low cost' solutions to nanolithography is nanoimprint lithography (NIL). In this technique, a prepatterned mask with features down to a handful of nanometers is held in contact with the resist material, and the pattern transferred using heat or UV radiation. In Chap. 5, Christophe Peroz, Vincent Reboud, and Clivia Sotomayor Torres describe the principles, capabilities and parameters of NIL. Thermal and UV NIL are described, including the challenges of resist flow when the stamp is brought into contact, stamp deformation, anti-sticking coatings, demolding, and resist materials. Full-wafer and step-and-print strategies are discussed as well as creation of 3D patterns. Applications to a variety of areas are briefly summarized.

1.3 Deposition at the Nanoscale

In conventional microfabrication, deposition involves processes such as evaporation, sputter deposition, chemical vapor deposition, or electrochemical deposition to add layers of a variety of conducting, insulating, semiconducting and other functional materials to the substrate. With care, all of these techniques can still find application to fabricating nanoscale structures. However, a new set of techniques also emerges for adding nanostructured materials. These rely heavily on wet chemistry, rather than more traditional vapor deposition.

The first of these techniques is atomic layer deposition (ALD). While the deposited layers are not nanostructured, they are deposited with monolayer control which is clearly of tremendous value in nanofabrication. This performance is achieved by using two self-limiting adsorption reactions which, when used in sequence, result in the addition of a monolayer of film to the surface. Because the adsorption step is to saturation, it is possible to deposit conformal layers over extremely high aspect ratio and complex topography. In Chap. 6, Ken Cadien and Ali Foroughi-Abari explain these principles in detail and provide an overview of some capabilities and applications of ALD. The development of ALD is summarized, along with the fundamental principles, equipment and variations such as plasma ALD. A discussion of precursors is presented including in the context of various applications, including the coating of anodized alumina nanopores (see Chap. 13), the formation of high-k gate dielectrics and diffusion barriers, and the formation of nanoparticles on surfaces and nanotubes.

Also in the domain of self-limiting monolayer deposition is the topic of surface functionalization. Here only one molecular layer is deposited, but it generally adds a highly specific affinity for a target biomolecule or other biologic function. This is an enabling capability for a whole spectrum of health and security related applications of nanotechnology. In Chap. 7, Adam Bergren and Richard McCreery discuss the importance of surfaces in the nanoscale, and how they can be tailored to provide a specific function. Aspects of the surface that can be controlled include wettability (by water and other materials), coverage and morphology, chemical reactivity, passivation, electronic activity, and thermal stability. Monolayer formation techniques such as Langmuir-Blodgett and self-assembled monolayers (SAM's) on metal, silicon and carbon surfaces are discussed, as are techniques to form multilayer nanomolecular coatings.

The immiscibility of certain polymers can be exploited to take advantage of the highly regular and reproducible 2D and 3D structures that can be produced using block co-polymers. In Chap. 8, Bhoje Gowd, Mallikarjuna Shroff Rama and Manfred Stamm present the various structures that can emerge including spheres, cylinders and lamellae and the principles and phase diagrams that govern their formation in bulk solutions and in thin films. Techniques for orienting structures in films are presented. Application of block copolymer films as templates and supramolecular assembly are also discussed.

In Chap. 9, the topic is the specialized one of epitaxial electrodeposition of metals on semiconductors. This is a low-cost technique to form arrays of metal crystals using the underlying substrate as a template, and is highly relevant to nanofabrication both for the exquisite nanostructures it can produce, as well as an important technique for forming electrical contact to semiconductor materials for both electronic and sensing applications. Karen Kavanagh summarizes the techniques of electrodeposition including the impact of various process parameters. Details are provided for both silicon and GaAs substrates. Application of the technique to form Cu and Au nanowires is presented.

1.4 Nanoscale Etching and Patterning

The final part of this book deals with the use of etching and other removal techniques to provide nanoscale structures or control. These range from techniques for producing the atomically flat surfaces necessary as a starting point for many nanoscale structures, to laser and ion bombardment to machine or modify highly localized areas, to anodization of 'valve metals' to produce highly controlled nanoporous materials both for their own right or as templates for the formation of even more exotic materials.

Chemical mechanical polish (CMP) has been used for decades as a final step to polish silicon wafers, but its use as a fabrication step is more recent, first for dielectric planarization of SiO_2, then for tungsten plug and copper damascene formation in modern microelectronics structures. The technique is useful beyond microelectronics, however, as many fabrication steps involving nanoscale control will also require a starting surface which is very nearly atomically flat. CMP offers this level of controllability through a combination of chemical surface softening and mechanical abrasion. In Chap. 10, Lucy Nolan and Ken Cadien discuss the main principles, parameters, and requirements to achieve the level of control required to enable many examples of nanofabrication.

In Chap. 11, we return to highly focused helium ion beams. However, this time they are not modifying the structure and solubility of resist materials. Rather, they are providing the energy to initiate a highly localized etching, deposition, or other chemical reaction involving reactive gases injected near the substrate. In this chapter, Paul Alkemade and Emile van Veldhoven explore the capabilities and limitations of ion induced deposition and etching as the ultimate nanomachining tool. The necessary equipment and techniques are described. The role of precursor gas injection, transport and decomposition, ion sputtering, proximity, secondary electron generation, and their governing relations are discussed. Rate, structure and composition of deposited films are presented.

We continue the nanomachining concept in Chap. 12, but now using laser pulses. As discussed by Robert Fedosejevs, Ying Tsui, Zhijiang Chen and Shyama Banerjee, the main benefit here is that the pulses can be varied from femtosecond duration to continuous wave, providing control over interaction times and resulting energy transport lengthscales down to below the nanoscale. This allows, for example, ablation or recrystallization with near monolayer levels of control. Laser induced forward transfer also presents opportunities for deposition control on similar scales. Frequency multiplication, microlenses and near field optical techniques are also possible, yielding nanoscale lateral control, as well. In addition, multiphoton adsorption provides an approach to 3D writing, adding to the versatility of laser patterning.

Finally, in Chap. 13, Karthik Shankar discusses the use of anodization of aluminum and titanium. Under the correct conditions, the nonlinear behavior of this process results in the production of highly regular, evenly spaced nanoscale pores that can extend microns or even millimeters deep. Of themselves, these can be useful for certain examples of nanoscale behavior. Further, however, they can serve as templates for the growth of nanorods and nanowires of a variety of materials that cannot easily be produced any other way. Further, the nanopore arrays can serve as a mask for etching and deposition reactions, providing another pathway for nanostructuring other materials. This chapter explores the key parameters, the main approaches to applying the nanopores, and some of the application areas of the overall technology.

1.5 Summary

Nanofabrication requires an extension of the set of microfabrication tools beyond traditional optical lithography, vapor deposition, and reactive ion etching. Through a set of focused chapters written by international experts, this book discusses the key technologies that are available to meet these requirements and enable the fabrication of complex systems and materials structured at the nanoscale.

Acknowledgements The authors are grateful for the support of the Natural Sciences and Engineering Research Council of Canada, the National Institute for Nanotechnology, and the University of Alberta NanoFab. As editors, we are also grateful for the hard work, vast expertise and professional manner of the contributing authors who have taken time out of their very busy schedules to write an excellent set of chapters on a fascinating topic.

References

1. Roco MC, Mirkin CA, Hersam MC. Nanotechnology research directions for societal needs in 2020, retrospective and outlook. Heidelberg Dordrecht London New York: Springer; 2011. ISBN 978-94-007-1167-9.
2. Roco M, Presentation on the national nanotechnology initiative 2. Washington, DC; 2010.
3. Cui Z. Nanofabrication: principles, capabilities and limits. Berlin: Springer; 2008. ISBN 0387755764.
4. Ampere A. Nanofabrication: fundamentals and applications. Singapore: World Scientific Publishing; 2008. ISBN 9812700765.
5. Wiederrecht G, editor. Handbook of nanofabrication. Amsterdam: Elsevier; 2010. ISBN 0123751764.
6. Ohtsu M, editor. Nanophotonics and nanofabrication. Darmstadt: Wiley-VCH; 2009. ISBN 9783527321216.
7. Ostrikov K, Xu S. Plasma-aided nanofabrication: from plasma sources to nanoassembly. Darmstadt: Wiley-VCH; 2007. ISBN 9783527406333.
8. Wang M, editor. Lithography. Vukovar: Intech; 2010. ISBN 9789533070643.
9. Kumar Challa, editor. Nanostructured thin films and surfaces. Darmstadt: Wiley-VCH; 2010. ISBN 9783527321551.

Part II

Nanolithography

Fundamentals of Electron Beam Exposure and Development

2

Mohammad Ali Mohammad, Mustafa Muhammad, Steven K. Dew, and Maria Stepanova

Abstract

Electron Beam Lithography (EBL) is a fundamental technique of nanofabrication, allowing not only the direct writing of structures down to sub-10 nm dimensions, but also enabling high volume nanoscale patterning technologies such as (DUV and EUV) optical lithography and nanoimprint lithography through the formation of masks and templates. This chapter summarizes the key principles of EBL and explores some of the complex interactions between relevant parameters and their effects on the quality of the resulting lithographic structures. The use of low energy exposure and cold development is discussed, along with their impacts on processing windows. Applications of EBL are explored for the fabrication of very small isolated bridge structures and for high density master masks for nanoimprint lithography. Strategies for using both positive and negative tone resists are explored.

2.1 Introduction

With its ability to form arbitrary two-dimensional patterns down to the nanometer scale, electron beam lithography (EBL) is one of the most important techniques in nanofabrication. In short, it involves the exposure by a highly focused electron

M.A. Mohammad (✉) • M. Muhammad • S.K. Dew
Department of Electrical and Computer Engineering, University of Alberta, Edmonton, AB, Canada
e-mail: M.A.Mohammad@ualberta.net

M. Stepanova
National Institute for Nanotechnology, National Research Council, and Department of Electrical and Computer Engineering, University of Alberta, Edmonton, AB, Canada

M. Stepanova and S. Dew (eds.), *Nanofabrication*,
DOI 10.1007/978-3-7091-0424-8_2, © Springer-Verlag/Wien 2012

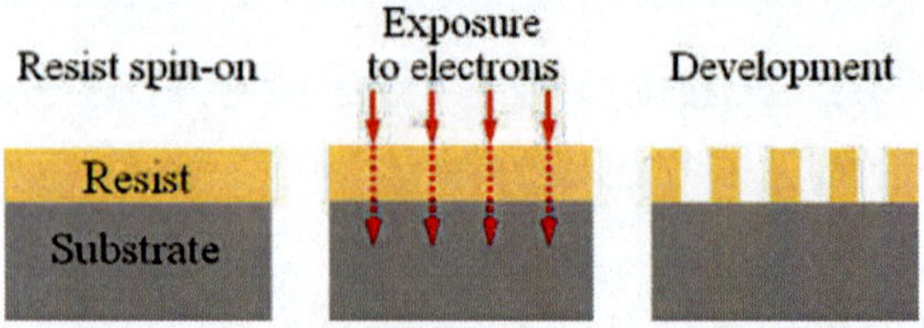

Fig. 2.1 Outline of EBL process steps to form a nanoscale pattern in a positive-tone resist layer

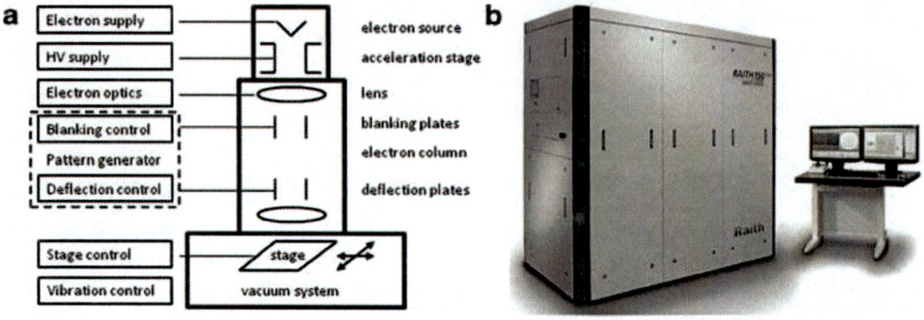

Fig. 2.2 Electron beam exposures systems: (**a**) schematic diagram, (**b**) Raith 150^{TWO} commercial EBL system (used with permission [7])

beam to dramatically modify the solubility of a resist material during a subsequent development step, see Fig. 2.1.

EBL was originally developed using scanning electron microscopes to which a pattern generator and beam blanker was added to control which areas of the viewing field are exposed [1–3] (see Fig. 2.2a for a schematic description of an EBL system). Modern EBL tools are fully dedicated patterning systems (Fig. 2.2b) that employ high brightness electron sources for faster throughput and high resolution mechanical stages to be able to expose step-by-step large substrates under the relatively narrow field of focus of the electron beam. These direct write systems have the advantage of extremely high resolution and the ability to create arbitrary patterns without a mask. Their disadvantage is the long times taken to write large, complex patterns. Efforts to overcome this challenge include projection EBL [4, 5] and the use of massively parallel beams [6]. Due to the developmental stage of these latter techniques, however, this chapter will focus on the single beam, direct write EBL technique.

The key objectives of EBL writing are to achieve arbitrary patterns in the resist with high resolution, high density, high sensitivity and high reliability. These characteristics are interrelated in a complex fashion. The key determinants will be the quality of the electron optics (e.g., the ability to create a finely focused spot), the choice of resist, substrate and developer, and the process conditions: electron beam energy and dose, and development time and temperature. Factors that complicate these objectives are delocalization of electrons due to forward and backscattering (proximity effects), collapse of the pattern due to swelling and capillarity forces, and fluctuations in the sizes of features (line edge roughness).

2.1.1 Electron Transport

The first requirement for a high quality beam is a stable, high brightness electron source, such as one employing thermal field emission. The quality of the spot is determined by the electron optics and degree of focus. It is necessary to have high positional accuracy with limited astigmatism and small spot size [8]. The electron column is under vacuum to reduce gas scattering of the beam, but divergence of the beam does result from mutual electrostatic repulsion by the electrons. This effect is more pronounced at higher currents and lower energies. Notwithstanding this, commercial EBL systems can usually deliver a spot size of just a few nanometers [7, 9, 10]. Unfortunately, other factors such as scattering will usually limit the final resist pattern to a larger size.

As the electrons enter the resist, they begin a series of low energy elastic collisions, each of which will deflect the electron slightly. This forward scattering broadens the beam by an amount that increases with thickness, and this effect is more pronounced at low incident energies [11, 12] (see also Fig. 2.3).

In addition to forward scattering, there is also backscattering to consider [13]. Typically, most of the electrons pass entirely through the resist and penetrate deeply into the substrate. Some fraction of those electrons will eventually undergo enough large angle collisions to re-emerge into the resist at some distance from the point at which they left it (see Fig. 2.4). At higher energies, these backscattered electrons may cause exposure microns away [14, 15] from where the beam entered. This leads to the so-called proximity effect [16–18] where electrons writing a feature at one location increase the exposure at a nearby feature, causing pattern distortion and overexposure. The density of features becomes an important factor in determining necessary exposure levels. Backscattering can be minimized by exposing on a thin membrane substrate.

Another electron transport consideration is secondary electrons [1]. These are low energy (a few to a few tens of eV) electrons produced by ionizations resulting from inelastic collisions by the primary incident electrons [19]. Because of their

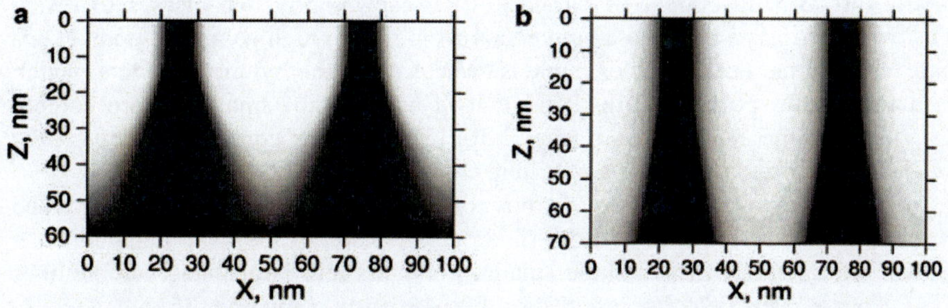

Fig. 2.3 Electron beam broadening due to forward scattering in the resist at (**a**) 3 keV and (**b**) 10 keV of incident energy. Shown is a predicted cross-section of resist exposure for two parallel lines

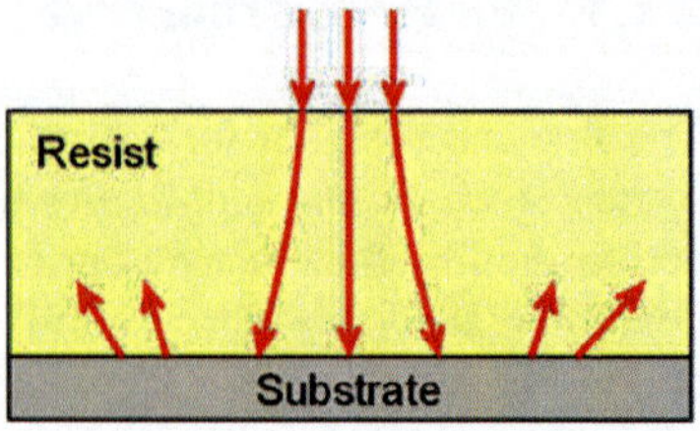

Fig. 2.4 Forward and backscattering of electrons in the resist and substrate leading to beam broadening and proximity effects

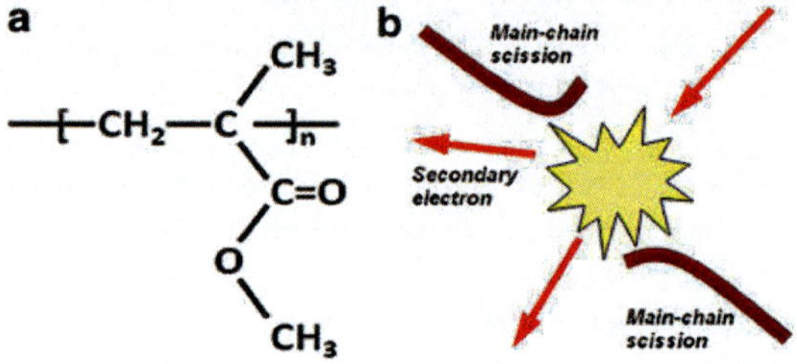

Fig. 2.5 (**a**) Polymer subunit of poly (methyl methacrylate), and (**b**) scission of the polymer chain during EBL exposure

energy, secondary electrons have short range (several nanometers) but may ultimately limit the resolution possible with EBL.

A final transport issue is electrostatic charging, particularly if writing onto an insulating substrate. If there exists no pathway for the absorbed electrons to dissipate, charge will build up and defocus the electron beam. In such cases, a thin metal [1] or conductive polymer [20] layer is required above or below the resist.

2.1.2 EBL Resists

Inelastic collisions of electrons with the resist result in ionization (secondary electron generation), which is accompanied by physico-chemical changes in the resist. Just as with optical lithography, EBL can employ two classes of resist. Positive tone resists undergo a conversion from low to high solubility upon exposure to electrons. The classic example is PMMA (poly-methyl methacrylate) which is a long chain polymer (Fig. 2.5a) that is broken into smaller, more soluble fragments by the electron beam (Fig. 2.5b) [21]. Another common positive resist is ZEP 520 which also consists of a long chain polymer [22, 23].

In a negative tone resist, the electrons convert the material to low solubility. The prime example here would be HSQ (hydrogen silsesquioxane) which undergoes a cross-linking reaction to combine smaller polymers into larger, less soluble ones [24]. Several other negative resists have been recently compared [25].

The most common positive resist, PMMA, consists of very long polymer chains with masses of 496 and 950 kDa being common. With such long chains, it takes many scission events before the resulting fragments become significantly soluble.

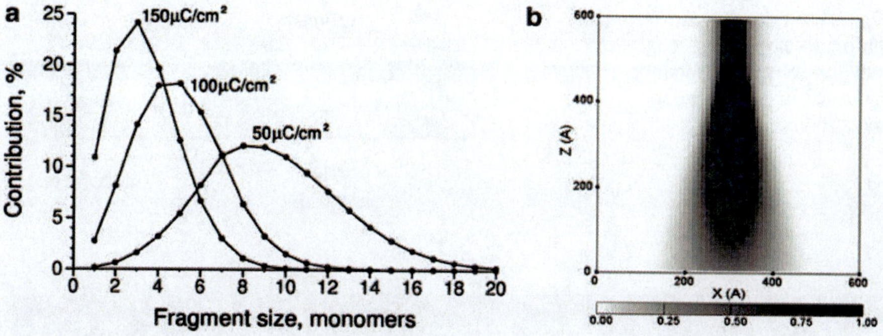

Fig. 2.6 (a) Calculated distribution of PMMA fragment sizes for various doses at 10 keV [26]. (b) Spatial distribution of the volume fraction of small (less than 10 monomers) fragments within the resist due to exposure at a single point

Hence, the distribution of fragment sizes becomes an important factor in understanding the relationship between exposure dose and development behavior. Figure 2.6a shows the distribution of PMMA fragment sizes and exposure dose [26]. As dose increases, the average fragment size decreases and solubility in the developer increases. Of course, dose varies spatially because of scattering, so the 3D distribution of scission events (Fig. 2.6b) becomes an important component to this total picture. Similar considerations also apply to other resists – positive or negative.

2.1.3 Resist Development

After exposure, the resist is typically immersed in a liquid developer to dissolve the fragments (positive tone) or non-crosslinked molecules (negative tone). Temperature and duration become important parameters here as the hotter/longer the development, the farther along the continuum of solubility the dissolution extends. For example, cold development of PMMA (discussed below) freezes out the development of all but the very smallest fragments, resulting in very high resolution as much of the scattered electrons cause insufficient exposure to reach this threshold.

During development, the solvent penetrates into the polymer matrix and starts to surround the fragments. As the molecules start to interact, a gel is formed (see Fig. 2.7). The thickness of the gel layer will depend on the amount of fragmentation and the strength of the solvent. Swelling of the polymer can also result. Once completely surrounded by solvent, the fragments detach from the matrix and diffuse into the solvent. Longer fragments are less mobile and more strongly bound to the matrix and take longer to dissolve [27–29]. More powerful solvents will remove longer fragments which is not always desirable if high resolution is required. Exposure and development are interrelated as short exposure with long or aggressive development can be equivalent to heavier exposure with short development.

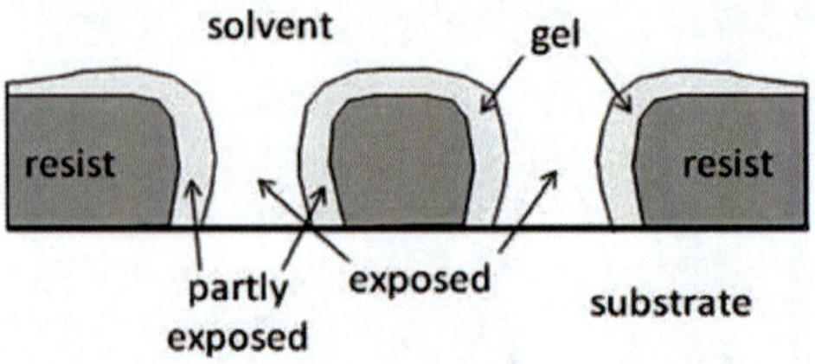

Fig. 2.7 Positive resist during development. Polymer-solvent interactions can result in gel formation and swelling

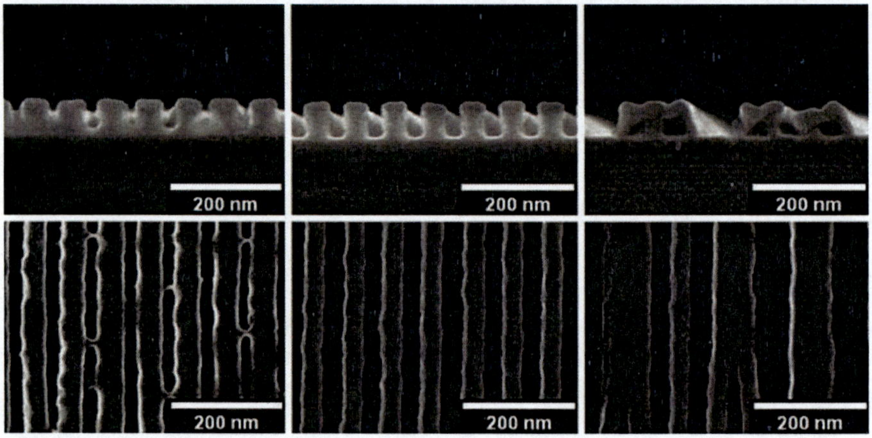

Fig. 2.8 PMMA grating structures in cross-section (*top*) and plan view (*bottom*). Shown are underexposed/underdeveloped (*left*), quality structures (*middle*) and collapsed patterns (*right*) [31]

This can lead to some ambiguity between the terms underdeveloped and underexposed or overdeveloped and overexposed. As will be discussed further below, the kinetics of the dissolution process becomes quite important to the optimization of EBL necessitating understanding of these factors at a detailed level. Often a mix of solvents (such as 1:3 methyl isobutyl ketone with isopropyl alcohol for PMMA) is used to tailor dissolution behavior.

An issue that arises if the resist is overdeveloped is the degradation of the resist-substrate bond and the capillarity forces that result as the solvent is removed. These lead to the mechanical failure of the resist structure resulting in pattern collapse [30]. Adjacent linear features are particularly vulnerable to this problem, especially for thick resists. Figure 2.8 shows examples of pattern collapse, along with underexposed/underdeveloped structures in PMMA.

2.1.4 Process Parameters

As has been implied above, there is a large number of parameters affecting the EBL process in a complex, interacting fashion. A partial list is given in Table 2.1. This

Table 2.1 Parameters affecting the EBL process

Parameter	Process impact
Exposure energy	Resolution, sensitivity, proximity
Exposure dose	Pattern quality
Pattern density	Proximity, pattern quality
Resist material	Sensitivity, resolution, contrast
Resist thickness	Sensitivity, resolution, pattern quality
Developer	Sensitivity, resolution, development window
Development temperature	Sensitivity, resolution, exposure window
Development time	Sensitivity, resolution, exposure window

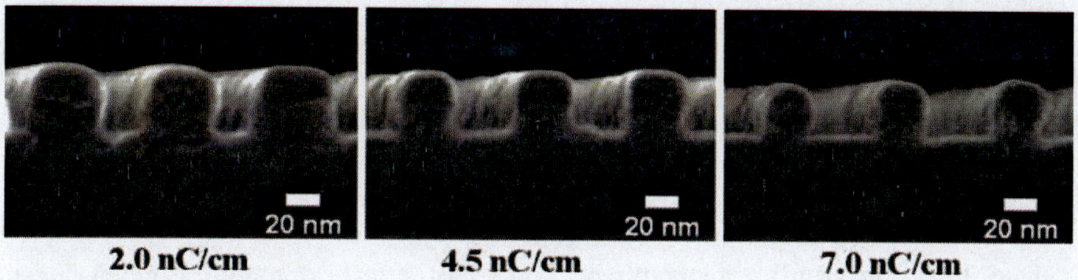

Fig. 2.9 Cross sectional profiles in 70 nm pitch PMMA gratings fabricated using 30 keV voltage and various line doses. The samples were developed at −15°C for 15 s, and the initial PMMA thickness was 55 nm [38]

excludes secondary factors such as resist polymer chain length which can impact sensitivity and contrast, or the use of techniques such as ultrasonic agitation [32–34] to reduce development times and improve clearance, or the use of critical point drying [35–37] to minimize pattern collapse. Of course, the objective of manipulating these parameters is to achieve a high resolution, high quality, high throughput result with large process windows to maximize yield and reproducibility.

An example of some of these process dependencies can be seen in Fig. 2.9 which shows the effect of dose on a single pixel line grating. While the structures are well-resolved at all three doses, the size of the final structures varies significantly. Similarly, Fig. 2.10 summarizes the aspect ratio (height:width) of the features resulting as a function of dose. The complexity of these interactions drives the need for better understanding and modelling techniques (see Chap. 3) for the optimization of EBL processes.

2.2 Process Windows with PMMA Resist

As requirements for lithography have progressed toward the sub-20 nm regime, major challenges have emerged for introducing controllable radiation-induced changes at molecular-size scales. Novel EBL processes that would extend capabilities of the technology significantly into the deep nanoscale regime entail

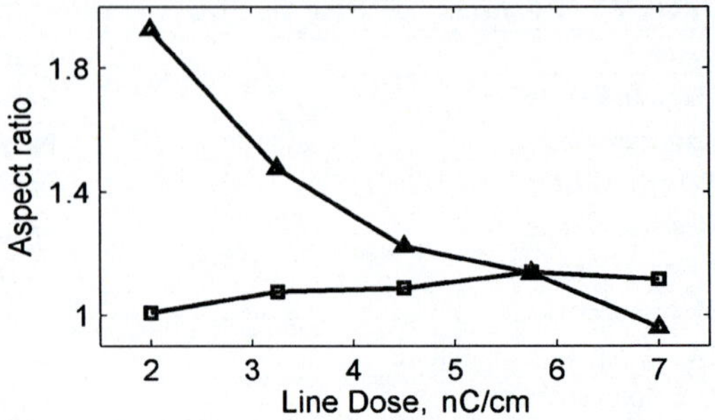

Fig. 2.10 Aspect ratios for the interline gaps (*triangles*) and PMMA lines (*squares*) as a function of the exposure dose, determined from cross-sectional SEM images for 70 nm pitch gratings fabricated using 30 keV electrons. Other process conditions were as in Fig. 2.9 [38]

new approaches to resist design, exposure strategies, and development techniques [31, 38–43]. To achieve this will require thorough, systematic understanding of the limiting factors [44] involved in both the electron-resist interaction and in the polymer dissolution (development), as well as the corresponding intricate interplay of the numerous process control parameters including the accelerating voltage, exposure dose, and development conditions.

Due to the proximity effect, this becomes particularly demanding when dense patterns with closely positioned features must be fabricated. Figure 2.11 presents an example of nanoscale morphologies after exposure and development of high-density gratings with various interline distances (pitches) in PMMA resist on a silicon wafer [38, 43]. From the figure, it is evident that the potential to fabricate a quality grating is strongly dependent on both the exposure dose and the interline distance. Thus, for 70 nm pitch, all the images present well developed gratings, with the exception of Fig. 2.11d that corresponds to the high area dose of 125 μC/cm^2. For 40 and 50 nm gratings, promising morphologies are seen in Fig. 2.11n, o, k for the doses from 50 to 75 μC/cm^2. For 30 nm interline distance, only Fig. 2.11m for 60 μC/cm^2 exhibits a promising morphology. The other images in Fig. 2.11 show various damaging influences. For example, underexposure, when the clearance depth of exposed lines is insufficient to generate a grating pattern, is manifested by a low contrast in Fig. 2.11m, q, r. The other extreme regime, overexposure, when the pattern is damaged by excessive clearance of PMMA, is evident in Fig. 2.11a, b. Mild signs of overexposure are also visible in Fig. 2.11c, e. Another common type of morphology damage is the collapse of interline resist walls. Examples of collapsed gratings are seen in Fig. 2.11d, f, g, and also to some extent in Fig. 2.11j, k. It is noteworthy that collapse only occurs in gratings with 40 nm and larger pitch. In 30 nm gratings the limiting mechanism is different as illustrated in Fig. 2.11e, i. In these cases, the grating is partly or entirely destroyed by redistribution of PMMA that tends to form islands at irregular locations. Formation of such globular islands or percolation networks is known to occur in immiscible liquids that undergo a phase transition. Since mixtures of PMMA fragments with most common EBL developers have regimes of limited miscibility, a phase separation can occur [45, 46].

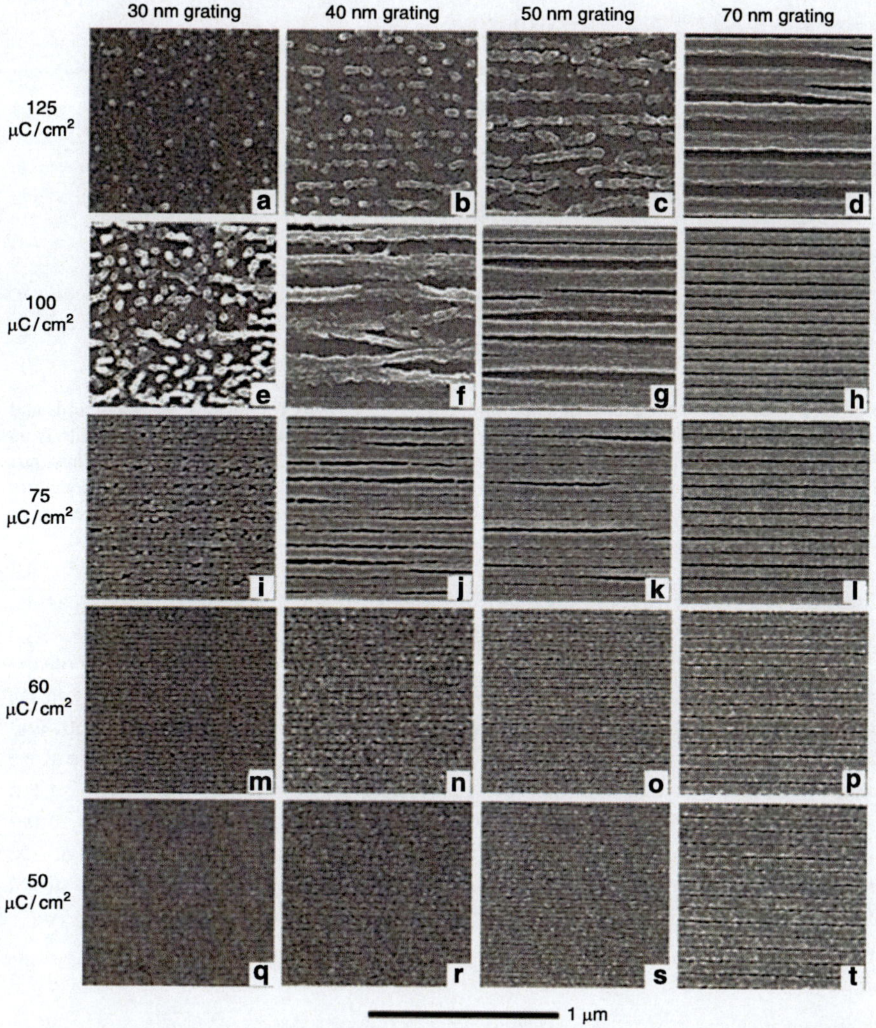

Fig. 2.11 SEM images of 30, 40, 50, and 70 nm gratings fabricated in a 65-nm thick PMMA layer on a silicon substrate, with 10 keV exposure at various area doses. The gratings were developed for 5 s in a 1:3 MIBK:IPA solution at room temperature. The lateral size of all images is 1 μm × 1 μm [43]. The average area doses are related with the line doses by $d_{area} = d_{line}/\lambda$, where λ is the interline distance (grating pitch)

Figure 2.12 summarizes the observed morphologic regimes for various grating periods and exposure doses. In the low dose regime, the limiting factor is underexposure, whereas at increased doses, the patterns degrade through phase separation or collapse. The densest gratings, with periods of 20 and 30 nm, degrade through phase separation, and the gratings with periods of 40 nm and larger rather tend to collapse. At even higher doses, gratings become overexposed. It can be seen

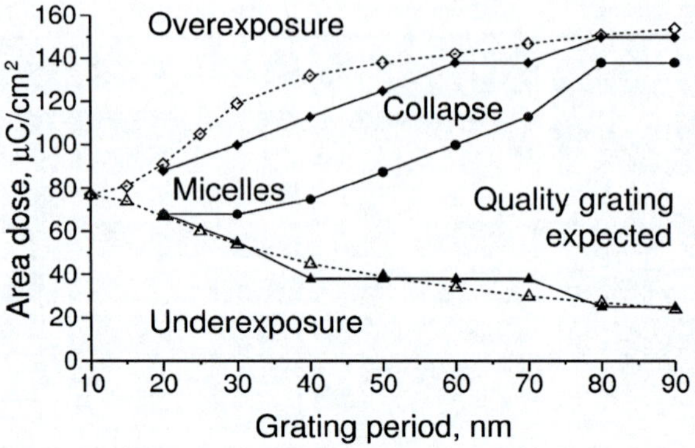

Fig. 2.12 Diagram of the characteristic morphologies in PMMA for various grating periods and area exposure doses using 10 keV voltage. Filled symbols represent experimental results from Fig. 2.9: *triangles* denote the boundary for underexposure (insufficient clearance); *diamonds* denote the boundary for overexposure (excessive clearance), and *circles* indicate the boundary for collapse or phase separation producing micellized patterns. Open symbols show the results of numerical modeling [43]

that the window of favourable exposure doses at which quality gratings can be expected decreases rapidly when the grating pitch decreases.

The width of the applicable dose window represents the robustness of the process [29, 38]. A large dose window means that the fabrication process has a better reproducibility, and also has greater controllability over the line width and aspect ratio. The minimum size of nanostructures for which a reasonable dose window is attained characterizes the resolution. For example, from Fig. 2.12 it follows that the highest resolution attainable at the experimental conditions considered corresponds to approximately 15 nm half-pitch representing the average line width. The minimum applicable doses can be associated with the sensitivity of the EBL process.

2.2.1 Temperature Dependence of Applicable Process Window

As already mentioned, the resist's development occurs through the removal of relatively low-weight fragments from the intensely exposed regions. This removal can be described as a kinetic diffusion-like process, with a molecular mobility represented by diffusivity, $D \sim n^{-\alpha} \exp(-U/kT)$, where U is the activation energy and the factor $n^{-\alpha}$ describes the mobility of fragments of size n in a medium whose properties are represented by power α. In most polymers, α varies from 1 in dilute solutions of small molecules to 2 for longer polymer chains in denser melts [27–29].

As noted in Fig. 2.6, in exposed PMMA, the average size of fragments $<n>$ is a function of both the exposure dose and location. For moderate exposure doses, the mean size of fragments $<n>$ is inversely proportional to the local probability of scissions, which in turn is related with exposure dose d, so that $<n> \sim 1/d$, and the diffusivity of fragments in exposed PMMA can be approximately described by,

$$D = cd^{\alpha} \exp(-A/kT) \tag{2.1}$$

where c is a location-dependent model coefficient of proportionality [29].

In Refs. [29, 31], for an example of a dense periodic grating in PMMA, it was demonstrated that the boundary applicable doses for quality nanofabrication, d_{min} and d_{max}, depend on temperature as follows,

$$d_{\mathrm{min,max}} = d^{ref}_{\mathrm{min,max}} \exp\left(-\frac{U}{\alpha k}\left(1/T - 1/T^{ref}\right)\right) \tag{2.2}$$

where '*ref*' indicates reference values for the minimum and maximum applicable doses d_{min} and d_{max}. In Fig. 2.13, (2.2) is compared with the experimental temperature dependencies for the doses d_{min} and d_{max} in a 70 nm pitch grating exposed with 10 keV electrons, and developed at various temperatures for 5s and 20s. The boundary doses determined experimentally for −15°C development were used for reference in this example, and the estimated value U/α = 0.22 eV was employed. It can be seen that (2.2) describes very well the broadening of the applicable dose window with the decrease of the development temperature. The lower boundary of the applicable dose window d_{min} (shown by solid lines in Fig. 2.13) can be interpreted as the dose at which PMMA fragments in the intensely exposed trenches are mobile enough to be removed over the time of development. The maximum applicable dose d_{max} (dashed lines) is, in turn, determined by molecular diffusion processes occurring in the walls. From Fig. 2.13, it is evident that both boundary doses d_{min} and d_{max} decrease with temperature, i.e. the process sensitivity increases at higher temperatures of development.

The regions between the solid and dashed lines in Fig. 2.13 represent the favorable dose windows where quality gratings may be fabricated. It can be seen that the width of the applicable dose window, $d_{max} - d_{min}$, increases strongly with

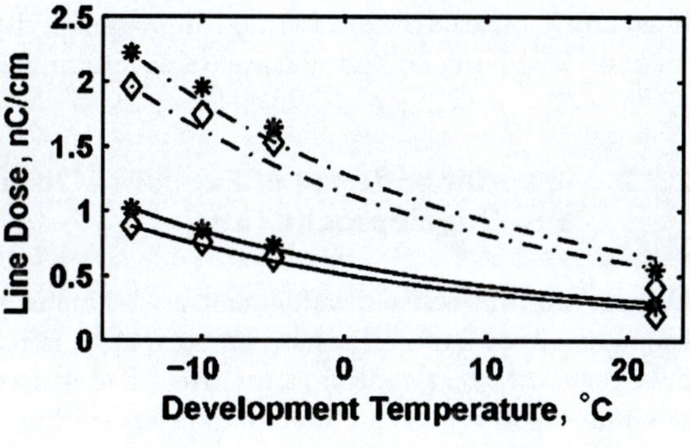

Fig. 2.13 Comparison of theoretical temperature dependencies for minimum (*solid lines*) and maximum (*dashed lines*) applicable exposure doses for a 70 nm pitch grating in PMMA exposed with 10 keV electrons. The symbols show the experimental data for development times of 5 s (*stars*) and 20 s (*diamonds*) [29]

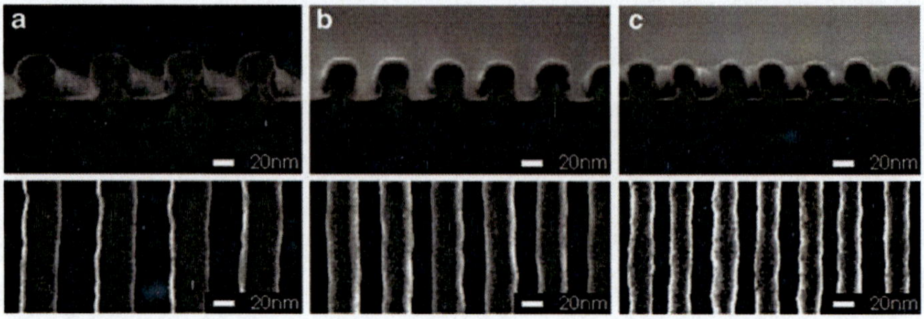

Fig. 2.14 Cross-section and top-view SEM micrographs showing examples of optimized dense nanoscale gratings fabricated in PMMA using a Raith 150 system with 10 keV voltage, and employing various development temperatures: (**a**) RT, 70 nm pitch; (**b**) −10°C, 50 nm pitch; and (**c**) −15°C, 40 nm pitch [29]

the *decrease* in the temperature of development. Thus, decreasing the development temperature from room temperature (RT) to −15°C results in an increase in the applicable line dose window by more than five times. This indicates a strong improvement in the EBL process robustness when decreasing the development temperature from RT to −15°C.

As an example of this increase in nanoscale resolution with decreasing temperature of development, Fig. 2.14 shows the highest resolution grating structures obtained in Ref. [29] by exposing 47–55 nm thick layers of PMMA to 10 keV electrons, for various development temperatures. It can be seen that RT development yields 33 ± 2 nm wide trench lines in a grating with a 70 nm pitch (a). With −10°C development, 20 ± 2 nm lines in a 50 nm pitch grating could be fabricated (b). This line width was further improved to 15 ± 2 nm in a 40 nm pitch grating at −15°C temperature (c).

One can conclude that the minimal resolved feature size significantly decreases with decrease of the development temperature from RT down to −15°C. This is consistent with the observed trend of broadening of the applicable dose windows, and can be explained by the kinetics of fragment mobility during resist dissolution. However, this advantage is accompanied by a drop in the process sensitivity. Reaching increased resolution and high sensitivity at the same time requires a trade-off involving co-optimization of several process conditions.

2.2.2 Interdependence of Exposure Dose and Development Time

Considering the resist development as a kinetic process involving diffusion of fragments from exposed resist into solvent, it is natural to expect the duration of development to be a control factor. The trend is illustrated by Fig. 2.15a, where the minimum (solid lines) and maximum (dashed lines) applicable doses, determined experimentally in a 50 nm grating pattern in PMMA, are shown as functions of time

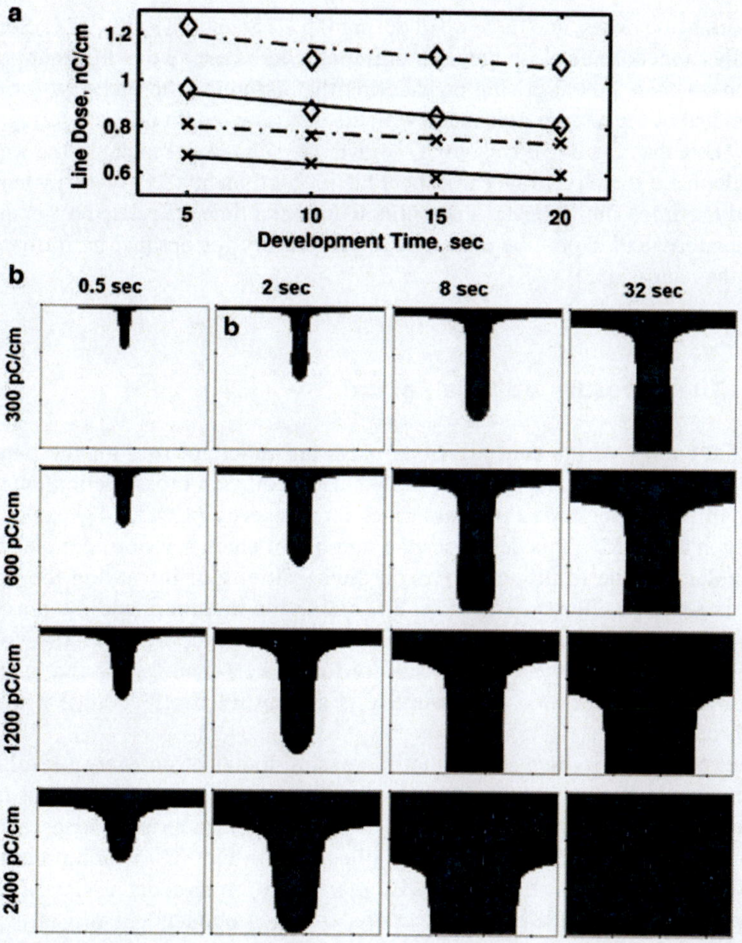

Fig. 2.15 (**a**) The applicable dose windows for 50 nm pitch gratings in PMMA showing minimum (*solid lines*) and maximum (*dashed lines*) line doses for quality patterning. The symbols indicate the temperature of development −5°C (*crosses*) and −15°C (*diamonds*) [29]. (**b**) Computed resist clearance profiles in a periodic grating with a 70 nm pitch exposed with 10 keV voltage with various line doses, and developed at −15°C for various durations. The width of all boxes is 70 nm, and the height is 60 nm. White denotes undissolved PMMA, and black denotes clearance [47]

of development at various temperatures. It can be seen that both the minimum and maximum boundaries show a moderate decrease with the development time. A minor to moderate decrease of the applicable dose window with the development time was also reported [38].

Figure 2.15b clarifies the interdependence of the exposure dose and development time, employing the results of recent kinetic modeling [47]. The figure presents a set of computed cross-sectional profiles in a grating with a 70 nm pitch, exposed with various line doses, and developed during 0.5, 2, 8, and 32 s at −15°C. Some of the profiles appear to be close in terms of trench width despite the different process conditions. This is not surprising considering that assuming the Fick law for diffusion, removal of the resist is associated with the diffusion length of PMMA fragments $(Dt)^{1/2}$, where the effective diffusivity D is given by (2.1). As a result, for the width of the development trench Δx one can expect the proportionality $\Delta x \sim d\,t^{1/2}$, where d is dose and t is time. This indicates that optimal dose and time are related by an inverse proportionality, and should be selected simultaneously for optimal performance of EBL at the nanoscale.

2.2.3 The Exposure Voltage Factor

The initial energy of incident electrons plays an important role for the exposure process due to several factors. First, the inelastic collision cross-section decreases roughly in proportion to an increase in electron energy [19, 26, 48]. In positive resists such as PMMA, this decreases the number of chain scissions per electron at higher voltages. The resulting impact on the sensitivity of increasing the voltage from 3 to 30 keV is illustrated in Fig. 2.16. Both the minimum and the maximum applicable exposure doses increase approximately in proportion to the voltage, resulting in a significantly lower sensitivity for 30 keV than for 10 and 3 keV. In a general case, the decrease in sensitivity is an unwanted effect since it leads to lower throughput.

However, Fig. 2.16 also shows that increasing the electron energy results in a dramatic increase of the applicable dose windows. The reason for this is that electrons with higher energies undergo less forward scattering and as a result, the beam broadens less [11, 31]. This is illustrated in Fig. 2.3, which presents the computed distributions of small (1–12 monomers) fragments in PMMA with exposed parallel lines [38]. Cross-sectional profiles obtained in resists exposed with different voltages are presented in Fig. 2.17. The grating exposed with

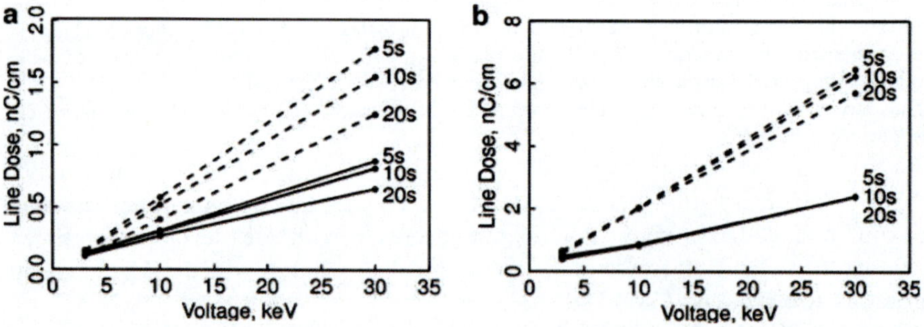

Fig. 2.16 Applicable line dose windows for 70 nm pitch gratings in for 3, 10, and 30 keV exposure voltages and different durations of development at RT (**a**) and at −15°C (**b**) [38]. The initial PMMA thickness was 55 nm. The meaning of the solid and dashed lines is as in Fig. 2.15

Fig. 2.17 SEM images of cross- sectional profiles for 70 nm pitch gratings fabricated at 3 keV (**a**), 10 keV (**b**), and 30 keV (**c**) voltages [31]

3 keV shows pronounced undercuts because of strong forward scattering of electrons, whereas 30 keV exposures produce almost straight interline walls. The low voltage tapered structures are more susceptible to pattern collapse, leading to reduced dose window. However, strong forward scattering of low energy electrons, which is usually believed to be the major resolution-limiting factor, may alternatively be employed to create nanoscale three-dimensional profiles in the resist [49] as is also demonstrated in the next section.

One more aspect to be mentioned in relation to the voltage, is its impact on the proximity effect. High energy electrons penetrate deeper into the substrate and can spread laterally due to backscattering. This gives rise to a significant proximity effect. The higher doses required at high energies compound this problem. In contrast, ultra low voltage electrons in the 1–3 keV regime deposit most of their energy within the resist, decreasing dramatically the proximity effect and also resulting in less substrate damage [11, 48].

2.3 Optimized EBL Processes: Examples

2.3.1 Low-kV, Cold Development PMMA Processing

As described in Sect. 2.2.3, 3 keV and lower voltage EBL has the advantages of higher sensitivity, lower substrate damage and proximity effect [11, 31, 48]. Furthermore, the strong forward scattering resulting in resist undercuts such as seen in Fig. 2.17a can be turned in an important processing advantage for metallization and lift-off patterning [49]. Conventionally, bi-layer resist schemes were used for metallization and lift-off at deep nanoscale [50–52]. A lower resolution resist layer would be coated first, followed by a higher resolution resist layer on top. When patterned, the resolution difference would yield a comparatively wider opening on the bottom resist layer as illustrated in Fig. 2.18. After subsequent metallization, this scheme would allow the solvent to access all areas of the trench and lift-off the PMMA resist layers without leaving any resist scum sticking to the substrate or deposited metal.

In contrast to using a bi-layer scheme, one can use a simpler single resist scheme by exploiting the re-entrant profiles created by scattering of low energy electrons. Using a single resist layer has two obvious advantages: (a) a thinner resist layer

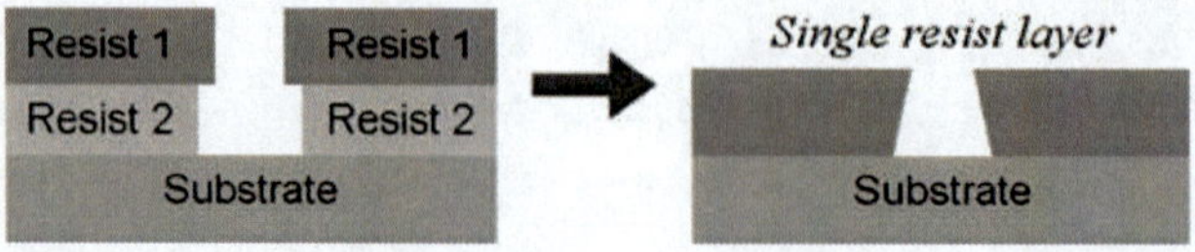

Fig. 2.18 Undercuts created in PMMA by using low-kV EBL enables replacement of bi-layer resist schemes (*left*) with a simpler single-resist layout (*right*)

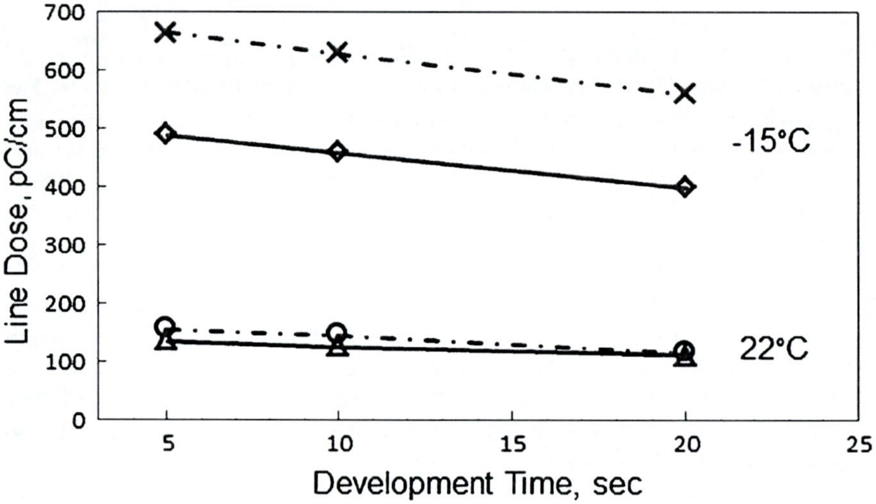

Fig. 2.19 Comparison of dose windows for 3 keV exposure voltages developed at room temperature (*triangles and circles*) vs. at −15°C (*diamonds and crosses*). The grating pitch is 70 nm and resist is 55 nm

allows a higher resolution since the aspect ratio requirement isn't as demanding, and (b) co-optimization of exposure doses and development conditions is simplified.

The advantages of using low voltage EBL are overcast by a major limitation: generally, the dose window for high-resolution, high-density gratings is very narrow for even thin resist layers. This property greatly affects the robustness of the EBL nanofabrication process i.e., small variations may lead to failure of the process. A useful work-around to this limitation is the use of cold development [38]. Figure 2.19 compares the dose windows for a 70 nm pitch grating in a 55 nm thick PMMA layer developed at room temperature and at −15°C. Cold development results in a dramatic increase in the applicable dose window by approximately an order of magnitude; however, this advantage is accompanied by a drop in the exposure sensitivity.

Using low-voltage exposure of PMMA in conjunction with cold development provides an extremely efficient and straightforward process of nanofabrication [29, 31, 38]. For example, at a PMMA thickness of 55 nm, the smallest grating pitch

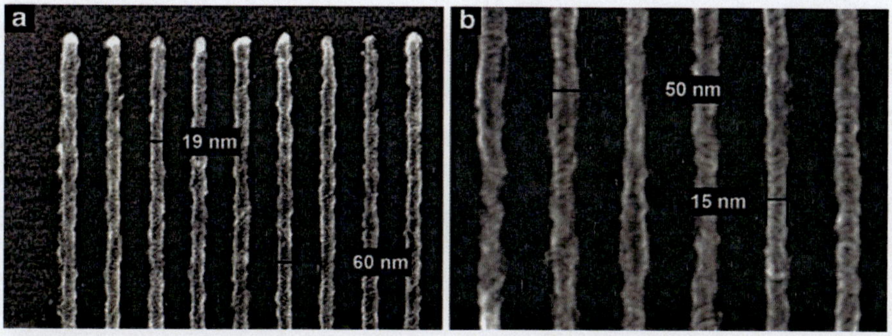

Fig. 2.20 (**a**) 20 nm wide Cr lines in a 60 nm pitch grating, and (**b**) 15 nm wide Cr lines in a 50 nm pitch grating, both fabricated by 3 keV exposure of a single layer of PMMA 950 K resist, and developed at −15°C

that our group could fabricate using 3 keV EBL at room temperature development was 70 nm. However, by using −15°C cold development, 50 nm pitch gratings could be realized with sub-20 nm line features after metallization and lift-off. Figure 2.20a, b shows sub-20 nm wide Cr lines in 60 and 50 nm pitch gratings, respectively, fabricated by deposition of a 12 nm thick layer of chromium on patterned PMMA features and subsequent lifted-off in acetone with ultrasonic agitation [53].

As described previously, one of the advantages of low-voltage EBL is the reduction of proximity effect [11, 48]. To illustrate this, Fig. 2.20a shows a corner of a Cr metallized grating array with uniform features at the edges of the pattern. This demonstrates the ability to fabricate high-resolution structures using low-voltage EBL in conjunction with cold development without the use of any proximity effect correction (PEC) routines [54–56]. The usage of low development temperature also improves the resistance of nano-patterned PMMA against post-exposure degradation such as wall collapse and fusion. In conclusion, cold development provides a significant number of advantages to a low-voltage EBL process such as higher resolution, larger dose window, pattern robustness, etc., at the expense of some loss in resist sensitivity.

2.3.2 Taking Control of PMMA: Sub-20 nm Wide Bridge Designs

As an example of optimized, PMMA-based ultrahigh resolution lithographic device fabrication, in this section we describe a recent improvement of existing silicon carbon nitride (SiCN) bridge resonator fabrication technology [57, 58] employing a low-voltage, cold development EBL process [53].

Figure 2.21 summarizes the improved process flow. In brief, an initial plasma enhanced chemical vapour deposition step is used to deposit a 50 nm thick SiCN layer followed by an anneal which sets the film to a tensile stress. A 45 nm thick PMMA film is then spun-cast, exposed at 3 keV, and developed at −15°C in MIBK: IPA 1:3. A 12 nm thick Cr layer is deposited by e-beam evaporation followed by an

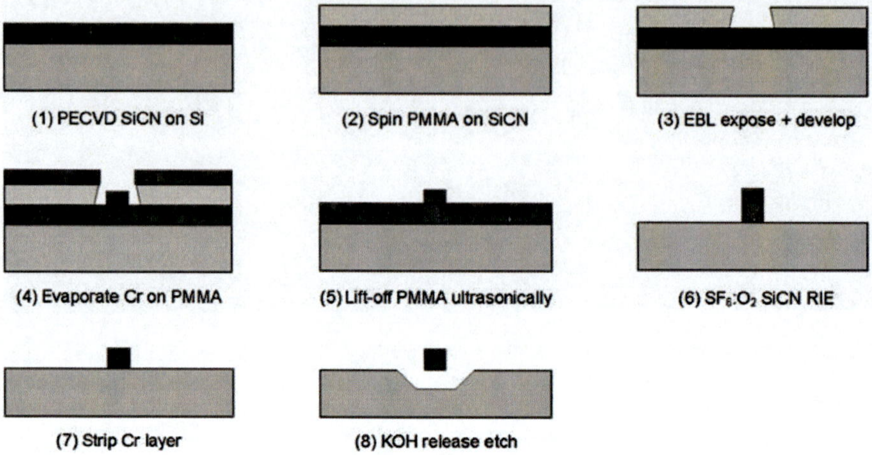

Fig. 2.21 Process flow outlining the steps used to fabricate doubly-clamped SiCN bridge resonators using a low voltage EBL and cold development process

ultrasonically assisted lift-off in acetone for 3 min. The Cr layer is used as an etch mask for the SiCN reactive ion etch (RIE) which is conducted with a 4:1 SF_6:O_2 recipe. Finally, after stripping the Cr layer, a wet release etch is conducted using hot KOH. Further processing details can be found in Ref. [53]

As a result of the above process, large arrays of nanoscale SiCN double-clamped bridge resonators could be fabricated with lengths between 1 and 20 μm. Figure 2.22 shows a representative micrograph of a 5 μm long bridge structure with a thickness of 50 nm and a width of 16 nm. In general, the minimum achievable linewidth of the bridge depended upon the desired length. The narrowest bridges with sub-15 nm widths could only be fabricated for lengths of up to 2 μm; bridges with widths between 14 and 18 nm could only be fabricated for lengths of up to 10 μm; and beyond that widths of 20–28 nm were required. Bridges with more demanding length to width aspect ratios failed due to mechanical fracture.

In Fig. 2.23 it can be seen that the bridge widths can be very finely controlled by applied electron beam single pixel line (SPL) dose for 1 μm long resonators. As the dose is reduced, the bridge width decreases: for doses of 2.0, 1.6, and 1.5 nC/cm, 16 $\pm$ 2 nm wide, 13 $\pm$ 3 nm wide, and 11 $\pm$ 5 nm wide bridges are fabricated, respectively. As the bridge widths decrease and approach 10 nm widths, the relative width non-uniformity strongly increases. At the smallest widths, small etch variations adversely affect the edge roughness.

An even more precise process control over the most critical part of the device, the clamping point, can be achieved with the usage of numerical modeling. The clamping point is a major factor in determining mechanical losses, so its fabrication is a key step to ensuring high resonator performance. In particular, any overhang or rounding at the corner is to be minimized. Considering the complexity of the interplay of many factors contributing to the EBL technique, being able to simulate

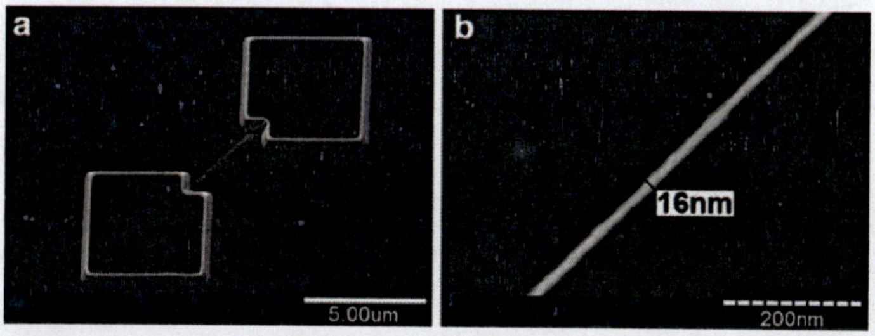

Fig. 2.22 A sub-20 nm wide, 5 μm long doubly-clamped SiCN resonator (**a**), and (**b**) a magnified image of the bridge showing a width of 16 ± 2 nm [53]

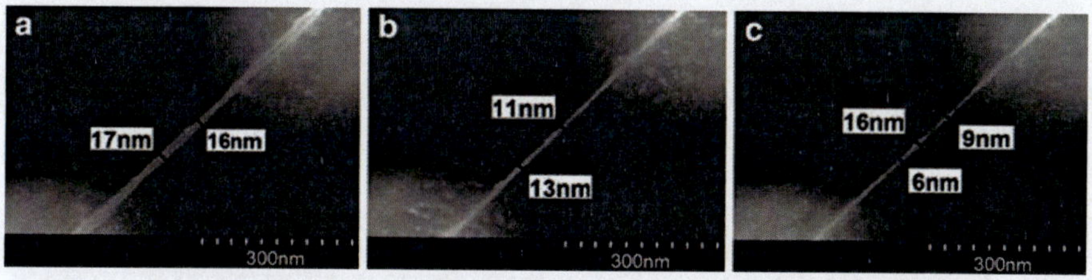

Fig. 2.23 Demonstrating control over the bridge widths: 1 μm long doubly-clamped SiCN bridges measuring (**a**) 16 ± 2 nm exposed at 2.0 nC/cm, (**b**) 13 ± 3 nm exposed at 1.6 nC/cm, and (**c**) 11 ± 5 nm exposed at 1.5 nC/cm [53]

all or some stages of the EBL process replaces time consuming and expensive cut-and-try experiments with a more efficient and rational, in-silico aided process design.

Figure 2.24 shows the user interface of the modeling tool and outlines the simulation procedure that was used to optimize the design of the clamping point in the resonator. This EBL simulator allows visualization of the electron beam exposure, fragmentation, and development of exposed positive tone resist, such as PMMA, on conductive substrates. In particular, 3D spatial maps of PMMA main chain scissions and 3D resist clearance profiles for given development conditions (duration, temperature) can be generated. More details on the EBL simulator can be found in Ref. [47].

Figure 2.25 demonstrates the results of the simulations and compares them to experimentally obtained results. A typical resonator clamping point is sketched in Fig. 2.25a and the results of simulation of the low-voltage exposure (3 keV) and cold development (−15°C) are presented in Fig. 2.25b, c, respectively. It can be seen that the numerically obtained results in Fig. 2.5c very accurately resemble the developed PMMA resist profile shown in Fig. 2.5d. The rounding of the clamping point seen in Fig. 2.25c, d arises because of the forward scattering of primary electrons.

The clamping point rounding results in an increase the overhanging area where undercuts occur after the final release etch. In Fig. 2.25e, the area where the SiCN

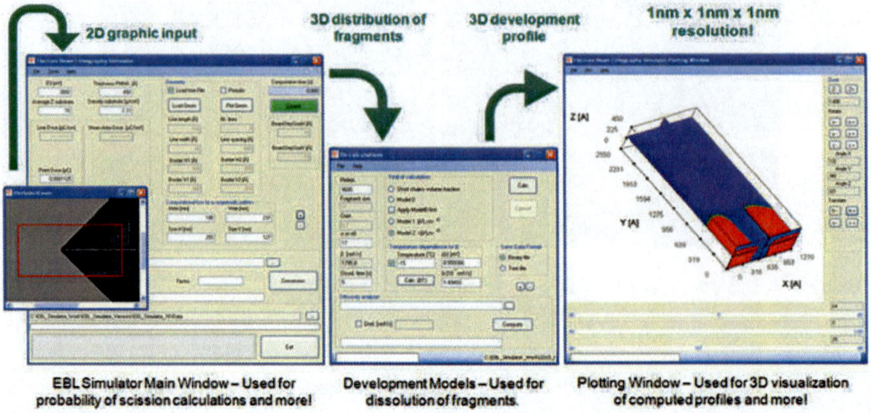

Fig. 2.24 Screen-shots of the EBL simulator user interface demonstrating the procedures of graphic input, simulation of the exposure (*chain scission*) and development, and visualization of the results. Further details on the simulation tool can be found in Ref. [47]

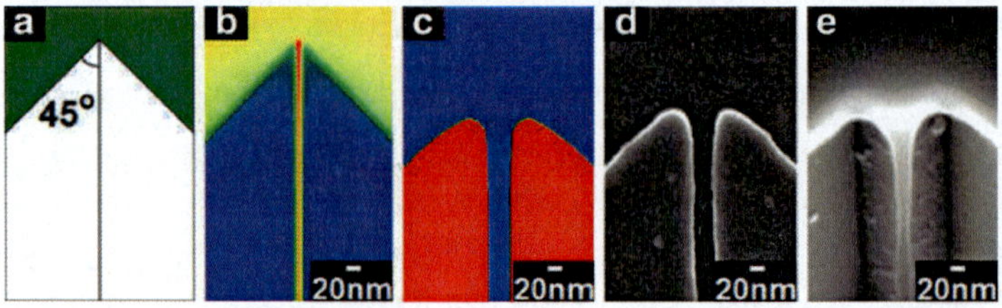

Fig. 2.25 (**a**) Sketch of a typical resonator clamping point design, (**b**) the computed yield of scission (*exposure plot*), (**c**) the computed dissolution profile (*development plot*), (**d**) a SEM image of the developed PMMA resist, and (**e**) a SEM image of the released SiCN resonator clamping point [53]

overhangs appears as the brighter parts around the clamping point. Generally, such overhangs contribute towards unwanted clamping losses and need to be avoided. To reduce the overhang around the clamping point, a number of alternative clamping designs were modelled and tested with the help of the EBL simulation tool. Designing and testing alternative clamping geometries provides greater flexibility than simple proximity effect correction (PEC) approaches, which are generally used to optimize the doses when complex structures involving multiple length scales are being fabricated.

Figures 2.26 and 2.27 illustrate two of the most successful designs. The first design (Fig. 2.26a) aims to sharpen the clamping point by defining a gap between the pad and resonator line at the design stage. This design exploits the movement of the resist-developer interface (the dissolution front) at the development stage such that the pad and resonator *just connect* yielding a sharp clamping point, provided the pad-resonator gap is optimized. An optimized gap of 170 nm was

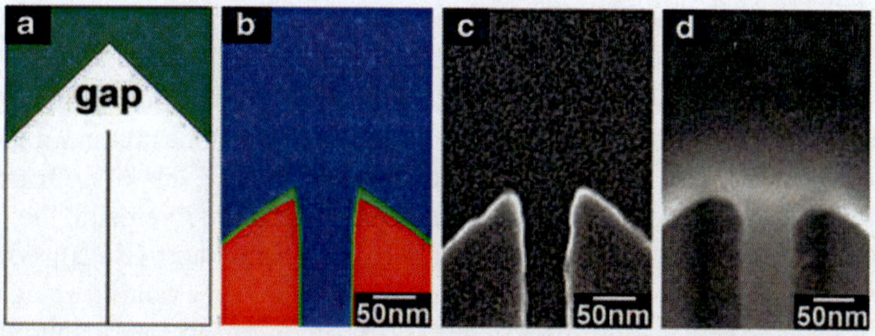

Fig. 2.26 (**a**) Diagram of an alternative clamping point design with an optimized gap of 170 nm between resonator and pad, (**b**) the final dissolution profile compared with (**c**) a SEM image of the developed PMMA resist showing sharper corners compared to Fig. 2.25d, and (**d**) a SEM image of the released SiCN resonator clamping point [53]

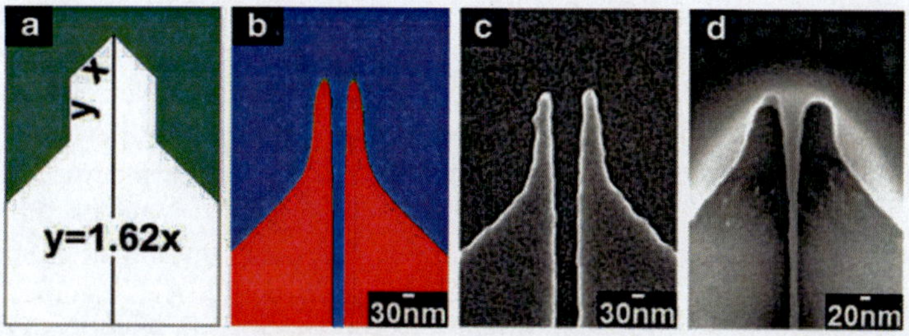

Fig. 2.27 (**a**) Diagram of an alternative clamping point design with an optimized side width of x = 165 nm, (**b**) the final dissolution profile compared with (**c**) a SEM image of the developed PMMA resist, and (**d**) a SEM image of the released SiCN resonator clamping point [53]

obtained by modelling, as shown in Fig. 2.26b, and experimentally realized, see Fig. 2.26c. After the release-etch, there still is some rounding in the device layer (Fig. 2.26d) as compared to the sharper corners seen in the resist layer in Fig. 2.26c; however, the result is better in comparison with Fig. 2.25e. One can hypothesize that the remaining rounding is due to the etchant contact angle and surface area minimization effects at the release etch stage. The second design (Fig. 2.26a) aims to isolate the resonator from the overhanging mass of the clamping point, and thus overcome the issue of the clamping point rounding. This design allows tuning the lengths of two sides such that an empty area is obtained around the beam. The two sides are related by the relation $y = 1.618x$. Through the help of the EBL simulator, an optimized value of $x = 165$ nm was obtained as shown in Fig. 2.26b and fabricated, see Fig. 2.26g. In this case after the release-etch, the overhanging area (the higher contrast and brighter part in Fig. 2.6d) is clearly disconnected from the resonator bridge.

One can conclude that employing low-voltage, cold development PMMA-based EBL process, complemented by in-silico aided optimization of the design, can be extremely efficient for ultrahigh resolution device fabrication.

2.3.3 Sub-10 nm Processing with HSQ

Employing polymer resist such as PMMA to fabricate sub-10 nm features in dense array configurations, although possible in principle, approaches the limit related to the size of the molecules, which challenges the attainable levels of uniformity and reproducibility. In the last decade, there has been significant interest in the usage of an alternative inorganic EBL resist hydrogen silsesquioxane (HSQ), which has shown considerable potential at the 10-nm scale (see for example review paper [59]). HSQ is a negative tone resist which cross-links to form an insoluble silica-like structure, although at significantly higher doses than required to process positive tone PMMA.

Many HSQ developers have been used such as TMAH – $(CH_3)_4NOH$ [60–66], NaOH [62, 66, 67], KOH [66, 68] and LiOH [66]. All of the above developer solutions are hydroxides i.e., they are basic solutions. Some of the optimization strategies included increasing developer strength from 2.38% TMAH to 25% TMAH, increasing the development time to 1 min [60], increasing the TMAH developer temperature to 50°C [63], adding NaCl solutions to NaOH [62, 67, 69] adding various salt solutions to all the above developers etc. [66]. The aforementioned optimization strategies have all yielded contrast improvements.

Apparently, TMAH-based development simultaneously provides the highest resolution while offering the greatest number of testable optimization strategies. Table 2.2 features three development recipes based on TMAH. Recipe (A) is a standard HSQ development recipe utilizing room temperature 25% TMAH developer diluted in water. Recipe (B) uses hot 50°C 25% TMAH developer [61–65]. Recipe (C) is a modified three step development recipe [70] in which a dilute hydrofluoric acid dip is incorporated between two hot TMAH development stages. Recipe (C) will be described in more detail later.

Figure 2.28 compares results for 50 nm pitch HSQ gratings exposed and developed using two different strategies. The gratings shown in Fig. 2.28a were exposed at 10 keV using a dose of 1.25 nC/cm and developed at room temperature. The gratings shown in Fig. 2.28b were exposed at 30 keV using a dose of 4.2 nC/cm and developed at 50°C. The images show that (a) the line resolution is superior to what has been illustrated in Sect. 2.3.1 for PMMA, and (b) using a higher voltage and hot development allows access to a sub-10 nm resolution at the cost of a significant reduction in sensitivity. One can conclude that a higher (30 keV) exposure and hot (50°C) development option appears as the baseline optimized process to achieve sub-10 nm resolution for HSQ [71].

Table 2.2 Examples of development schemes useful for HSQ

Recipe	Developer formulation
(A)	75 s dip in 25% TMAH
(B)	75 s dip in 50°C 25% TMAH
(C)	75 s dip in 50°C 25% TMAH
	+30–60 s dip in 2000:1 H_2O:BOE
	+75 s dip in 50°C 25% TMAH

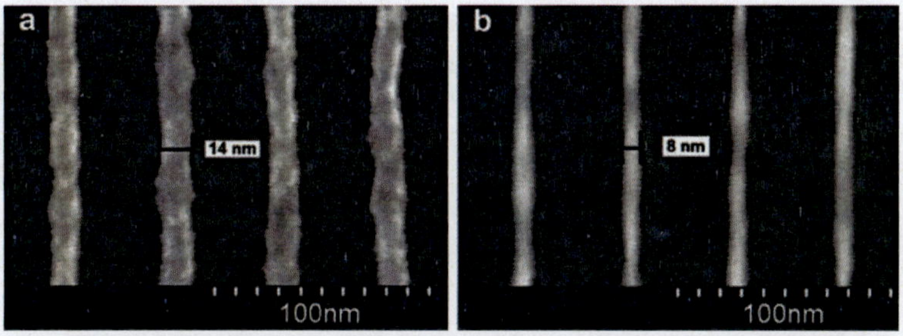

Fig. 2.28 Comparison of a 50 nm pitch HSQ grating developed at (**a**) room temperature, and at (**b**) 50°C, in 25% TMAH

Fig. 2.29 Arbitrary area patterns (**a**) showing EBL rastering, and (**b**) magnified image of the rastering showing 10 nm diameter dots

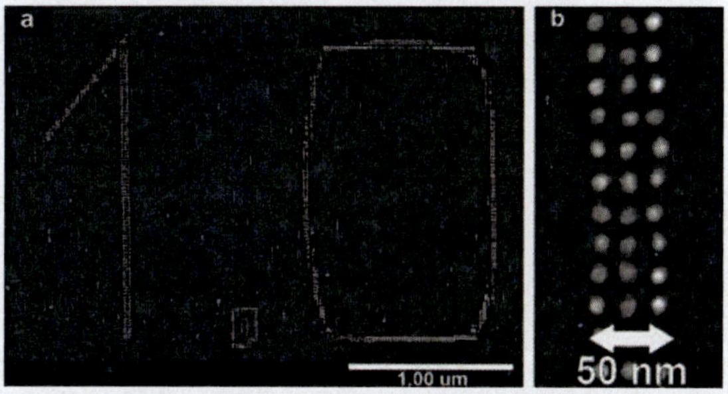

While optimizing HSQ process, care must be taken regarding a number of conditions during the exposure and development stages. Thus at the exposure stage, beam step size is important. Since the resolution HSQ is superior to that of PMMA, both the single pixel line (SPL) and area step sizes must be chosen less than for PMMA. Figure 2.29a shows an arbitrary text written with a Raith 150^{TWO} EBL system using an area step size of 20 nm. As shown in the magnified image provided in Fig. 2.29b, the rastering spots approx. 10 nm in diameter are visible. A continuous exposure can be achieved with decreased step sizes of 2–10 nm.

After choosing a small enough beam step size, exposure dose selection is the next step. Figure 2.30 shows an area dose test pattern with squares exposed at successively higher doses. The dose factors are marked above each square. At the extremes, either the dose is too low for pattern delineation, or the dose is too high and the pattern is severely distorted. In the middle, a window of applicable doses can be seen. In contrast to positive resists, selecting the minimum cross-linking dose would not work if the pattern is intended to be used as a masking layer for subsequent etching. In that case, the cross-link density may not be high enough to be a suitable etch mask. Thus with negative tome resists, the preferred dose is often the largest possible cross-linking dose before pattern distortion, as opposed to selecting the minimum clearance dose for positive tone resists.

The selection of the largest possible cross-linking dose, however, has a downside of the process of cross-linking becoming extremely sensitive to exposure by

Fig. 2.30 Dose test for HSQ showing various exposure regimes

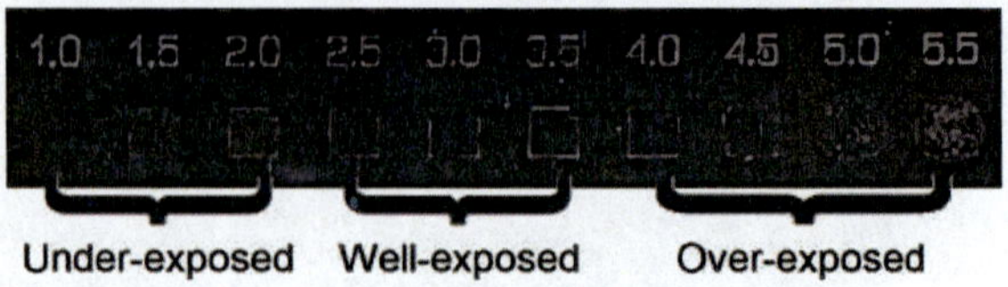

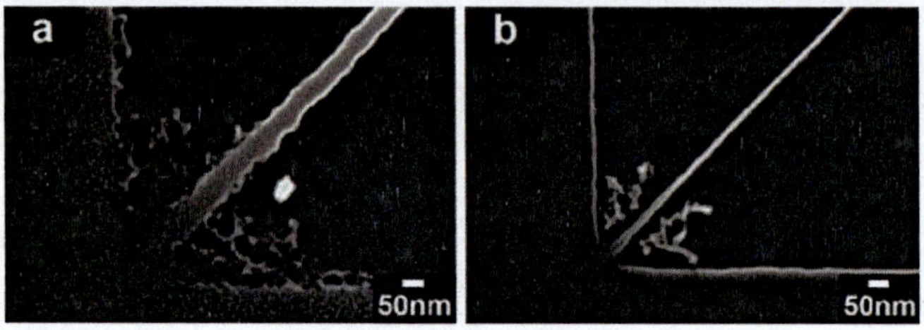

Fig. 2.31 SEM images of un-released SiCN masking layer at the pad-resonator interface developed using (**a**) standard TMAH recipe (A) with centre width 48 ± 5 nm, and (**b**) with modified hot TMAH-BOE-TMAH recipe (C) with centre width 15 ± 2 nm [71]

secondary and backscattered electrons. These proximity effects may cause unwanted partly-exposed siloxane-like scum [70–72] around pattern sidewalls and dense features as shown in Fig. 2.31a. Since the siloxane-like scum consists of an oxide, $HSiO_x$, it can be etched away using dilute HF rinse [70] which is also beneficial for the resist contrast as it prevents the development step from saturating [70] and also trims the patterns which enhances the resolution even further [72].

Additionally, hot development has also been reported to remove these partly cross-linked structures [63]. To solve the described issues, an alternative development scheme has been developed recently [71], see recipe (C) in Table 2.2. In this recipe, hot development is combined with HF-trimming or multi-step TMAH-HF-TMAH development. This combination helps to clean up the unwanted siloxane-like scum and also provide pattern trimming as shown in Fig. 2.31b. The trimming is estimated to be as high as 20 nm/min.

2.3.4 HSQ Resist as Etching Mask: 8 nm Wide Bridge

Exploiting the optimized HSQ based nano-patterning techniques discussed in Sect. 2.3.3, the SiCN resonator fabrication process discussed in Sect. 2.3.2 has been further improved [71]. By using HSQ both as the resist layer and the masking layer for the SiCN RIE, the resonator fabrication process has been simplified by eliminating the chromium metal layer, and its resolution has been enhanced to achieve sub-10 nm wide, microns long bridges. The highly optimized SiCN resonator fabrication process [71] is summarized in Fig. 2.32.

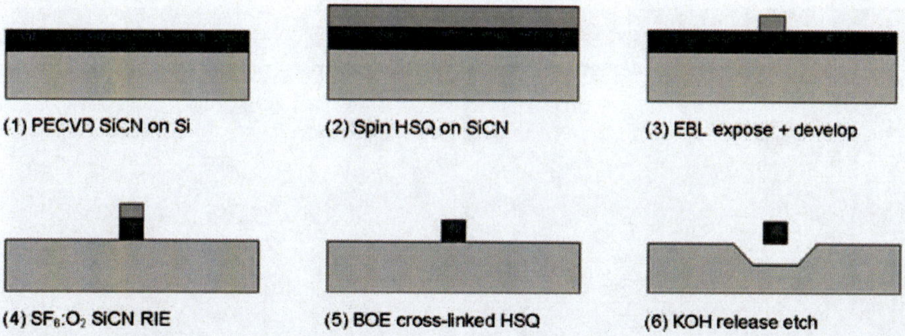

Fig. 2.32 Process flow outlining the steps used to fabricate doubly-clamped SiCN resonators using a HSQ mask (Adapted from Ref. [71])

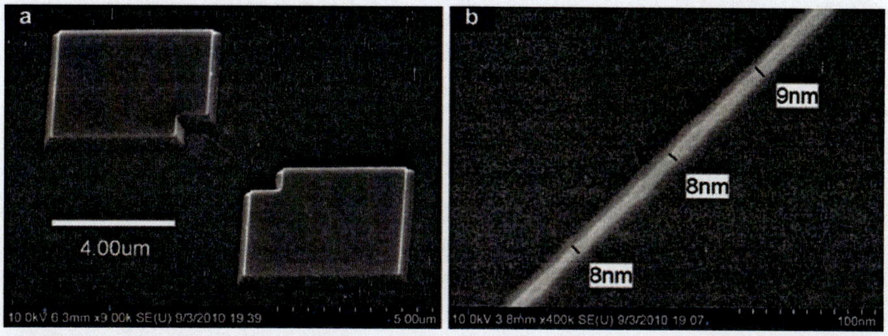

Fig. 2.33 A sub-10 nm wide, 5 μm long doubly-clamped SiCN resonator: (**a**) overview image, and (**b**) a magnified image of the bridge

After depositing and annealing a 50 nm thick SiCN layer on Si, a 25–30 nm thick HSQ layer is spin on top. The HSQ layer is exposed by 30 keV EBL and developed using the multi-step hot TMAH-HF-TMAH recipe (C) from Table 2.2 as described in Sect. 2.3.3. The cross-linked HSQ layer is used as an etch mask for the SiCN RIE and then is stripped off using a 30 s BOE. Finally the resonator is released in a 75°C 28.3% KOH bath saturated with IPA for short etch durations ranging from 30 to 45 s. Further details can be found in Ref. [71].

Figure 2.33a presents a SEM image of a typical sub-10 nm wide, 5 μm long doubly-clamped SiCN resonator with SiCN layer thickness of 50 nm. The EBL area and line doses used were 2.5 mC/cm^2 and 9 nC/cm respectively. Figure 2.33b is a magnified top-view SEM image of the bridge. The SEM image is taken from the centre of the resonator shown in Fig. 2.33a. The measured width of the bridge is 9 ± 1 nm. Bridges of similar resolution can also be obtained using single step room temperature recipe (A) or hot development recipe (B); however, this is inferior because a higher dose is required, the bridges are less uniform, and a presence of siloxane-like scum in the clamping area.

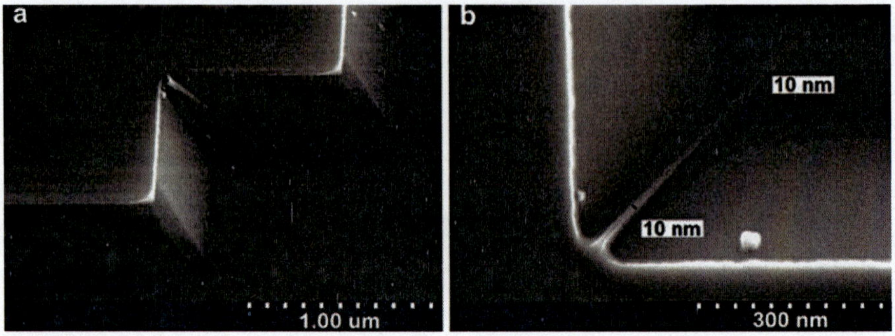

Fig. 2.34 SEM images of the clamping point: (**a**) tilt-view showing clamping features and release, and (**b**) top-view showing width uniformity

Figure 2.34a, b presents magnified SEM images of the resonator clamping point fabricated using recipe C from Table 2.2. Both figures show that little or no residual scum is visible at the clamping point and the surfaces and edges are very clean. The areas around the clamping point show only a minor overhanging. The bridges are suspended 400–600 nm above the silicon surface. Finally, both Figs. 2.33b and 2.34b show that the bridge width is uniform and less than 10 nm along its entire length. This demonstrates that by co-optimization of exposure and development conditions, proximity effects can be avoided or compensated without the use of complicated correction algorithms.

2.4 Insulating Substrates

Application areas such photonics, nanoelectronics, and bio-nano-electromechanical systems (bio-NEMS) often require fabrication of large, macroscopic-scale arrays of nanoscale features. Nanoimprint lithography has emerged as an efficient and low cost solution [73]. Ultraviolet nanoimprint lithography (UV-NIL), which involves an optical exposure, requires fabrication of transparent nanostructured masters. Dielectric materials appropriate for this purpose present a difficulty when employing EBL for their nanostructuring. Unlike with conducting and semiconducting substrates, polymeric resist layers such as PMMA on top of insulators accumulate charge during EBL exposure, which deflects the beam and distorts the pattern [74–76].

The solutions revolve largely around usage of conducting anti-charging layers. It has been demonstrated that thin (from 5 nm) overlayers of light metals such as Al, Cr, or Cu, can be functional when fabricating periodic grating patterns with a pitch up to 150–200 nm [77]. Such coatings, however, result in electron beam scattering in the metal layer and subsequent broadening of the exposure profile in the resist, which limits the resolution at the deep nanoscale. The technique also tends to decrease the process sensitivity [77].

An example of an alternative process where a metallic anti-charging layer is located underneath the resist is given in Fig. 2.35 [78]. In brief, a UV transparent fused silica (FS) substrates is coated with a sputtered 30 nm Al film, followed by spin coating a 60 nm layer of PMMA resist. In this PMMA/Al/FS scheme, the Al layer works as a charge conduction layer during e-beam exposure, and later, works as hard mask for pattern transfer into the FS substrate via glass etching. After 10 keV EBL exposure with a Raith 150 system and development of the PMMA, reactive ion etch (RIE) is carried out to transfer the pattern to the underlying Al layer.

As the next step, The EBL features from the Al layer are transferred to the fused silica substrate by a RIE process. The Al mask is then removed from the substrate. Figure 2.36a, b shows the examples of an Al grating pattern with composed of ~40 nm lines with a ~60 nm interline distance, and a corresponding grating pattern in fused silica, respectively. An edge roughness is visible that occurs because nanocrystalline grains are formed during Al deposition by sputtering. Although

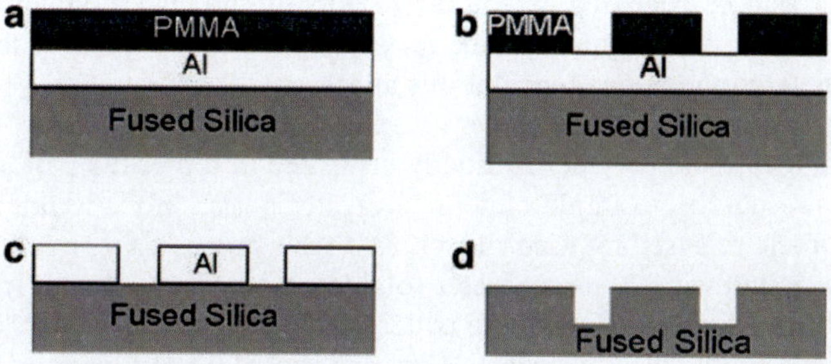

Fig. 2.35 NIL mask fabrication schemes: (**a**) glass substrate is coated with Al and PMMA. (**b**) Patterns generated by e-beam exposure are realized in PMMA, (**c**) which is then transferred to the Al layer using dry etch with PMMA acting as the etch mask. (**d**) Patterns are then transferred to the glass substrate via glass etch process with the Al layer acting as hard mask. Finally the Al mask is removed from the fused silica surface

Fig. 2.36 100 nm pitch gratings with less than 40 nm wide lines realized in the Al layer (**a**), and released in a fused silica substrate (**b**). An edge roughness resulting from metal grains is visible

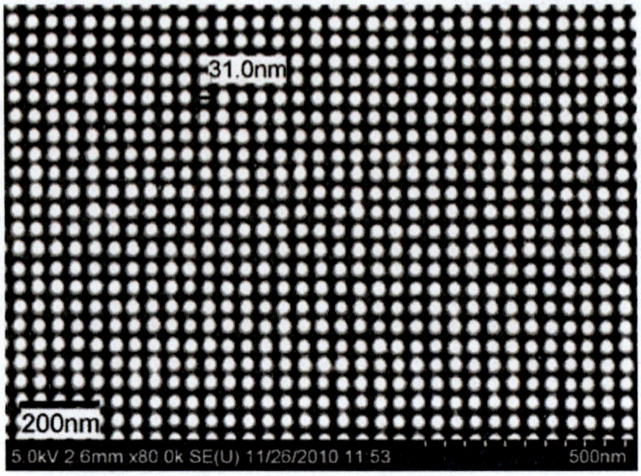

Fig. 2.37 Array of posts with 50 nm × 50 nm pitch fabricated in a fused silica substrate employing a conducting polymer aquaSAVE [81] of top of PMMA as an anti-charging layer

the grain size can be deceased by carefully selecting the conditions of deposition and the metal [79], the roughness in the released FS pattern caused by the metallic grains is a quite common challenge of this approach.

A different solution is to use a layer of a conducting polymer instead of a metal layer, in which case the polymer is usually deposited of top of the EBL resist [80]. Figure 2.37 presents an example of dense arrays of posts, only ~30 nm across and with 50 nm pitch, released in fused silica [78]. In this process, a 90 nm thick film of PMMA was coated with 70 nm of water soluble, conducting polymer (aquaSAVE from Mitsubishi Rayon [81]), and the pattern was generated using 30 kV e-beam exposure with a Raith 150^{TWO} system. After removal of the anti-charging layer and development of PMMA, an 8 nm film of Cr was sputter deposited and then lifted-off in an ultrasonic bath. The patterns realized in Cr were then transferred to the FS substrate by RIE. In comparison with metallic layers deposited on top of the resist, polymers introduce less beam broadening, offering a stronger potential in an increase of resolution at the deep nanoscale. The lift-off stage, however, turned out to be somewhat pattern sensitive, with the quality of the released pattern of same nominal size varied depending on the geometry [78]. Clearly, further process optimization effort, including the insight from numerical modeling, is needed to fully realize the potential of EBL on insulating substrates.

2.5 Summary

EBL is a complex process with many interacting parameters that affect the quality of the resulting nanofabricated structures and the robustness of the process. An approach has been presented for mapping regimes within which successful results can be achieved, with a particular emphasis on PMMA as a model resist. An analytic model of exposure and development has been presented as a tool to aid this analysis. Use of low energy EBL has been explored to improve sensitivity and reduce proximity effects. Cold development has been shown to improve process windows and resolution, particularly in concert with low energy exposure. Application of these tools and techniques to the improvement of an isolated nanoscale bridge structure and a dense, insulating nanoimprint lithography master have been presented.

Acknowledgements The authors are grateful for the support of the Natural Sciences and Engineering Research Council of Canada, the National Institute for Nanotechnology, the Alberta Ingenuity Fund, Raith GmbH, and the University of Alberta NanoFab.

References

1. McCord MA, Rooks M. Electron beam lithography. In: Rai-Choudury P, editor. Handbook of microlithography, micromachining and microfabrication, vol. 1. Bellingham: SPIE; 1997. ISBN 978-081-942-378-8.
2. Nabity J, Compbell LA, Zhu M, Zhou W. E-beam nanolithography integrated with scanning electron microscope. In: Zhou W, Wang ZhL, editors. Scanning microscopy for nanotechnology: techniques and applications. 1st ed. New York: Springer; 2006. ISBN 978-144-192-209-0.
3. Wu CS, Makiuchi Y, Chen CD, In: Wang M, editors. Lithography. High-energy electron beam lithography for nanoscale fabrication. Croatia: InTech; 2010, ISBN 978-953-307-064-3.
4. Liddle JA, Berger SD. Proc SPIE. 2014;2014:66–76.
5. Pfeiffer HC, Stickel W. Future Fab Intl. 2002;12:187.
6. Mapper Lithography, Delft, www.mapperlithography.com
7. Raith GmbH, Dortmund, www.raith.com
8. Goldstein J, Newbury DE, Joy DC, Lyman CE, Echlin P, Lifshin E, Sawyer L, Michael JR. Scanning electron microscopy and X-ray microanalysis. 3rd ed. New York: Springer; 2003. ISBN 978-030-647-292-3.
9. Jeol electron beam lithography, Tokyo, www.jeol.com/PRODUCTS/SemiconductorEquipment/ElectronBeamLithography/tabid/99/Default.aspx
10. Vistec electron beam GmbH, Jena, www.vistec-semi.com
11. Lee YH, Browing R, Maluf N, Owen G, Pease RFW. J Vac Sci Technol B. 1992;10:3094–8.
12. Yang H, Fan L, Jin A, Luo Q, Gu C, Cui Z. In: Proc. 1st IEEE Intl. Conf. Nano/Micro Engg. & Molec. Sys. Jan 2006, Zhuhai, p. 391–4.
13. Kyser DF, Viswanathan NS. J Vac Sci Technol. 1975;12:1305–8.
14. Brewer G, editor. Electron-Beam Technology in Microelectronic Fabrication. New York: Academic; 1980. 978-012-133-550-2.
15. Kamp M, Emmerling M, Kuhn S, Forchel A. J Vac Sci Technol B. 1999;17:86–9.
16. Chang THP. J Vac Sci Technol. 1975;12:1271–5.
17. Lo CW, Rooks MJ, Lo WK, Isaacson M, Craighead HG. J Vac Sci Technol B. 1995;13:812–20.
18. Mun LK, Drouin D, Lavallée E, Beauvais J. Microsc Microanal. 2004;10:804–9.
19. Wu B, Neureuther AR. J Vac Sci Technol B. 2001;19:2508–11.
20. Showa Denko ESPACER, www.showadenko.us
21. Hatzakis M. J Electrochem Soc. 1969;116:1033–7.
22. ZEONREX electronic chemicals, Japan http://www.zeon.co.jp/index_e.html
23. Nishida T, Notomi M, Iga R, Tamamura T. Jpn J Appl Phys. 1992;31:4508–14.
24. Olynick DL, Cord B, Schipotinin A, Ogletree DF, Schuck PJ. J Vac Sci Technol B. 2010;28:581–7.
25. Bilenberg B, Schøler M, Shi P, Schmidt MS, Bøggild P, Fink M, Schuster C, Reuther F, Gruetzner C, Kristensen A. J Vac Sci Technol B. 2006;24:1776–9.

26. Aktary M, Stepanova M, Dew SK. J Vac Sci Technol B. 2006;24:768–79.
27. Masaro L, Zhu XX. Prog Polym Sci. 1999;24:731–75.
28. Miller-Chou BA, Koenig JL. Prog Polym Sci. 2003;28:1223–70.
29. Mohammad MA, Fito T, Chen J, Aktary M, Stepanova M, Dew SK. J Vac Sci Technol B. 2010;28:L1–4.
30. Tanaka T, Morigami M, Atoda N. Jpn J Appl Phys. 1993;32:6059–64.
31. Mohammad MA, Fito T, Chen J, Buswell S, Aktary M, Stepanova M, Dew SK. Micr Eng. 2010;87:1104–7.
32. Lee K, Bucchignano J, Gelorme J, Viswanathan R. J Vac Sci Technol B. 1997;15:2621–6.
33. Yasin S, Hasko D, Ahmed H. J Vac Sci Technol B. 1999;17:3390–3.
34. Kupper D, Kupper D, Wahlbrink T, Bolten J, Lemme M, Georgiev Y, Kurz H. J Vac Sci Technol B. 2006;24:1827–32.
35. Namatsu H, Yamazaki K, Kurihara K. J Vac Sci Technol B. 2000;18:780–4.
36. Goldfarb D, de Pablo J, Nealey P, Simons J, Moreau W, Angelpoulos M. J Vac Sci Technol B. 2000;18:3313–7.
37. Wahlbrink T, Kupper D, Georgiev Y, Bolten J, Moller M, Kupper D, Lemme M, Kurz H. Microelectron Eng. 2006;83:1124–7.
38. Mohammad MA, Fito T, Chen J, Buswell S, Aktary M, Dew SK, Stepanova M. In Wang M editors, Lithography. The interdependence of exposure and development conditions when optimizing low-energy EBL for nano-scale resolution. Croatia: InTech; 2010, ISBN 978-953-307-064-3.
39. Ocola LE, Stein A. J Vac Sci Technol B. 2006;24:3061–5.
40. Häffner M, Heeren A, Fleischer M, Kern DP, Schmidt G, Molenkamp LW. Microelectron Eng. 2007;84:937–9.
41. Cord B, Lutkenhaus J, Berggren KK. J Vac Sci Technol B. 2007;25:2013–6.
42. Yan M, Choi S, Subramanian KRV, Adesida I. J Vac Sci Technol B. 2008;26:2306–10.
43. Mohammad MA, Dew SK, Westra K, Li P, Aktary M, Lauw Y, Kovalenko A, Stepanova M. J Vac Sci Technol B. 2007;25:745–53.
44. Cord B, Yang J, Duan H, Joy DC, Klingfus J, Berggren KK. J Vac Sci Technol B. 2009;27:2616–21.
45. Hasko DG, Yasin S, Mumatz A. J Vac Sci Technol B. 2000;18:3441–4.
46. Yasin S, Hasko DG, Khalid MN, Weaver DJ, Ahmed H. J Vac Sci Technol B. 2004;22:574–8.
47. Stepanova M, Fito T, Szabó Zs, Alti K, Adeyenuwo AP, Koshelev K, Aktary M, Dew SK. J Vac Sci Technol B. 2010;28:C6C48–57.
48. Schock K-D, Prins FE, Strähle FES, Kern DP. J Vac Sci Technol B. 1997;15:2323–6.
49. Brünger W, Kley EB, Schnabel B, Stolberg I, Zierbock M, Plontke M. Microelectron Eng. 1995;27:135–8.
50. An L, Zheng Y, Li K, Luo P, Wu Y. J Vac Sci Technol B. 2005;23:1603–6.
51. Cord B, Dames C, Berggren KK, Aumentado J. J Vac Sci Technol B. 2006;24:3139–43.
52. Yang H, Jin A, Luo Q, Li J, Gu C, Cui Z. Microelectron Eng. 2008;85:814–7.
53. Mohammad MA, Guthy C, Evoy S, Dew SK, Stepanova M. J Vac Sci Technol B. 2010;28: C6P36–41.
54. Anbumony K, Lee S. J Vac Sci Technol B. 2006;24:3115–20.
55. Leunissen L, Jonckheere R, Hofmann U, Unal N, Kalus C. J Vac Sci Technol B. 2004;22:2943–7.
56. Ogino K, Hoshino H, Machida Y, Osawa M, Arimoto H, Maruyama T, Kawamura E. Jpn J Appl Phys. 2004;43:3762–6.
57. Fischer LM, Wilding LMN, Gel M, Evoy S. J Vac Sci Technol B. 2007;25:33–7.
58. Fischer LM, Wright VA, Guthy Cz, Yang N, McDermott MT, Buriak JM, Evoy S. Sens Actuators B. 2008;134:613–7.
59. Grigorescu AE, Hagen CW. Nanotechnology. 2009;20:292001.
60. Fruleux-Cornu F, Penaud J, Dubois E, Francois M, Muller M. Mater Sci Eng C. 2005;26:893–7.

61. Chen Y, Yang H, Cui Z. Microelectron Eng. 2006;83:1119–23.
62. Yang H, Jin A, Luo O, Gu C, Cui Z. Microelectron Eng. 2007;84:1109–12.
63. Haffner M, Haug A, Heeren A, Fleischer M, Peisert H, Chasse T, Kern DP. J Vac Sci Technol B. 2007;25:2045–8.
64. Choi S, Jin N, Kumar V, Adesida I, Shannon M. J Vac Sci Technol B. 2007;25:2085–8.
65. Ocola LE, Tirumala VR. J Vac Sci Technol B. 2008;26:2632–5.
66. Kim J, Chao W, Griedel B, Liang X, Lewis M, Hilken D, Olynick D. J Vac Sci Technol B. 2009;27:2628–34.
67. Yan M, Lee J, Ofuonye B, Choi S, Jang JH, Adesida I. J Vac Sci Technol B. 2020;28:C6S23–7.
68. Lauvernier D, Vilcot J-P, Francois M, Decoster D. Microelectron Eng. 2004;75:177–82.
69. Yang JKW, Berggren KK. J Vac Sci Technol B. 2007;25:2025–9.
70. Lee H-S, Wi J-S, Nam S-W, Kim H-M, Kim K-B. Vac Sci Technol B. 2009;25:188–92.
71. Mohammad MA, Dew SK, Evoy S, Stepanova M. Microelectron Eng. 2011;88:2338–41.
72. Tiron R, Mollard L, Louveau O, Lajoinie E. J Vac Sci Technol B. 2007;25:1147–51.
73. Schift H. J Vac Sci Technol B. 2008;26:458–80.
74. Liu W, Ingino J, Pease RF. J Vac Sci Technol B. 1995;13:1979–83.
75. Satyalakshmi KM, Olkhovets A, Metzler MG, Harnett CK, Tanenbaum DM, Craighead HG. J Vac Sci Technol B. 2000;18:3122–5.
76. Joo J, Jun K, Jacobson JM. J Vac Sci Technol B. 2007;5:2407–11.
77. Samantaray CB, Hastings JT. J Vac Sci Technol B. 2008;26:2300–5.
78. Muhammad M, Buswell SC, Dew SK, Stepanova M. J Vac Sci Technol B. 2011;29:06F304.
79. Bhuiyan A, Dew SK, Stepanova M. Comput Commun Phys. 2011;9:49–67.
80. Dylevwicz R, Lis S, De La Rue RM, Rahman F. J Vac Sci Technol B. 2010;28:817–22.
81. aquaSAVE Electronic Conductor, Mitsubishi Rayon America Inc., New York, http://www.mrany.com/data/HTML/20.htm

Simulation of Electron Beam Exposure and Resist Processing for Nano-Patterning

3

Ioannis Raptis and George P. Patsis

Abstract

Electron beam lithography simulation is a powerful tool for the prediction of the resist profile after a complete lithographic process and the process optimization towards ultimate resolution and increased process latitude. In this chapter the structure of such a simulation tool is presented and all core modules are explained in detail along with all major approaches developed for each of them (exposure, development, thermal treatment). Finally, two particular examples are given from simulation of complex patterns in the deep sub-100 nm regime along with experimental data that confirm the accuracy and capabilities of the modern simulation tools.

3.1 Introduction

Electron beam lithography (EBL) is a very powerful lithographic technology able to resolve in a repeatable and reliable way, sub-10 nm patterns over large areas. Even though several other methodologies have been developed offering the same performance, EBL remains the choice for mask making applications and for prototyping. EBL is also used in the production of particular components such as Read-Write hard disk heads and is considered as a competitor for the next generation of ASIC's and Multi Project Wafers, replacing optical lithography. Even though EBL has been around for decades, with the first vector scan tool developed by IBM at the early 70s, EBL tools are very expensive and their writing speed is still low or moderate if patterning very fine complex and dense structures. For those

I. Raptis (✉)
Institute of Microelectronics, NCSR "Demokritos", Athens, Greece
e-mail: raptis@imel.demokritos.gr

G.P. Patsis
Department of Electronics, TEI of Athens, Athens, Greece

M. Stepanova and S. Dew (eds.), *Nanofabrication*,
DOI 10.1007/978-3-7091-0424-8_3, © Springer-Verlag/Wien 2012

reasons, EBL machine time is in great demand and process optimization turn-around time should be minimized in order to decrease the overall cost. EBL is quite complicated, and includes several physical and chemical processing steps: electron beam – matter interaction, thermal processing steps (in the Chemically Amplified resists) and resist development. In order to decrease the process optimization time and cost, fast and accurate holistic simulation tools are needed based on accurate physical and chemical models describing all phenomena taking place during the various processing steps. Simulation outcomes should be multi-dimensional with detailed and accurate 3D information on resist profile, actual feature dimension and Line Edge Roughness (LER). Pattern dimension and LER define at a great percentage the device electrical characteristics and the overall IC size. All simulation features described above are common for all lithographic technologies. Additionally in EBL, the proximity effect sets significant obstacles in the realization of very fine and dense features. Proximity effect is the unwanted re-exposure of the resist film from the backscattered electrons at distances several microns from the point of initial incidence. This effect is one of the limiting factors for the minimum dense features that could be produced with EBL. Furthermore simulation should be able to process the original layout in terms of pattern dimensions and exposure dose in such a way to produce the desired resist patterns. At the early days of EBL simulation, its application was bound by the shortage of CPU power and by the fact that critical structures in the layout were of moderate size and complexity. However, nowadays layouts mastered by EBL are of high density, complexity and are full of features with critical dimensions in the nanoscale. Thus, simulation should be performed in reasonable times and offer very high accuracy.

Lithography simulation has progressed a lot since its infancy at the end of 80s. Nowadays, lithography simulation is considered of great importance for the process optimization and the development of the technologies required for the implementation of next generation lithography for the realization of the devices for the forthcoming technology nodes, Fig. 3.1. The simulation software should predict the resist profile and provide accurate information on dimensions and line edge roughness. Furthermore, complete software should include the capability of comparison with experimental data (e.g. top-down and cross-section SEM images) for automatic assessment of simulation results.

EBL simulation consists of several parts, each one of them dedicated to a particular task. The major simulation components are

(a) Calculation of energy deposition initially from a point beam and, after convolution with beam size and type, the energy deposition for the whole layout
(b) Simulation of thermal processing effect in the case of chemically amplified resists

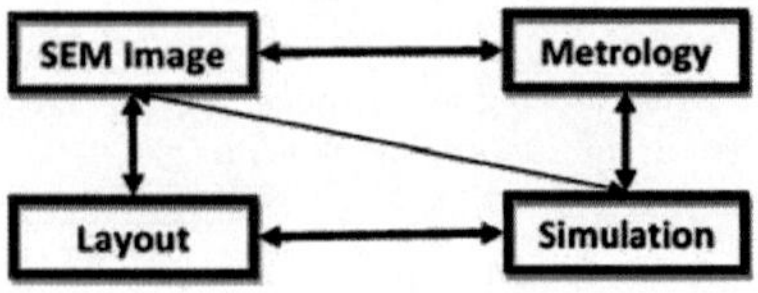

Fig. 3.1 Core modules of a complete lithography simulation tool

(c) Simulation of resist profile by taking into account the results from the previous steps and the resist development algorithm
(d) Comparison with experimental results.

For the simulation needs, a series of information should be provided by the user, such as layout (in GDS or CIF formats), electron beam energy, beam size and type, exposure dose, layer stack details, and resist dissolution characteristics.

Overall the simulator should be able to

(a) Predict the resist profile in 3D with moderate CPU resources and
(b) Apply proximity effect correction through tuning of exposure dose (fractioning) and if necessary layout adjustment.

3.2 Simulation Flow Chart

Due to the high cost of electron beam lithography in terms of machine price, running costs, etc., it is necessary to optimize the entire process aiming at highest possible resolution and process latitude to be achieved in short exposure times. This quite challenging goal, especially in the nanoscale regime, can be achieved only through fast, accurate, and user-friendly software tools. Several tools have been developed by academia over the last decades and some of them matured to commercial products, as will be presented later. In all cases, the simulation software is structured into a set of modular components that allows the intermediate use of particular results. This way the modules needing intense CPU power could be executed separately, and the results can be used several times. The first software that offered complete simulation of EBL exposure and development steps was SAMPLE, developed at University of California Berkeley, that initially was written for 2D simulation and later on was extended to 3D topology and lithography simulation. In recent years, this simulation tool is offered in the Web through the LAVA (Lithography Analysis through Virtual Access) platform [1]. A typical simulation sequence is illustrated in Fig. 3.2.

In the first module (Energy Deposition), the energy deposition (Energy Deposition Function, EDF) due to an ideal "point" beam (with negligible beam diameter, a delta function) is calculated. Even though electrons travel in three directions (x, y, z), due to cylindrical symmetry, the calculation can be reduced to 2D (r, z), resulting in the function EDF(r, z). Input for this module is (a) film stack i.e. the composition in terms of density, mean atomic number and mean atomic weight and film thicknesses and (b) beam energy. Interestingly, EDF(r, z) is independent of beam type (Gaussian, Rectangular, Variable Shape) and size as well as writing strategy and the layout to be written. Therefore the output of this module can be used for all subsequent simulations where the stack of films and beam energy are employed and for any layout and beam type/size. It should be noted that the module's accuracy is a dominant parameter for the resist profile prediction accuracy. The EDF(r, z) module accuracy is primarily dependent on the physical models employed for the electron trajectory simulation. In certain cases, feedback from experimental results could considerably improve prediction accuracy. However, in

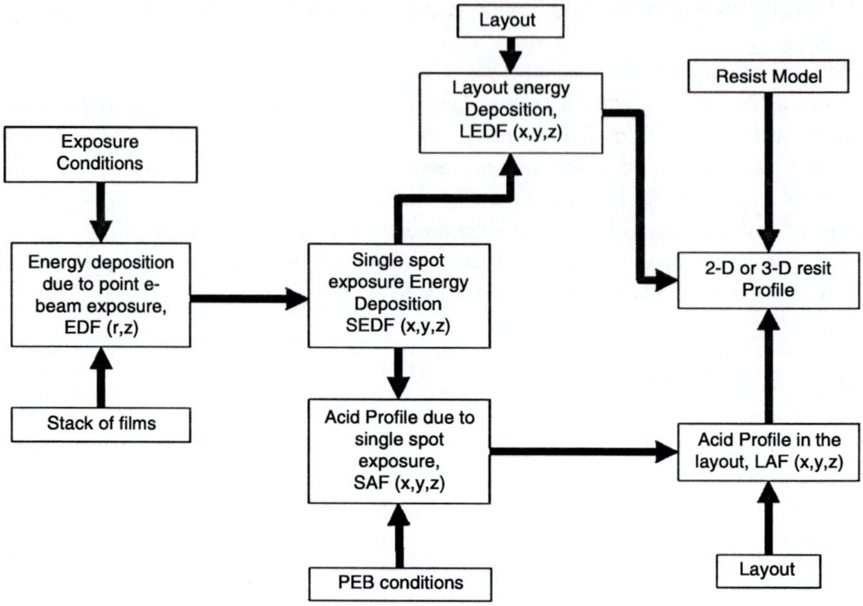

Fig. 3.2 Complete simulation sequence incorporating all lithography steps

this simulation module certain exposure characteristics, such as sample charging and heating are difficult to simulate and for that reason are not taken into account. Such phenomena are considered by subsequent modules.

In the second module (Spot Energy Deposition), the EDF(r, z) is used as input along with beam type and size, and the Spot Energy Deposition Function, SEDF(x, y, z), is calculated. Even though the cylindrical symmetry is applicable for Gaussian beams and SEDF could be expressed in radius and depth, usually it is selected that SEDF to be calculated in Cartesian coordinates to allow for easy coupling with the layout and the next simulation steps. In general, SEDF(x, y, z) calculation is performed through EDF(r, z) convolution with the beam profile, B(r). SEDF(x, y, z) at the resist/substrate interface is the most convenient quantity for comparison with experimental data as was shown in numerous studies e.g. [2, 3]. In this module, the shot noise effect, which is related to the beam quality [4] should be taken into account in order to allow for a reliable transfer to the layout level simulation. However, certain beam shape issues such as resist heating, electron–electron interaction in the beam and repelling electrical field due charging are very difficult to be included. Such phenomena are usually studied with the use of specialized software tools e.g. TEMPTATION [5] from abeam technologies [6].

At this point, the simulation path differentiates depending on the resist class – conventional (e.g. PMMA) or Chemically Amplified (CA) resists. This differentiation comes from the different effect of absorbed energy in conventional and CA resists.

In the Layout Energy Deposition module and for the conventional resist case, SEDF(x, y, z) is used as input along with the layout and finally the Layout Energy Deposition Function, LEDF(x, y, z), is calculated, i.e. the energy absorbed in every pixel in 3D. In order to accomplish this task, SEDF(x, y, z) and LEDF(x, y, z) grids are matched by taking into account the nominal pattern size in the layout. This way LEDF(x, y, z) is totally layout dependent, i.e. for 0.2 μm dense lines and 0.1 μm isolated rectangles layouts, the same SEDF(x, y, z) can be used if the stack of films and beam energy/shape are identical. This module consumes huge CPU power especially for fine grids and complex layouts.

The next step in the case of conventional resists is the resist dissolution simulation. This step is performed in 2D or preferably in 3D, by taking into account LEDF (x, y, z) and resist dissolution rate R(E), where E is the local absorbed energy. By applying this procedure, excellent simulation results have been obtained for complex layouts with critical dimension in the nanoscale and with very good agreement with experimental ones [7]. This simulation module is of critical importance for the final resist profile. Thus, very accurate models on resist behaviors are necessary which is very difficult to obtain and correlate with absorbed energy.

In the case of Chemically Amplified (CA) resists, processing is more complicated. In the CA resist processing, acids released by the exposure radiation diffuse during the Post Exposure Bake (PEB) step. These acids render surrounding polymer soluble in developer. A single acid molecule can catalyze many such "deprotections"; hence, fewer photons or electrons are needed. Acid diffusion is important not only to increase photoresist sensitivity and throughput, but also to limit Line Edge Roughness due to shot noise statistics. However, the acid diffusion length is itself a potential resolution limiter. In addition, too much diffusion reduces chemical contrast, leading again to more roughness. During the e-beam processing of commercial CA resists, the following reaction [8] takes place:

$$e^- + \text{photoacid generator} \rightarrow e^- + \text{acid cation} + \text{sulfonate anion}$$

The e^- represents an electron that may react with other constituents of the solution. It typically travels a distance on the order of many nanometers before being contained. This parasitic exposure would degrade the resolution of the photoresist.

Due to this extra and at the same time critical processing step, simulation also is more complicated and as in the case of the resist dissolution module is very difficult to measure experimentally. On the other hand this step affects dramatically overall accuracy. The PEB effect on pattern dimensions and resist profile is either experimentally calculated [9] or simulated by solving a two non-linear equation system for the chemical reaction and the acid diffusion [10]. The solution of the system requires knowledge of parameters related to the chemical reaction (activation energy, reaction order) and the diffusion coefficient, e.g. [11]. These parameters are obtained by various experimental methodologies including post-lithographic measurements (single pixels [3], lines) and/or physicochemical and optical measurements. In order to decrease the CPU power needed, the acid concentration calculation is done on SEDF(x, y, z) rather than LEDF(x, y, z). Then, by taking into account the layout of interest, the total acid concentration in every pixel of the grid is calculated, LAF(x, y, z). This function later on is used in the resist development module along with the development model to predict the actual 2D or 3D resist profile.

3.3 E-Beam Exposure Simulation Module

One of the critical modules for the accurate prediction of the final output of the electron beam lithography process is the simulation of the electron beam exposure step. During the exposure step, the focused electron beam interacts with the sample; both with the resist film but also with the layers underneath, if any, and the substrate, Fig. 3.3. During this step the interaction volume of electrons with matter is computed which significantly defines the final feature dimensions.

Due to the significance of electron beam – matter interaction in numerous applications, this issue has been the subject of international research from the early seventies, when the first electron beam tools were developed. At the same time similar electron beam energies are used also in Scanning Electron Microscopes (SEM), thus electron beam-matter interaction simulation finds application in this technology also.

The electron beam – sample interaction is the core module for all e-beam lithography simulators. In this module the energy deposited in the resist film is calculated by taking into account all the necessary information about the sample and the actual beam to be used. The electron beam can be of various shapes with

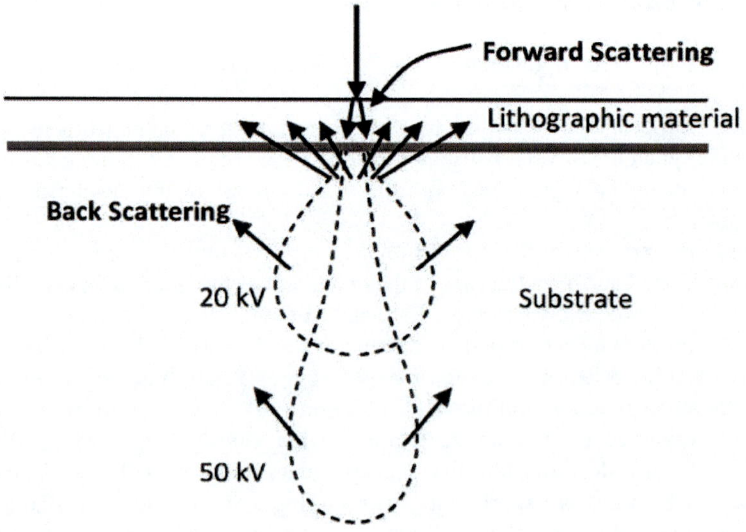

Fig. 3.3 Electron beam – matter interaction cross section

Gaussian beams used most for R&D purposes while shaped beams are the choice for production. All beams can be simulated as the convolution of a point beam with the actual beam shape [12] i.e. the electron beam interaction with the sample is reduced to the calculation of the interaction of the point beam with the sample. Then the 2D or 3D energy profile is convolved with the actual beam profile and all beam effects, such as noise, FWHM are taken into account in detail.

The sample in electron beam lithography consists of the resist film on the substrate. In the simplest case, the substrate is a silicon wafer; however, there are numerous applications where (a) several films exist between the resist film and the substrate, or (b) the resist film consists of two or even three layers with different dissolution properties. Furthermore, as will be explained later, in certain cases the substrate is a thin film (membrane) in order to improve further the patterning resolution.

3.3.1 Modeling Approaches

Numerous articles have been published during the last decades targeting the simulation of the point beam interaction with the substrate. In the vast majority of these publications, the sample consists of PMMA film deposited on bulk substrate and the beam energy is in the 10–50 keV range e.g. [13–22]. This trend can be explained by the fact that PMMA was the material of choice for high resolution patterning for several decades and the beam energy range was the one offered by most of the available tools. During the last two decades, simulation approaches covering beam energies of 100 keV or higher and multilayer substrates were developed, e.g. [7, 23–26].

The simulation approaches that have been developed for the modeling of the energy dissipation during the electron trajectory are based on two principles: (a) Monte Carlo and (b) analytical. The vast majority of the simulation tools are based on Monte Carlo methods and only a small percentage are based on analytical methods.

3.3.1.1 Monte Carlo Method

Monte Carlo methodology was the first approach adopted for the modeling of the electron path in matter. The force that dominates in the electron trajectory is the Coulomb force between the electron and the particles in the sample. The electron deflects due to elastic scattering with nuclei and loses energy due to numerous mechanisms such as ionization, Compton scattering, bremsstrahlung, etc. [27].

In most Monte Carlo methodologies, electron trajectories are handled by applying the following two assumptions [28]:

(a) Electron's trajectory is defined by elastic scattering events only (scattering angles in the 5–180° range) since inelastic scattering events cause deflections at much smaller angles (<2°).
(b) Electron's energy loss is approximated by utilizing the Continuous Slowing Down Approximation (CSDA). In this approximation, the energy lost per unit

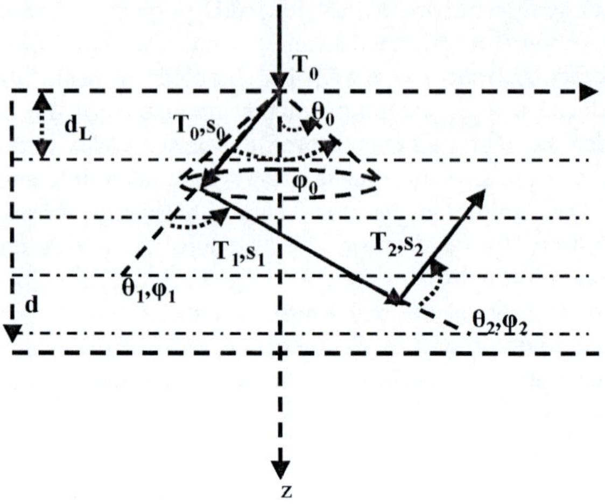

Fig. 3.4 Coordinate system used in the electron track simulation. Between successive scattering events, energy loss is calculated from the continuous slowing down approximation

path length (dT/ds) is a function of the energy and the atomic number and the angular deflection and can be calculated by applying a differential scattering cross section.

The core element of the Monte Carlo methodology is the description of a scattering event. The coordinate system used in the electron track scattering simulation is shown in Fig. 3.4. Initially, the electron's direction is described by two angles (θ_n, φ_n) and the electron has kinetic energy of T_n. When an elastic event occurs, the electron is scattered to new angles (θ_{n+1}, φ_{n+1}) and the electron continues through the target, ionizing atoms and losing energy until it undergoes a new elastic scattering event. The distance between two elastic scattering events can be chosen to be the mean free path length (λ) or by an exponential distribution of path lengths:

$$s = -\lambda \ln(\mathrm{RND}) \tag{3.1}$$

where RND is a random number in the [0–1] range. The mean free path length (λ) is given by:

$$\lambda = \frac{1.02\beta(1+\beta)AT^2}{Z(Z+1)\rho}(\mu m) \tag{3.2}$$

where Z is the atomic number, A the atomic mass number, and ρ the density. Bethe [29] has derived the following expression, which gives the kinetic energy lost by a nonrelativistic electron as it travels a path of length ds in matter

$$\frac{dT}{ds} = -7.83\left(\frac{\rho Z}{AT}\right)\ln\left(\frac{174T}{Z}\right)\left(\frac{keV}{\mu m}\right) \tag{3.3}$$

where *T(keV)* is the electronic kinetic energy, $\rho(g/cm^3)$ the density of the target material, *A(g)* the atomic weight of the target and *Z* the atomic number of the target. The penetrating electrons are subject to a screened Coulomb potential

$$V(r) = \frac{Zq^2}{r} e^{-r/a}, \quad a = \frac{a_0}{Z^{1/3}} \tag{3.4}$$

where q is the electron charge, r the distance between the colliding electron and the nucleus, and α_0 the Bohr radius of hydrogen. The α parameter in the exponential factor approximately represents the screening of the nucleus by the orbital electrons. This leads to the following shielded Rutherford cross-section

$$\frac{d\sigma}{d\Omega} = \frac{Z(Z+1)e^4}{p^2 \upsilon^2} \frac{1}{(1-\cos\theta + 2\beta)^2}, \quad \beta = 0.25\left(\frac{1.12Z^{1/3}\hbar}{0.885pa_0}\right)^2 \tag{3.5}$$

where $p = m\ \upsilon$ is the electron momentum, and angle θ represents the scattering angle. The total screened cross-section, σ_T is found by integrating the above expression over all solid angles.

An electron is followed in its trajectory into the solid target until its energy becomes lower than the mean ionization energy J calculated in eV by:

$$J = \left(9.76 + \frac{58.8}{Z^{1.19}}\right) Z \tag{3.6}$$

The expressions given so far are the fundamental ones for the Monte Carlo calculation. A typical flow diagram of a Monte Carlo simulator is shown in Fig. 3.5.

The calculated electron trajectories in the case of 20 keV electrons incident on 200 nm PMMA film over bulk Si are shown in Fig. 3.6. A number of electrons undergo either an elastic scattering event of large angle or a series of scattering events of smaller angles and at the end are backscattered and re-expose the resist film. The percentage of backscattered electrons is a typical criterion for the Monte Carlo algorithm accuracy. In Fig. 3.7, the calculated backscattering coefficient is compared with experimental data [30] for two cases: (a) for different substrates covering a wide range of atomic numbers, and (b) for films of Ag and Mo and for a wide range of film thickness. In all these cases, simulated backscattering coefficient values are in very good agreement with experimental ones which is clear indication of the particular Monte Carlo accuracy.

In general, the major disadvantage of the Monte Carlo approach is that it requires much CPU time in order to obtain a result with sufficiently small statistical fluctuations (typically 5,000–100,000 trajectories are used). In order to overcome this drawback, special tricks are used (e.g. varying cell dimensions). This problem becomes more intense in the case of multilayer substrates where a lot of boundary conditions are involved and in surfaces with topographic discontinuities.

Fig. 3.5 Monte Carlo flow diagram

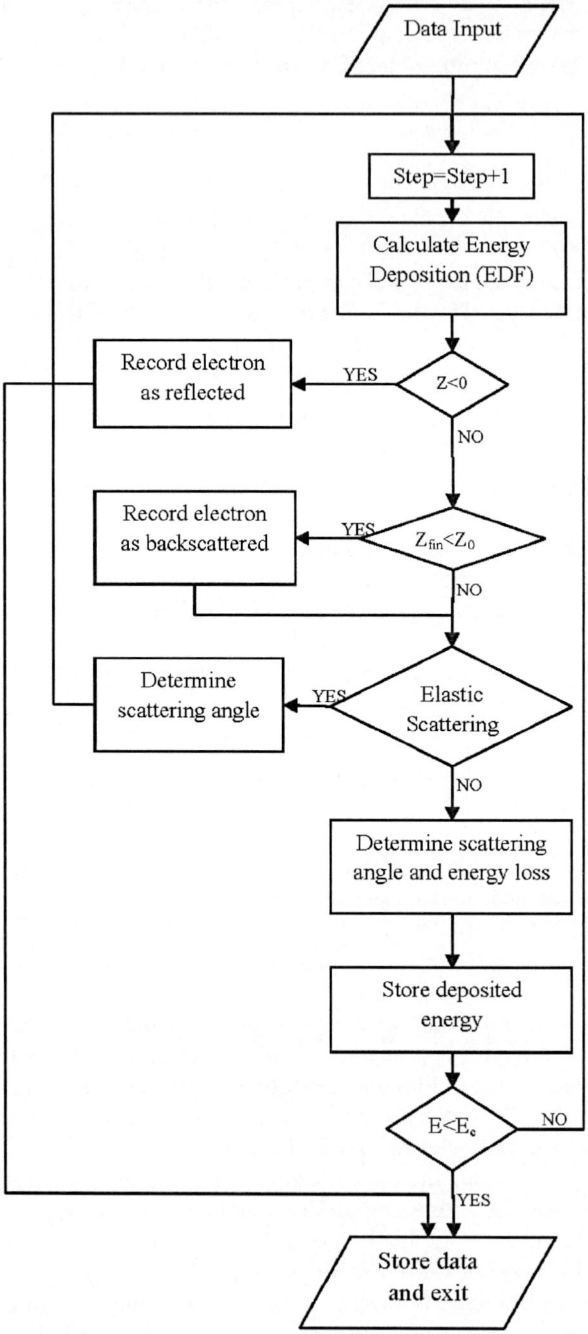

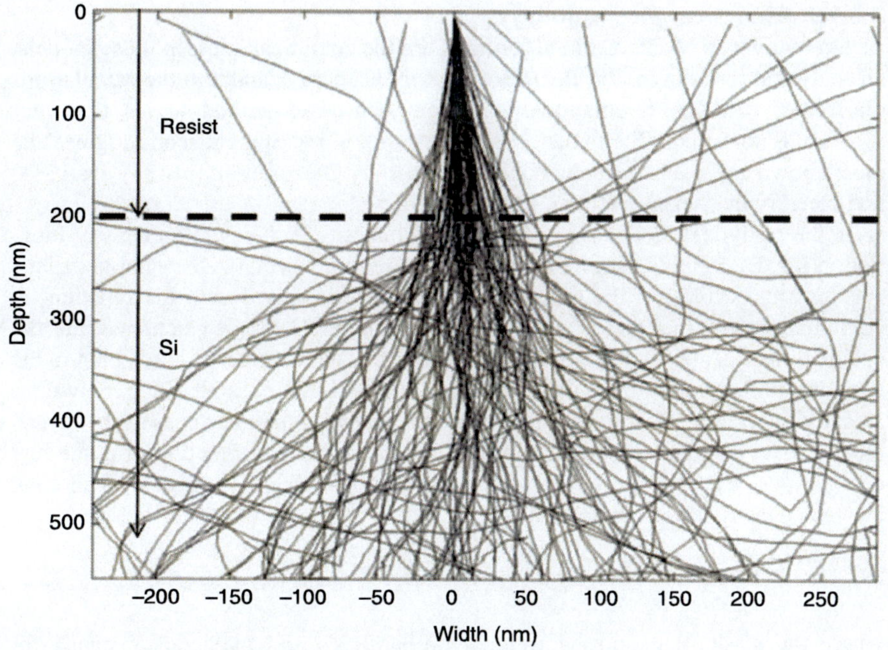

Fig. 3.6 Electron trajectories in 200 nm PMMA on bulk Si substrate

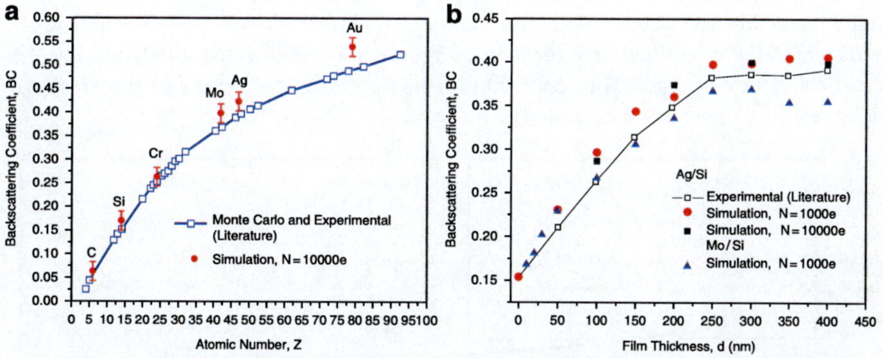

Fig. 3.7 (**a**) Electron backscattering coefficient vs. atomic number of bulk target material. Points connected by line are both experimental and Monte Carlo data [28], for $T_0 = 10$ keV. Simulations with $N = 10^5$ and 10^4 electrons show the agreement with the published data. Backscattering coefficient as a function of (**a**) atomic number of the target, (**b**) electron backscattering coefficient vs. increasing Ag and Mo film thickness on top of Si bulk material. Points connected by line are experimental data for $T_0 = 20$ keV. Simulations with $N = 10^5$ and even 10^3 electron tracks, show the agreement with the published data (*squares* and *circles*)

3.3.1.2 Analytical Methodology

As alternatives to Monte Carlo algorithms are the analytical ones. In those models, an analytical formula for the electron trajectory is applied, and thus the calculations are free of statistical fluctuations. However, with those methodologies, it is very difficult, if not impossible to handle surfaces with topographic discontinuities. The most known approaches are the ones developed by Glezos et al. [31–33], Paul [34] and recently by Stepanova et al. [22, 35].

In the analytical models, the fundamental quantity is the electron density function, ρ(r, z, E). In this function, z denotes the direction normal to the sample surface and inwards, while r is the radial distance from this axis. From this function, in combination with the Bethe energy loss function, the EDF(r) is calculated. For the calculation of ρ(r, z, E), the Boltzmann transport equation must be solved but in the case of multilayer substrates the analytical solution is not possible since it involves an additional set of complicated boundary conditions at the interfaces of the layers. Due to these problems, the electron density calculated as a function of depth and energy, ρ(z, E), and the lateral distribution is calculated separately. The total electron density is written as:

$$\rho(r,z,E) = \rho(z,E)[\rho_f(r|z,E) + \rho_{bd}(r|z,E)+] + \rho_{bs}(r,z,E) \qquad (3.7)$$

where the slash (|) denotes correlated probabilities and the index *f* stands for forward scattered, *bd* for backscattered due to a series of small angle scattering events and *bs* for backscattered due to large angle scattering events.

Therefore, the energy deposition within the resist film consists of I_f, I_{bd}, and I_{bs}, where the indices have the same meaning as in the electron density equation. In Fig. 3.8a, these partial contributions are shown in the case of 0.5 μm PMMA resist over bulk Si substrate for 50 keV and in Fig. 3.8b for 10 keV. These two energies have been selected due to their importance for mask making (10 keV) and direct writing (50 keV) applications. From both figures, it is obvious that the I_f part has the smallest lateral spread. This contribution has a Gaussian form and the standard

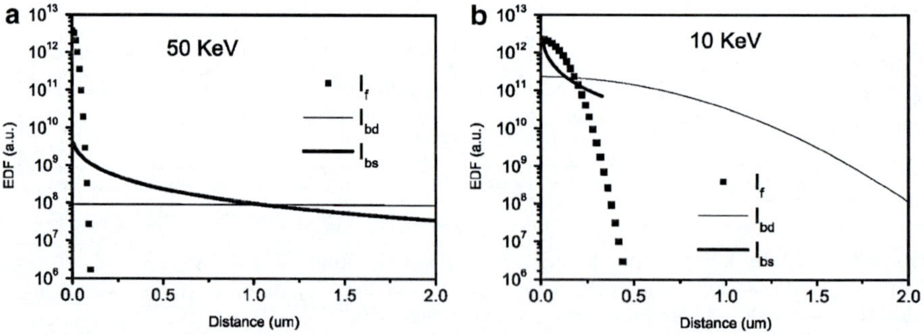

Fig. 3.8 Partial energy deposition contributions in the case of 0.4 μm PMMA over bulk Si substrates. In both cases, the Gaussian form of I_{bd}, I_f is apparent. Exposure energies shown are (**a**) 50 keV and (**b**) 10 keV [7]

deviation is the well-known proximity parameter α. On the other hand I_{bd} has very wide lateral spread and its standard deviation is parameter-β. The third contribution (I_{bs}) cannot be fitted with a Gaussian form, but it is possible to fit it with a simple function.

It should be noted that very few publications have combined the study of both exposure and resist dissolution. Only very recently, Stepanova and coworkers [22] have published a work on the simulation of exposure and PMMA development suitable for patterning simulation in the deep nano-size regime. In that work, the exposure model employs kinetic transport theory to determine the distribution of forward, back scattering and secondary electrons. First, generation and transport of secondary, and higher electrons, produced by a point beam with energy E_P moving along a given direction, is described by the Boltzmann transport equation. Iterative numerical solution of the Boltzmann equation produces a distribution function of secondary electrons moving with energy E at a distance ρ from the primary beam, $f_S(E_p, \rho, E)$. The entire model system comprising both the primary beam and the secondary electrons is given by the distribution function,

$$f(E_n, \rho, E) = f_n \delta(\rho)\delta(E - E_P) + f_S(E_n, \rho, E) \tag{3.8}$$

where f_p is a flux constant, and δ is the Dirac delta function. The last equation is further employed to compute the corresponding rate of scissions of the C-C bonds in the main chain in PMMA,

$$Y(E_p, \rho) = \int f(E_P, \rho, E) v\, \mu^{tot}_{c-c}(E) dE \tag{3.9}$$

where v is the electron velocity and μ^{tot}_{c-c} is the total cross-section of inelastic collisions with valence electrons involved in backbone C-C bonds in PMMA.

The propagation of primary electrons has been factorized in such a way that inelastic collisions decrease the energy of primary electrons in accordance with the stopping power form, whereas elastic collisions change only the direction of motion of the electrons, but not their energy. Broadening of the primary electron beam is described through the classic diffusion approximation [33, 34]. For a point beam traveling a distance z, the lateral broadening is given by

$$P_P(\rho, z)\rho d\rho = \frac{3\lambda}{(z_{\max} - z)^3} \exp\left(-\frac{3\lambda\rho^2}{2(z_{\max} - z)^3}\right)\rho d\rho \tag{3.10}$$

where z is depth ($z = 0$ corresponds to the bottom of the resist interfacing the substrate), z_{max} is the thickness of the resist, and λ is the depth dependent elastic transport mean free path. The resulting depth distribution $P_P(\rho, z)$ is then convolved with the function $w(E_p, \rho)$, which produces a depth dependent radial distribution of the yield of scissions in a planar layer of PMMA exposed to a point electron beam, $w_P(\rho, z)$. For the distribution of backscattered electrons over the emission angle θ, the dependence cos(θ), has been employed which is in reasonable agreement with numerical and experimental results.

Adding together the local yields of scissions generated by forward and backscattered electrons, $w_P(\rho, z)$ and $w_B(\rho, z)$, provides the total depth dependent radial distribution of the scissions of the main chain in a planar layer of PMMA exposed by a point beam, $w(\rho, z) = w_P(\rho, z) + w_B(\rho, z)$. This distribution replaces the conventional energy deposition function (EDF). The corresponding cross-sections are intimately related to the molecular mechanisms involved, allowing for clearer interpretations and more direct paths toward further improvement of the model. Also the usage of the kinetic approaches favorably affects the efficiency of the approach in comparison with direct Monte-Carlo simulations, particularly when computing the contribution from backscattered electrons, at the same time allowing for nano-scale resolution for the contribution from the forward beam.

3.3.2 Energy Deposition Function

Both simulation methodologies are capable of calculating the energy deposition within the resist film. Due to the cylindrical symmetry of the electron trajectory, the energy deposition can be expressed as EDF(r, z) instead of EDF(x, y, z), where r is the radial distance from the point of incidence and z the depth in the resist film. EDF (r, z) depends on numerous parameters such as electron energy, atomic number, atomic mass and density of all layers in the sample. In Fig. 3.9, the energy deposited as a function of radial distance for three different energies is illustrated when the sample consists of a thin PMMA film on bulk Si wafer. Clearly the energy deposited is higher in the short range for lower e-beam energy. This increased energy

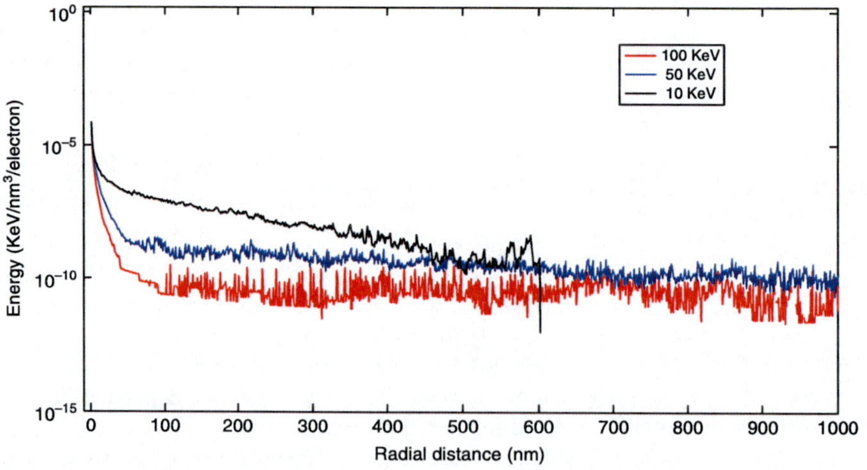

Fig. 3.9 EDF(r) at the resist/substrate interface for 10, 50 and 100 keV energies. Substrate considered was bulk Si

deposition is a killing factor for high resolution dense patterns and the application of proximity effect corrections algorithms is necessary. On the other hand, energy deposited is lower at long range for the low e-beam energies. In the case of 100 keV the backscattering contribution remains almost flat even for few tens of micrometers range. This uniform distribution of backscattered contribution allows for easier realization of dense high resolution features at the expense of higher exposure-longer exposure times needed. In Fig. 3.10 the parameter α is presented with respect to resist depth for 10 and 50 keV. Clearly, forward scattering is higher at 10 keV as well as its dependence on resist thickness. Therefore, for e-beam exposure with low energies, the resist thickness should be also small to avoid any problems with sloped resist profiles.

In Fig. 3.11, a comparison of simulated energy deposition with experimental results is presented. The experimental data are from the Single Pixel Exposure (SPE) method on PMMA resist over a Si substrate. The resulting pattern from a single pixel exposure is a hole (positive resist case) and its diameter depends strongly on the exposure dose (high dose results in large hole diameter, low dose results in a small hole diameter). The differences observed between these two curves are very small, despite the large range examined (0.1–8.0 μm). In order to perform this and the following comparisons with the experimental results, the beam

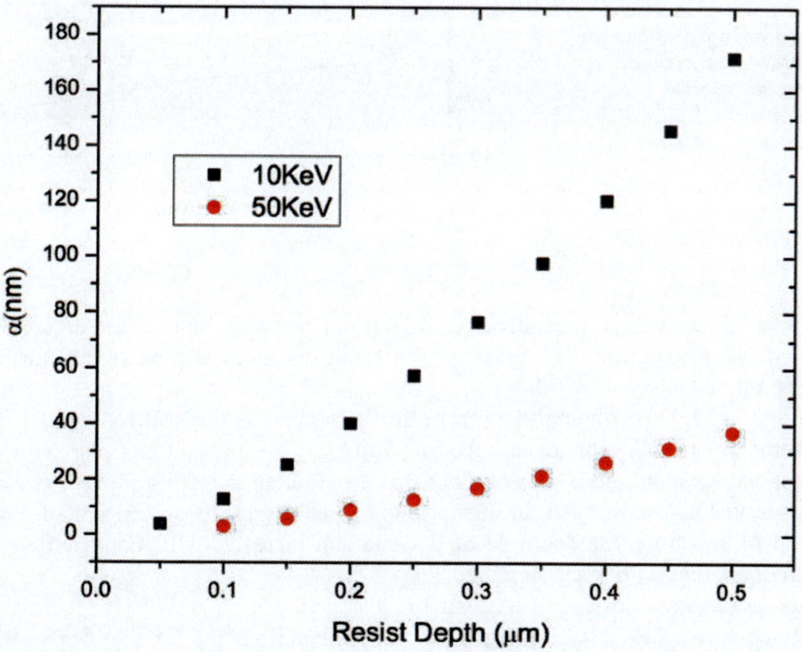

Fig. 3.10 Forward scattered standard deviation dependence on resist depth for 10 and 50 keV exposure. The resist material is 0.5 μm PMMA. Both series of simulation data could be fitted with a power function

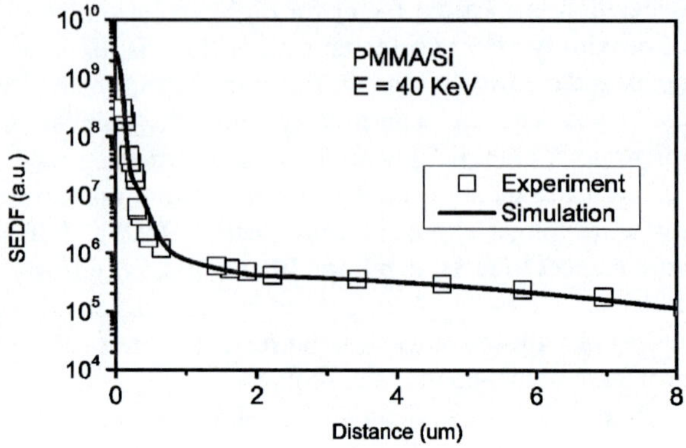

Fig. 3.11 Comparison of simulation results with experimental data for PMMA resist on bulk Si (single pixel exposure). Experimental data cover the range of radii 0.1–8.0 μm [7]

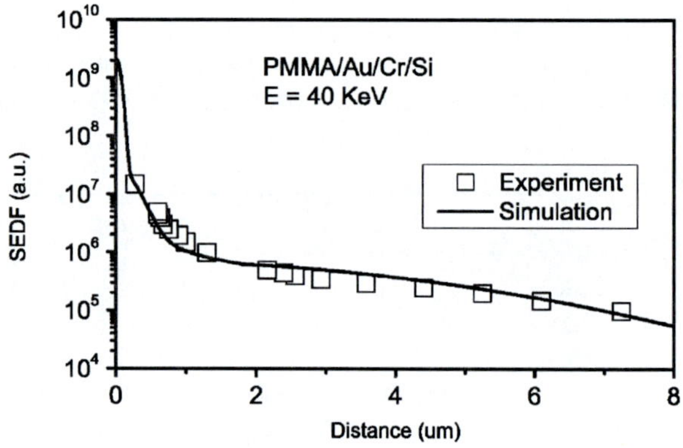

Fig. 3.12 Comparison of experimental and simulated EDF(r) for 0.1 μm Au/0.04 μm Cr/Si. The difference is very small in the 0.2–8.0 μm range. Measurement of smaller patterns was not possible due to the flow from excess heating during SEM measurement [7]

diameter has been taken into account (SEDF(r)). Otherwise, since beam diameter is comparable with α, the differences in the small distances regime (<0.5 μm) are significant.

In Fig. 3.12, experimental data from SPE experiments for PMMA over 0.1 μm Au film/0.04 μm Cr film over bulk Si (Au/Cr/Si) are presented along with the corresponding simulations. It is obvious that the simulation results are very close to the experimental ones. Also, in this case the SEDF(r) results are presented. From Figs. 3.11 and 3.12, the decrease of β value and increase of I_{bs} contribution are apparent for the multilayer case.

SPE tests have been applied also for a negative high resolution chemically amplified resist, EPR [36]. In Fig. 3.13, experimental results for thin films of Au and Ag on bulk Si are presented. Since EPR is a chemically amplified resist, the

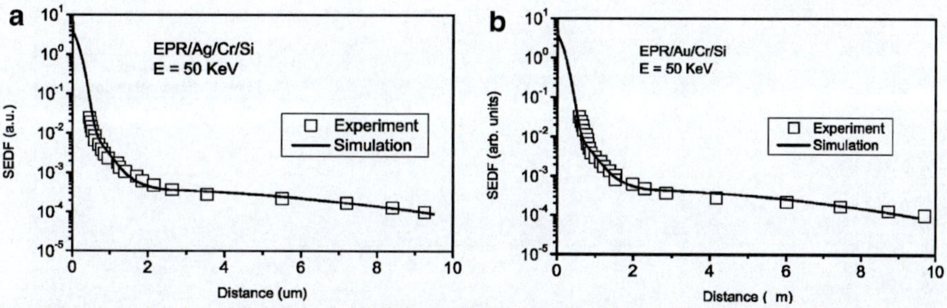

Fig. 3.13 Single Pixel Exposures on 0.5 μm EPR (negative CAR). Comparison of simulation and experiment where in the case of simulation results, PEB effect has been taken into account. **(a)** Substrate was 0.18 μm Ag over 0.04 μm Cr on bulk Si, **(b)** substrate was 0.10 μm Au over 0.04 μm Cr on bulk Si

acid diffusion effect that occurs during PEB (Post Exposure Bake) must be taken into account. From Fig. 3.13, it is clear that the simulation results are again in very good agreement with the corresponding experimental ones. Specifically, the simulated forward parameter α (since EPR is CAR, α contains the acid diffusion effect) is almost identical to the experimental one. Additionally, very good agreement exists for the large distance part where backscattering due to a series of small angle scattering events dominates. In this range, acid diffusion is very small due to the low level of acid concentration and the small space gradient.

In general, the EDF(r) for the Si substrate can be approximated very well with a sum of Gaussians and in most cases two or three Gaussians are enough. However, when the substrate is either bulk from a heavier material, or if a multilayer stack exists between the resist and the substrate, then a sum of Gaussians is not enough, and other functions such as exponentials should be also involved, e.g. [17].

3.3.3 Beam Profile Effect

Since a real beam is not a point beam, the Beam intensity is *B(r)* if the beam is Gaussian and B(x, y) for other beam shapes. In order to calculate the Spot Energy Deposition Function, *SEDF(x, y, z)*, the *EDF(r, z)* should be convolved with *B(r)*:

$$SEDF(x, y, z) = B(r) \otimes EDF(r, z) \tag{3.11}$$

In most cases, this operation is performed through a Fourier transform.

However, *B(r)* should describe the real beam intensity, i.e. the statistical variations that become very critical especially when low beam currents and fast resists are used. For that reason, for optimum results *B(r)* is calculated for each time conditions (beam current, exposure dose = number of electrons) by applying a random distribution of electron intensity that of course follows the Gaussian profile of the actual FWHM (Full Width Half Maximum). In Fig. 3.14, an ideal Gaussian

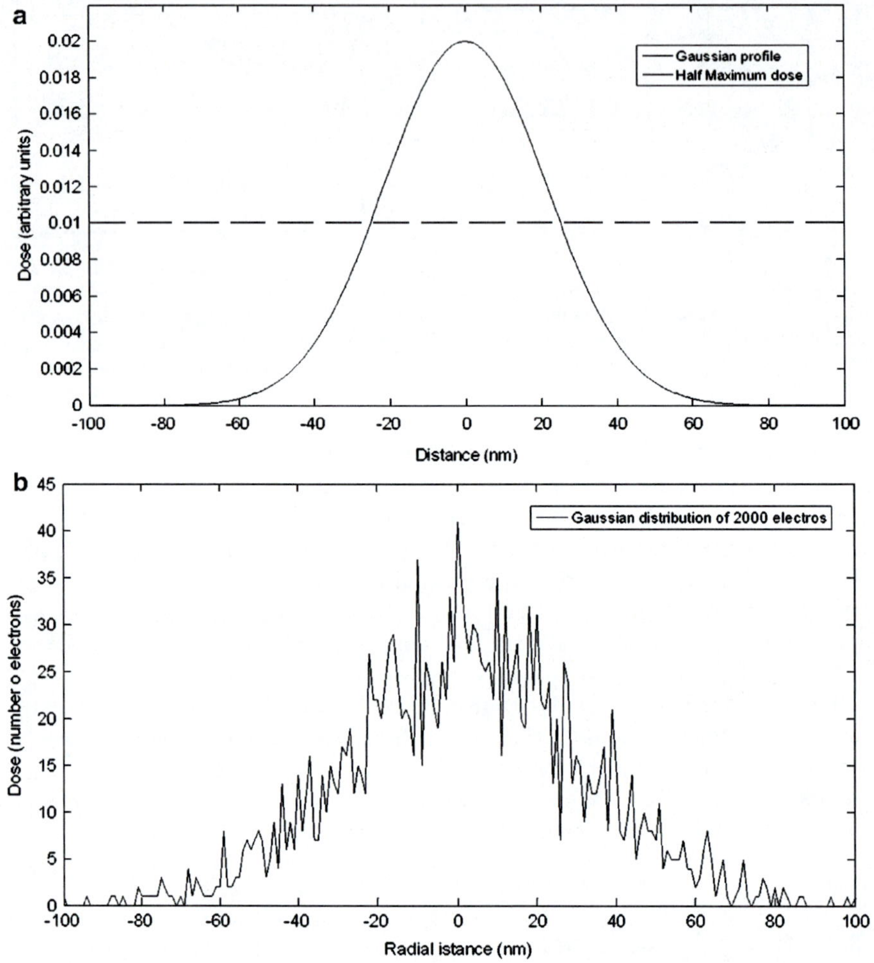

Fig. 3.14 (**a**) Ideal Gaussian beam with FWHM = 50 nm and (**b**) statistical representation of a Gaussian beam with FWHM and 2,000 electrons (12.8 μC/cm^2)

beam of FWHM = 50 nm and the corresponding one for the same FWHM and for 2,000 electrons is shown.

3.3.4 Layout Simulation

For the calculation of *LEDF(x, y, z)*, i.e. Layout Energy Deposition Function, the SEDF(x, y, z) and the layout is used. For optimum usability, standard formats such as CIF or GDS should be supported. In Fig. 3.15a, a simple layout of three

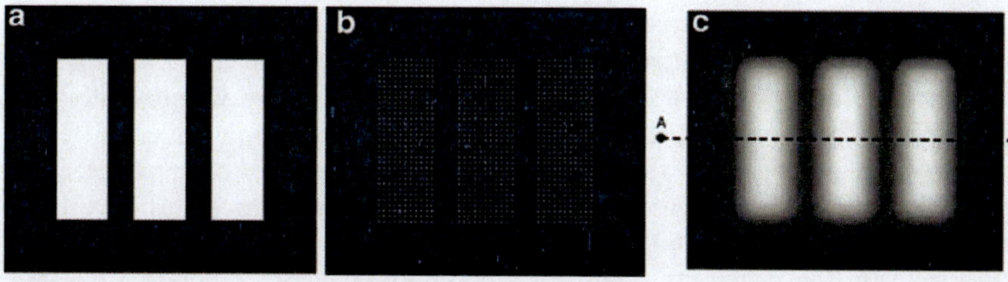

Fig. 3.15 (a) Layout consisting of three rectangles, (b) exposure shots, step =10 nm, (c) layout energy deposition at the resist/substrate interface for 100 keV

rectangles with 100 nm width, 150 nm pitch and 300 nm heights are shown. For the writing of this particular layout, the e-beam steps in both directions at a beam step s (s = 10 nm for the present study), Fig. 3.15b. Therefore, for the *LEDF* caclulation, the *SEDF* should be convolved with *L(x, y)* which is a logic function with logic value of 1 for all points representing a beam spot and 0 for all other points. i.e.

$$LEDF(x, y, z) = SEDF(x, y, z) \otimes L(x, y) \tag{3.12}$$

This simulation step is the most demanding from a CPU time and memory point of view, since the energy deposition should be calculated for the whole layout and for all resist depths. For that reason, several tricks are applied such as: (a) calculation only at layout regions with patterns of very small dimensions or very high density, (b) variable pixel sizes, (c) increase of pixel size in the resist height axis, or (d) decrease of radial distance of *EDF(r, z)*.

In general, the pixel size defines the resolution of the calculation. Increased pixel size relaxes considerably both CPU time and memory needs but deteriorates calculation accuracy. In Fig. 3.15c, the energy deposition *LEDF(x, y)* at the resist/substrate interface is shown in greyscale. The resist was 100 nm PMMA and electrons of 100 keV were used and the unwanted exposure from backscattered electrons is clearly seen.

3.3.5 Proximity Effect Correction

Proximity Effect (PE) is the unwanted exposure in a pattern area that mainly comes from the backscattered electrons [37]. Actually, PE is one of the major limiting resolution factors of electron beam lithography and is an issue of continuous research. The unwanted exposure appears in two ways: (a) energy deposition variation between patterns, usually named as inter-proximity effect and (b) energy deposition variation inside a pattern i.e. the energy deposited close to the edge is smaller than the energy deposited in the areas close to the center of a pattern, usually named as intra-proximity effect.

Due to the inter-proximity effect, in a smaller pattern the energy deposited is smaller than from a larger pattern, while the energy deposited in an isolated pattern is smaller than the energy deposited for identical dense patterns. Due to both types of proximity effect, the final patterns obtained after resist development are different from the nominal ones. This effect becomes stronger as layouts become denser and pattern dimensions shrink. For that reason, several software packages have been developed and are applied for the realization of patterns.

In Fig. 3.16, the proximity effect for a complex layout incorporating isolated features, dense patterns and large pads is shown. In particular, in Fig. 3.16a the layout as it was designed is shown, while in Fig. 3.16b the energy deposited is illustrated in grey scale. Finally, in Fig. 3.16c the iso-energy contours are plotted. In this last figure both types of proximity effect are clearly seen. In Fig. 3.17 experimental verification for the proximity effect is shown in PMMA resist exposed with 50 keV electrons. Due to the inter-proximity effect, the exposure energy in the areas in the center of the layout (dense lines with 500 nm pitch) is high and PMMA resist is totally dissolved. On the other hand, the deposited energy at the edges of the layout is smaller and fine trenches are resolved.

Due to the high importance of PE in realization of complex layouts with small critical dimension, numerous articles have been published from the mid 70s. The first in-depth study of the phenomenon and its possible solutions was published by M. Parikh at the late 70s [38–40].

In order to correct proximity effects, a solution should be found such that all exposure points receive the same amount of energy deposited, e.g. [41, 42]. This can be done either by changing the layout locally, i.e. change the shape and dimensions of the patterns contributing in PE or by changing the exposure dose in particular areas of each pattern. In the latter case, each pattern is fractioned to smaller patterns, and a different exposure dose is assigned to each of them. In certain cases both methodologies should be applied e.g. [43], and actually as critical dimension decreases and patterns become more dense, this is the preferred choice. It should be noted that in the vast majority of studies, proximity effect is corrected at the resist/substrate interface. However, as it was pointed out earlier in this chapter, the energy deposited function depends on the resist depth and this dependence becomes more severe for thicker resist films. For that reason, 3D Proximity Effect Correction is already studied, e.g. [44]. Due to the large computational needs for PEC, the use of neural networks has been also suggested [45]. In Fig. 3.18, Proximity Effect Correction in the very demanding Photonic Crystals area is illustrated [46].

3.4 Resist Simulation Module

The photopolymer material traditionally was attributed just a few physical parameters related mainly to exposure absorption (e.g. Dill parameters [47]) and various parameters related to its baking behavior (e.g. glass transition temperature). In most commercial simulators it is still considered as bulk and the mesh created to

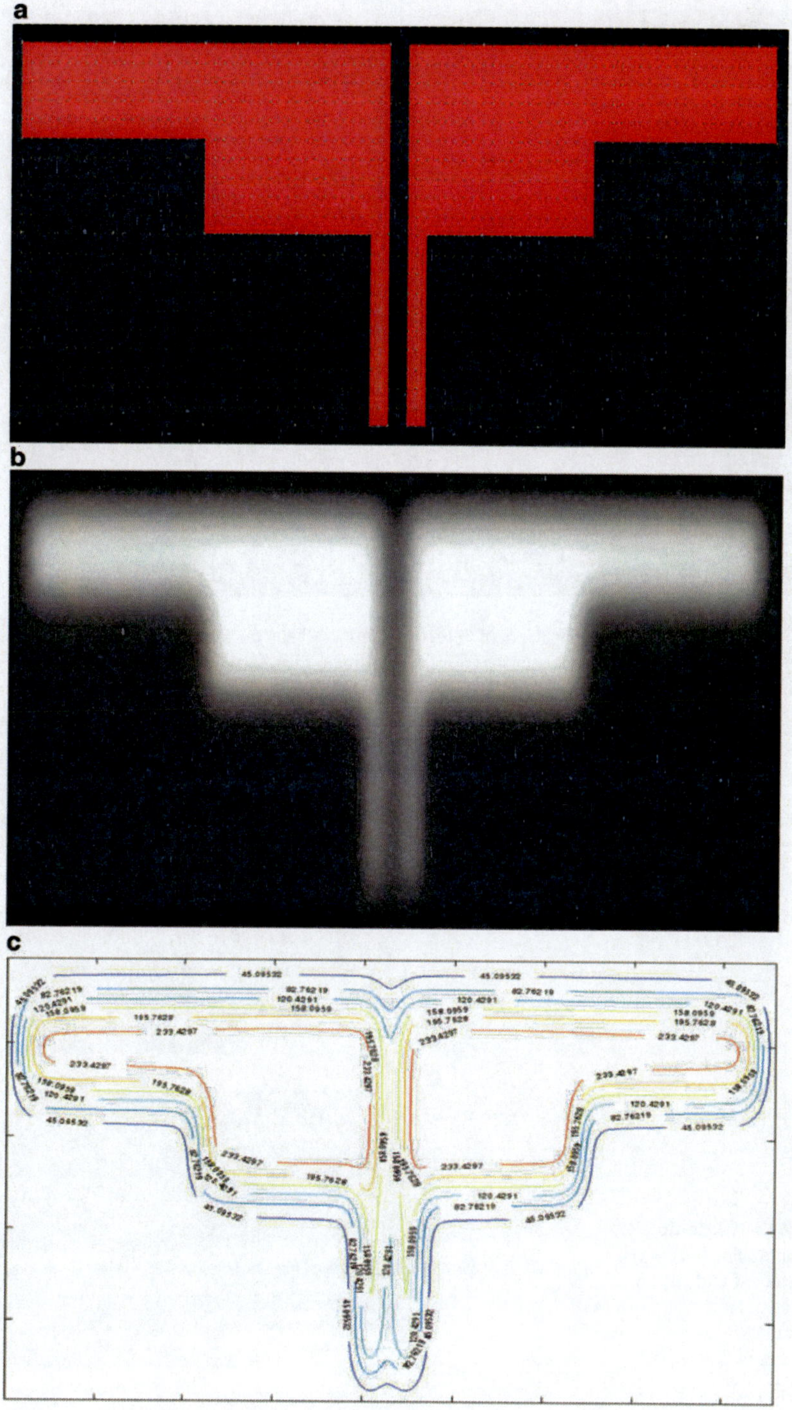

Fig. 3.16 Proximity effect: (**a**) layout consisting of large pads, dense fine patterns and fine patterns close to large pads, (**b**) energy deposited in grey scale (LEDF) at the resist/substrate interface and (**c**) iso-energy deposited contours

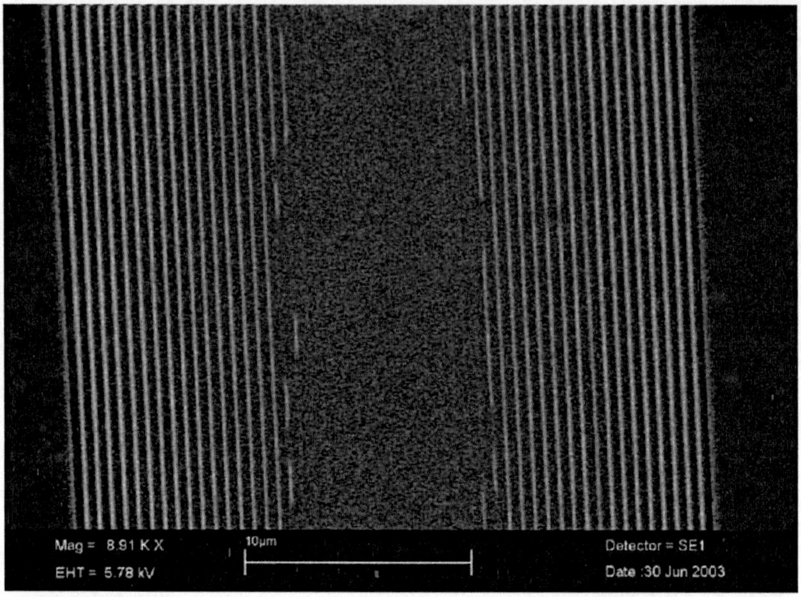

Fig. 3.17 Experimental verification of inter-proximity effect on PMMA resist patterned with 50 keV electrons. The layout consists of dense long lines with 500 nm pitch. Exposure dose was 550 μC/cm^2

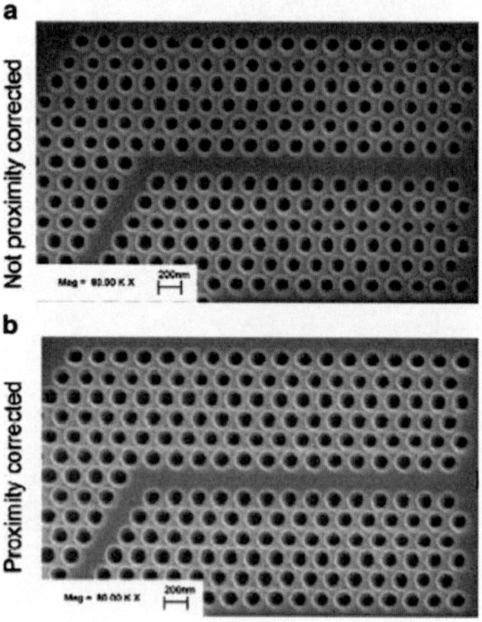

Fig. 3.18 Close-up micrographs of (**a**) uncorrected and (**b**) corrected bend structures written into 220 nm PMMA resist (From Wüest et al. 2003 [46])

solve the analytical models assumes that its composition is the same in each volume element occurring from the discretization process.

Modeling strategies for lithography generally make a fundamental assumption about the physical world being described: the continuum approximation. This approach worked well with existing exposure models which are indeed very advanced and take most of the burden of having a very accurate description, since after all, if the image to be printed in the resist is not optimal, the material certainly will not compensate for the problem. At the same time the corresponding dissolution algorithms performed the process of dissolving the material assuming average values of protected to deprotected sites within each discretized volume element of the computational lattice and through calibration were able to determine the required time step and finally the dissolution rate of the resist. In this way the profile of the printed structure on the resist was obtained, starting from the geometry of the shapes on the mask and the properties of the exposure source.

Even though light energy or electron energy is quantized and chemical concentrations are quantized into spatially-distributed molecules, the physical description of aerial and latent images so far have ignored the discrete nature of these fundamental units and instead use continuous mathematical functions. While in most cases the simulation volumes of interest are large enough in order not to worry about this distinction, when trying to understand lithography at the nanometer length scale, the continuum approximation begins to break down.

Figure 3.19 shows qualitatively the continuum handling of CA resists in (a–d) and the stochastic handling in (e–h). Initially, the deposited energy in the resist due to photons or electrons should be computed. Polymer dissolution during photoresist development is the last lithographic step for pattern formation prior to etching. Analytic models for resist dissolution [48–52] are unable to capture the microscopic details of the resist surface and describe the current problem of resist roughness which is of paramount importance for the patterning technologies of features with critical dimension <65 nm. Much more detail related to resist chemical structure and the processes occurring in the microscopic environment can be modeled in the stochastic or probabilistic approach.

Figure 3.20 shows an example of a simulated resist lattice (Fig. 3.20a), part of this lattice after PEB and development (Fig. 3.20b), and finally a magnification (Fig. 3.20c) of a portion near the edge of the simulated line profile. It is seen that with such modeling, roughness quantification could be taken into consideration. Line Edge Roughness (LER) simulation requires additional consideration of (a) the inherent stochastic nature of reactions during processing, and of (b) the size and shape of the resist polymers. These factors are taken into account by mesoscopic simulation models. Such models currently find applications especially in dissolution modeling [53] and as variants of the percolation dissolution i.e., the aggregate extraction dissolution model [54].

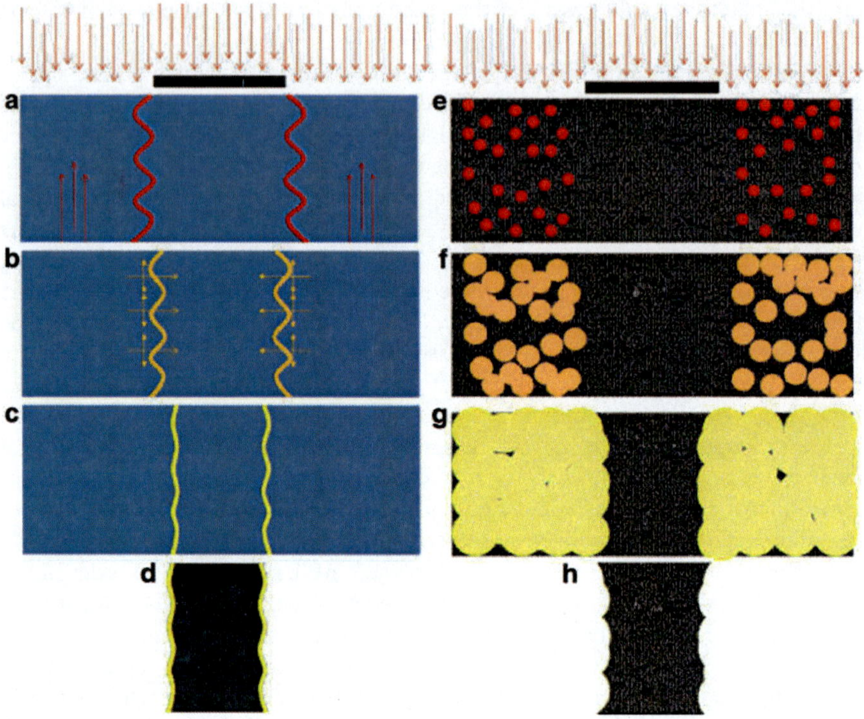

Fig. 3.19 Qualitative depiction of the continuum and stochastic description of the lithography (optical in this case) process of chemically amplified films: (**a**–**d**) continuous modeling, and (**e**–**h**) stochastic resist modeling

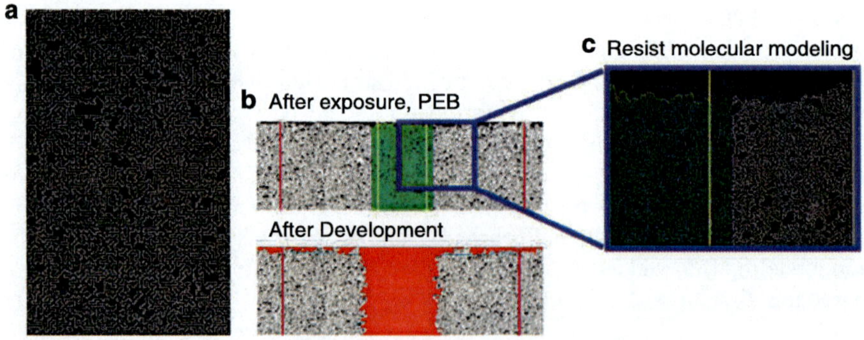

Fig. 3.20 Example of chain lattice and development process. Energy deposition used as a "template". The continuous energy distribution on polymer chains is "discretized" and line-edge roughness occurs

3.4.1 Macroscopic Photoresist Modeling Concepts

3.4.1.1 Exposure Modeling

The modeling of the **photoresist exposure** behavior is associated with great uncertainty because of the extreme complexity of the physical and chemical processes involved. Especially the distinct sensitivity of the resist performance to the actual chemistry and composition makes a predictive simulation very difficult. Often the only possibility to obtain reasonable results within a short time period and with an acceptable effort is to fit simple models to experimental data, and the IC manufacturers keep the fitting parameters strictly confidential.

As was presented earlier, photoresist in EBL is considered as a continuum within which electrons travel and its chemical composition is taken into account through the scattering probability in each scattering event. It is assumed that electrons initiate PAG (PhotoAcid Generator) molecules contained in the resist matrix in order to produce acid species. The reactions governing the bake effects have to be considered simultaneously with the diffusion of the involved species.

3.4.1.2 PEB Modeling

In a PEB model proposed by Ferguson et al. [55], two species are considered, namely **reactive sites** $m(\vec{r};t)$ not consumed by the acid and the **acid concentration** $h(\vec{r};t)$ produced during exposure (concentrations are normalized values). The concentration $m(\vec{r};t)$ is driven by its availability and the acid concentration $h(\vec{r};t)$ raised to a power n, and, secondly, the catalyzing acid $h(\vec{r};t)$ decreases with time according to its concentration because of possible loss mechanism such as neutralization or time-sharing with the deprotected material. These two reactions are completed by a diffusion term for the acid $h(\vec{r};t)$ yielding two coupled partial differential equations

$$\frac{\partial m(\vec{r};t)}{\partial t} = -k_{peb,1} m(\vec{r};t) h^n(\vec{r};t) \tag{3.13}$$

$$\frac{\partial h(\vec{r};t)}{\partial t} = -k_{peb,2} h(\vec{r};t) + \nabla \cdot [D_h(\vec{r};t) \nabla h(\vec{r};t)] \tag{3.14}$$

The reaction order n and the first rate parameter $k_{peb,1}$ characterize the chemical amplification, whereas the second rate parameter $k_{peb,2}$ describes potential loss mechanisms of the catalyst. The diffusion coefficient $D_h(\vec{r};t)$ of the acid depends on the already reacted, consumed sites $x(\vec{r};t) = 1 - m(\vec{r};t)$. Various models have been proposed to describe this dependence, e.g.

$$D_h(\vec{r};t) = D_{h,0} + D_{h,1}[1 - m(\vec{r};t)] \tag{3.15}$$

$$D_h(\vec{r};t) = D_{h,0} \exp[-w_h(1 - m(\vec{r};t))] \tag{3.16}$$

The linear relationship reflects possible use of deprotected sites as stepping stones, and the exponential dependence accounts for free volume effects [56]. All resist parameters typically exhibit an Arrhenius-type temperature behavior. At the beginning of the bake, no sites are activated and the acid concentration is obtained from the preceding exposure simulation.

The following **boundary conditions** are usually imposed:

- The resist/substrate interface can be assumed to be impermeable so that no diffusion occurs across it.
- At the lateral boundaries either periodic or homogeneous Neumann conditions are imposed, i.e., in the latter case a vanishing flux across the lateral boundaries is prescribed.
- The physically most important boundary segment is the resist surface, since acid diffusion into the atmosphere above the wafer occurs. The amount of this acid evaporation is a function of the acid size and the degree of its interaction with the resist polymer. In general, the evaporation is modeled by:

$$\frac{\partial h(\vec{r}_s;t)}{\partial t} = -k_{evap}[h(\vec{r}_s;t) - h_{air}(\vec{r}_s;t)] \tag{3.17}$$

whereby $h_{air}(\vec{r}_s;t)$ is the acid concentration in the atmosphere in the vicinity of the resist surface $\vec{r}_s$. Usually, PEB takes place in a reasonably open environment with enough air flow to eliminate any buildup of evaporated acid above the resist. Thus $h_{air}(\vec{r}_s;t)$ can be neglected. If k_{evap} is very small, then virtually no evaporation takes place. On the other hand, if k_{evap} is very large, the effect is to bring the acid surface concentration in the resist to zero.

3.4.1.3 Development Modeling

The next step in simulation is the modeling of photoresist development. A wide range of dissolution rate models ranging from purely empirical fits to physical based approaches are available with the most popular to be:

Dill's "E"- dissolution model. Along with the **"ABC"-model** describing the exposure/bleaching phenomenon, F. Dill also introduced a dissolution rate model [47]. The dependence of the local development rate $r(m)$ on the normalized local PAC concentration m is thereby given by the three-parameter equation:

$$r(m) = \exp(E_1 + E_2 m + E_3 m^2) \tag{3.18}$$

The three parameters E_1, E_2 and E_3 are obtained using least-square fitting to experimental data. Experience shows that the relationship (3.18) is capable of attaining good agreement with data at high values of the inhibitor concentration m. However, it does not properly characterize rates at low concentrations since it predicts an unphysical rate maximum in this regime. This can usually be ignored since the development rate for small m is so high that it can be considered infinite anyway [57]. Dill's "E"-model was the default model of early versions of the lithography simulator SAMPLE.

Kim's "R"-dissolution model. Kim et al. addressed in [50] the shortcomings of Dill's "E"-model. Two assumptions were made to obtain a more realistic expression to fit experimental rate data. Firstly, the chemical reaction at the resist/developer interface is postulated to be the rate-limiting process, which implies that the mass-transfer dynamics, i.e., the diffusion process, can be neglected. Secondly, the time required to dissolve a differential resist layer is assumed to consist of two additive terms: (a) the time required to dissolve all the PAC in the layer, and (b) the time required to dissolve the photo-produced acid and all other compounds. With these assumptions the resulting rate expression can be written as:

$$r(m) = \left[\frac{1 - me^{-R_3(1-m)}}{R_1} + \frac{me^{-R_3(1-m)}}{R_2}\right]^{-1} \tag{3.19}$$

where R_1 is the dissolution rate of a fully exposed resist ($m = 0$), R_2 of an unexposed resist ($m = 1$), and R_3 is a sensitivity parameter with the interpretation of a rate enhancement due to photoinduced acid.

Mack's "a"-dissolution model. Both models described so far provide mainly fitting formulae with parameters having minor or no physical meaning. Mack proposed a four parameter model based on simple kinetic considerations with the advantage that all involved parameters have physical significance [58]. The approach relies on the assumption that the development process involves two dominating mechanisms, namely the **diffusion of the developer from the bulk solution to the resist surface and the reaction of the developer with the resist.** Modeling the rates of both reactions with fundamental kinetic theory and equating them yields:

$$r(m) = r_{\max}\frac{(a+1)(1-m)^n}{a(1-m)^n} + r_{\min} \quad a = \frac{(n+1)}{(n-1)}(1-m_{th})^n \tag{3.20}$$

Here $r_{\min}$ is the development rate of the unexposed resist, $r_{\max}$ is the rate of the fully exposed resist, n is a selectivity parameter that describes the kinetic order of the dissolution reaction occurring at the resist surface, and m_{th} is a threshold PAC concentration corresponding to the concentration at which the development curve displays an inflection point. It can thus be interpreted as the concentration of a transition between fast and slow development regimes.

3.4.2 Mesoscopic Photoresist Modeling Concepts

The important characteristic in this approach is the detailed model of the resist film. Its molecular structure is considered in much more detail than in macroscopic models. Specifically, the film is considered as a collection of polymer chains or molecular resist molecules, PAG and PAG quencher sites. Each one of these is a distinct entity during all the stages of the simulation process. This way we can incorporate the effects of the resist molecular structure and make correlations of material properties with the measured line-edge roughness.

Molecular models, which incorporate the detailed microscopic processes through Monte Carlo simulations, are necessary for describing photoresist dissolution for LER quantification in stochastic lithography simulation [59, 60]. In this category, the percolation e.g. [61–65], and the critical ionization approaches [66–70] belong.

In order to get to such detail, the simulation flow stores and retrieves data in "lattice-instances", which are created and updated. In the first place, the chain lattice is created, where all polymer chains are placed as self-avoiding and mutually excluded random walks. Then the PAG sites are placed in a separate lattice. The exposure lattice is used as a template to perform the PAG initiation. Reaction/Diffusion of acid species is recorded also on a separate lattice, as well as the developer diffusion and material ionization/deprotection events. All lattice instances are in 1:1 correspondence with each other, and information among them is exchanged easily during the simulation. Figure 3.21 shows the order of creation of each lattice-instance.

The use of such a stochastic simulation framework can be used effectively for *LER* quantification in terms of material and process parameters. In order to get a better view of the whole simulation, 2D results are shown in Fig. 3.22. The layout (Fig. 3.22a) (without dimensional details) is convolved with an e-beam exposure module [71] and the deposited energy (Fig. 3.22b) is used to initiate PAG sites (Fig. 3.22c). During diffusion, the resist becomes deprotected in the exposed region, and finally the developer removes that portion of the resist film (Fig. 3.22d).

3.4.2.1 Film Representation in Stochastic Models

The photoresist in mesoscopic models is spatially discretized using a regular, cubic grid. Mesoscopic simulation requires the description of the local concentration of each photoresist component by the discrete number of molecules in each cell. Additional information required for changing from a continuous to discrete description is the absolute number of molecules corresponding to the relative concentration values in the macroscopic models. The number of molecules for each photoresist component is then randomly distributed on the grid, since no concentration gradients exist prior to exposure. The size of the photoresist components also has

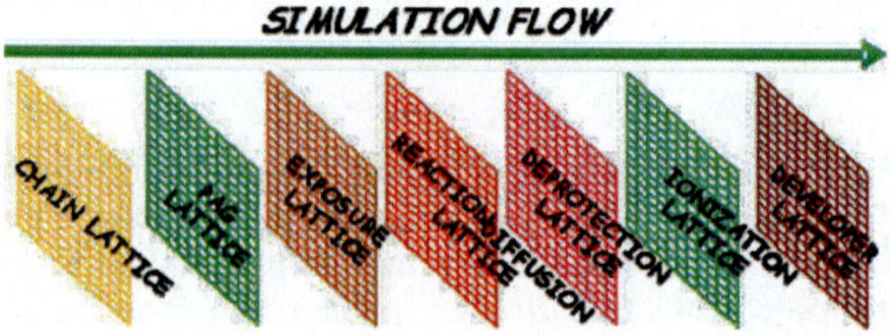

Fig. 3.21 The simulation flow stores and retrieves data in "lattice-instances", which are created and updated throughout the lithography simulation process

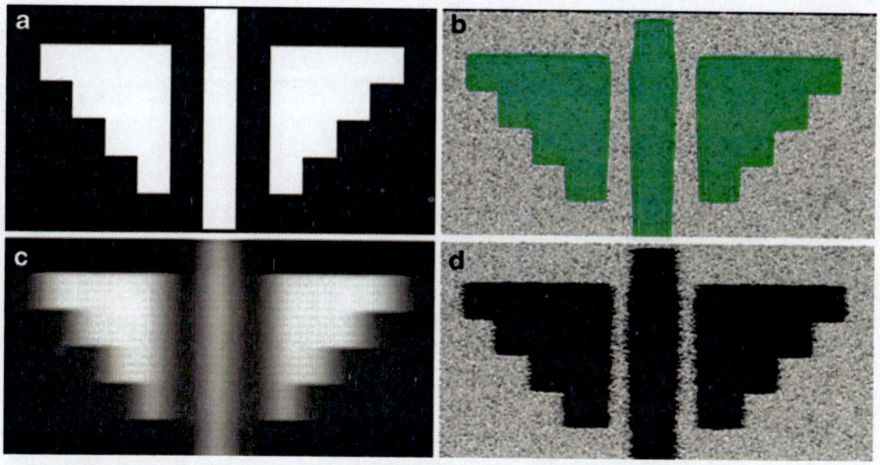

Fig. 3.22 (**a**) Drawn layout, (**b**) E-beam energy deposition, (**c**) deprotected sites, (**d**) final resist profile after dissolution

to be taken into account. Given a radius of gyration of up to 3–5 nm for typical photoresist polymer chains, individual molecules can have a significant effect on the structure of the final photoresist feature [70]. These effects are lost in a continuum description, irrespective of the grid size used. Figure 3.23 shows a typical 2D photoresist lattice representation. Each lattice point corresponds to a site where exposure events may be located. The event density is represented by lattice occupation probability, so the larger the dose is, the larger the lattice occupation probability will be. The occupation probability is computed using a Poisson distribution of the photons exposing the resist. In the case of a CAR, the absorption of two photons within the diffusion length of the acid will not increase the local dissolution rate, since it is already saturated.

Various methods for the description of the polymer chain geometry have been suggested in publications. They have been developed in conjunction with the CIM (Critical Ionization Model) [66–68, 70, 72] to analyze the impact of polymer chain architecture on the resulting LER of the photoresist. Application of CIM for development simulation requires an explicit representation of the polymer chains, for they represent the dissolution units. Patsis et al. [73, 74] proposed the self-avoiding random walk algorithm. In order to add a monomer unit to the polymer, the position is selected randomly from those neighboring cells that are not yet occupied by a polymer chain. Overlapping of polymer chains occurs only if all neighboring cells are already occupied. Patsis et al. proposed an additional variant for generating a polymer distribution, the randomly grafted chain algorithm. An unoccupied lattice cell is randomly selected for adding the first monomer unit of a polymer chain. All additional monomer units are sequentially placed as neighbors of any previously added monomer, if possible in an unoccupied lattice cell. The polymer distribution generated by the self-avoiding random walk and the randomly

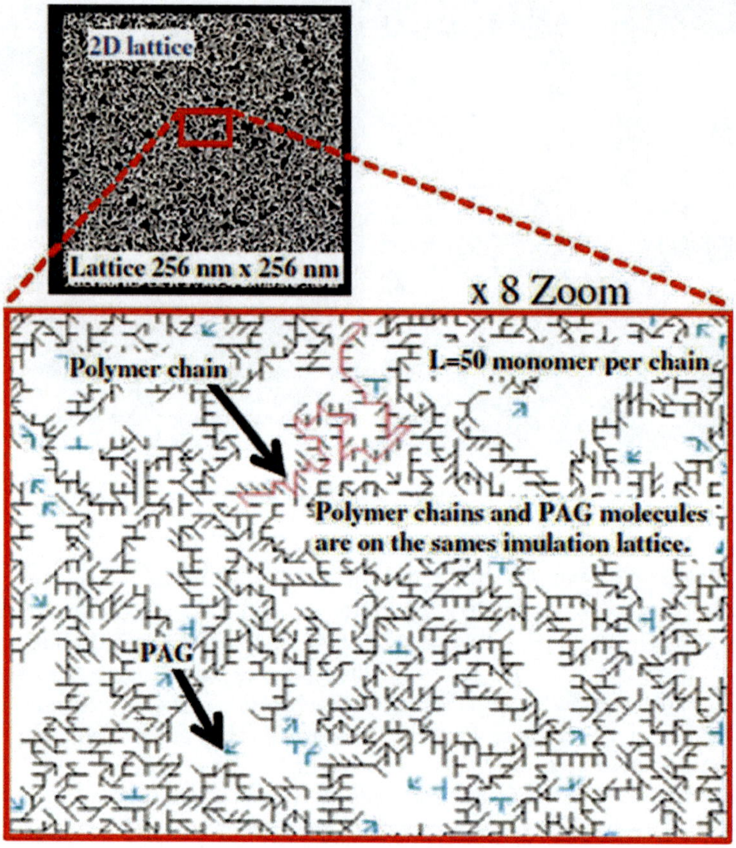

Fig. 3.23 Typical 2D resist lattice with magnification of a small region in order to present the distribution of polymer chains and PAG molecules. L is the average degree of polymerization of resist chains

grafted chain algorithms lead to more realistic polymer distributions. They partly incorporate volume restriction (i.e.,polymers may not overlap), resulting in only a low spatial overlap of monomer units.

As an example Fig. 3.24a, b shows a sample 2D 100^2 nm^2 square polymer lattice where linear polymer chains are clearly seen. Exposure through a mask and simulation of the dissolution process can deliver detailed SR and *LER* evolution profiles, as well analysis vs. resist thickness loss, and *LER* vs. edge depth. The same procedure could be applied in 3D modeling (Fig. 3.24c).

Each lattice site corresponds to a physical size comparable to the minimum molecular group considered on the material. E.g. if a polymer chain is comprised of monomers each of which is equal to 1 nm in radius, then this is the unit of length in the chain lattice. The material algorithmically is a structure of nodes, each one of which represents a monomer in the chain or a molecular group in the molecular

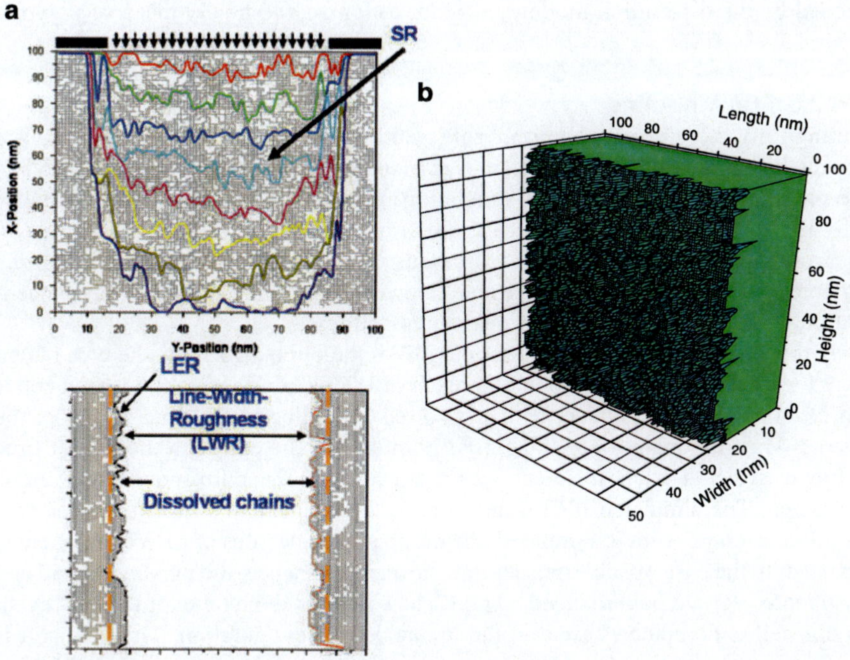

Fig. 3.24 Stochastic simulation of the dissolution process can deliver detailed SR and LER evolution profiles, as well analysis vs. resist thickness loss, and LER vs. edge depth: (**a**) 2D example, (**b**) 3D example

resist case. So if a monomer is a "cell", then a chain is considered a "collection of cells", and the chain-lattice, a collection of chains [62, 73, 74]. During chain-lattice creation, the free volume usually is considered ~10% and implemented through the self-avoidance and mutual exclusion of the molecules in the lattice. The different chain architectures result in different chain-lattice textures. This property has effects on *LER* exhibited by each material model.

3.4.2.2 Exposure Modeling

Exposure data are required in order to initiate PAG and create acid species. Energy deposition data could be obtained from exposure simulation e.g. from PROLITH [www.kla-tencor.com] in the case of optical lithography simulation or from Monte Carlo e-beam simulation. The amount of deposited energy film vs. position in the resist film is first normalized in order to be used as a template for PAG initiation. When a PAG site is inserted in the lattice, among others, it is assigned a random number which describes its initiation threshold. The normalized energy distribution is compared with this threshold and if higher, the specific PAG produces an acid site. The acid sites perform random walks in the lattice (during the simulation of the post-exposure bake), and when they encounter the protection sites on polymer chains, they change them to deprotected ones. This is the information needed by the dissolution module, in order to ionize and finally dissolve polymer chains.

3.4.2.3 PEB Modeling

During PEB, the polymer chains are sufficiently described by the attached inhibitor groups. The finite number of molecules is accounted for, but geometrical aspects of the particles are ignored in PEB simulation. As a result, polymer chain geometry only has to be taken into account for the simulation of the development process. Macroscopic models assume a continuous and deterministic evolution of the photoresist PEB. At the molecular level, however, photoresist component concentration values can only change in discrete quantities.

In the initialization of the mesoscopic PEB simulation algorithm, the type of and time for the next event is computed for each cell. This information about each cell is stored in a next event queue, which is sorted according to the time values of the events. Then the main loop of the algorithm starts. The cell with the lowest time value is selected from the next event queue, and the corresponding event is simulated. The simulated PEB time is set to the event time value of the selected cell. If a reaction event is simulated, the number of molecules of each component is updated. In the case of acid base neutralization, for instance, the number of acid and base molecules are both reduced by one. The type of and the time for the next event in this cell is computed based on the updated cell configuration. Then the cell is again inserted into the next event queue with the sum of the time computed for the next event plus the currently simulated PEB time. If a diffusion event is simulated, the same procedure as described above for a reaction event is additionally performed for the cell to which the molecule is diffused. In the case of a reaction, the next reaction time only has to be updated for one cell. In case of a diffusion event, two cells have to be updated. These steps of selecting the cell with the lowest time value from the next event queue and updating the cell in which the number of resist component molecules change, are iteratively performed until the specified PEB time has been simulated.

3.4.2.4 Development Modeling

As for the simulation of the development process in the mesoscopic framework, developed regions will expand outward to non-occupied regions as time goes on, so the development proceeds by removing the dissolving cells, leading to a front propagating through the material. In the development process, the polymer chains are the dissolution units and their geometry is represented additionally. Each lattice cell represents the volume of a single monomer unit. Free volume is represented by leaving a certain fraction of the lattice cells empty. All discrete polymer description schemes are based on an orthogonal, structured grid. This imposes some artificial constraints on the shape of the polymer chains.

The average degree of polymerization (i.e., the number of monomer units per polymer) and the standard deviation can be obtained from the molecular weight distribution of the photoresist polymers. The molecular weight distribution of the polymer chains is close to Gaussian [68] and can be measured by size exclusion chromatography [75]. Periodic boundary conditions are applied in lateral x and y directions to remove artificial effects due to the limited size of the simulation area. In the vertical z direction, reflective boundary conditions are assumed.

Macroscopic development rate models provide a description of the rate dependence on the average inhibitor concentration. They provide neither a description of the actual chemical processes nor of the size of the polymer chains. Thus, unlike exposure or PEB models, the macroscopic development models cannot be generalized to mesoscopic models. At the mesoscopic level, photoresist development is a binary process in which each polymer molecule either remains in the photoresist film or is dissolved in the developer liquid during the specified process time [70]. Based on the discrete photoresist description, mesoscopic development simulation requires a description of the dissolution behavior of each polymer. Polymer solubility is essentially governed by the degree of deprotected (i.e., ionizable) monomer units where the initially attached inhibitor group was decomposed during PEB. The primary reactions during the development process are [69, 76]:

- Diffusion of developer molecules to the photoresist film surface.
- Deprotonation (ionization) reaction of the developer molecules with the deprotected acidic phenolic polymer sites (generation of phenolate anions).
- Diffusion of the dissolved, ionized polymers into the bulk of the developer.

Polymer dissolution is a combined process involving mass transfer resistances and chemical reactions, and thus different rate-limiting steps can lead to different dissolution behavior. All chemical models for the dissolution of phenolic polymers in an aqueous base assume one of these processes as the rate limiting step and the others as occurring almost immediately [67].

The fundamental mechanisms that dominate dissolution behavior of the polymer chains, and thus the photoresist development rate behavior, have been the subject of considerable discussion. A universally accepted theory that accounts for all observed dissolution phenomena still does not exist. Two models are prevalent in the literature and have been applied for simulation: the percolation model (PM) [77] and the critical ionization (CIM) model [66–70].

Variations and improvements of the models have been presented by various groups e.g. [73], and efforts to use these models for LER predictions have shown promising results [77]. From a computational point of view, both the CIM and PM are one-parameter models (the parameters are the f_{crit} for CIM, and the diffusion probability for PM). However, CIM has difficulties in "completing dissolution" when the blocking fraction of the polymer (i.e., the fraction of the protected OH groups versus the total OH groups) increases. These happen because this model considers polymer dissolution as a surface process. In addition there are computational difficulties when increasing the length of the polymer chains. On the other hand experimental results suggest that the gel layer at the developer-polymer interface predicted by the PM is either very thin or does not exist [78].

PM and CIM have attracted attention in the last few years, although other models have also been discussed such as the aggregate model [54]. Application of the CIM for development simulation is based on a discrete polymer chain description.

The additional information required about the polymer chains is the fraction of the ionizable sites (i.e., monomer units). This can be obtained from exposure and PEB simulations that predict the local extent of inhibitor deprotection. Each decomposed inhibitor molecule represents an unblocked, ionizable site.

To sum up, the CIM provides a more chemically sound explanation of the development process than percolation theory or other suggested models. Simulations with the CIM model dissolution mechanism might be useful for qualitatively estimating the potential and limitations of new photoresist formulations or the optimization of existing ones. Despite several refinements of the simulation approach since the first publication of the CIM in 1997, it has not proven to be suitable for quantitatively predicting the development rate curve required for the accurate simulation of photoresist profile geometry.

3.5 Commercial Software for E-Beam Lithography Simulation

As an outcome of the worldwide research effort in lithography simulation over the last decades, several software tools were developed. These software tools were initially focused in the simulation of electron beam-matter interaction with the resist film and the substrate. Later on the center of gravity was shifted to the proximity effect simulation. The latest versions of the software tools are very powerful covering all aspects of the e-beam lithography process towards the so-called computational lithography.

3.5.1 Electron Beam Matter Interaction

The majority of the simulation tools that were offered as commercial products were based on algorithms developed in research labs. Most of these tools are based on Monte Carlo simulation of electron beam – matter interaction and support 2D and 3D simulation, and they can handle wide electron energy ranges as well as numerous materials and multilayer stacks. Furthermore, in several cases the energy deposition module is coupled with resist development modules providing thus integrated software tools for in-depth study of EBL process. With those simulation tools certain parameters can be tuned such as film stack and resist thickness, proximity effect correction, and beam voltage impact. The most known and used commercial simulation tools are:

Sceleton: that initially was offered by AISS, Germany and now is a product available by Synopsys [79]. This software tool takes a user-defined resist-substrate multilayer stack of components defined in a materials database and traces a specified number of electrons in the material. The electron trajectory is followed through a series of scattering events in the resist/substrate stack through a Monte Carlo algorithm. Elastic scattering events are described using the screened Rutherford formula. The backscattering range and the ratio of backscattered over forward scattered contributions is determined numerically through the calculation of the proximity function. Energy dissipation due to inelastic scattering is modeled by Bethe's energy loss formula in the continuously slowing down approximation. Typical calculation time was 11 h on a LINUX-based Xeon processor system in which 10^7 electrons were traced [80]. The resulting calculated radial energy can be used directly as input for other PEC tools.

ProBEAM: that was initially developed by Finle Technologies at the end of the 90s. This simulation tool has been also developed around a Monte Carlo code able to follow in 3D the electron path in the resist/substrate stack. The initial major ProBEAM application was for mask making. Monte Carlo simulations are combined with a beam shape to generate a single "pixel" energy distribution. This pixel is then used to write a pattern by controlling the dose of every pixel on an address grid [81]. The resulting dose pattern is used to expose and develop a resist, by implementing appropriate resist algorithms that were developed by the same company for optical lithography, to form finally a simulated 3D resist pattern. Nowadays, ProBEAM is offered by KLA-Tencor [82] and includes models for electron beam processes up to 100 keV [5].

LITHOS was initially developed at the Institute of Microelectronics NCSR "Demokritos" in the beginning of the 90s [83] and was commercialized by Sigma-C. LITHOS followed a different path to handle the long CPU times needed by Monte Carlo. In LITHOS, the electron trajectory is calculated in 2D by applying the Boltzmann transport equation that can be solved numerically, and after taking into account the same equations for elastic and inelastic scattering events the energy deposition can be calculated in CPU times at least one order of magnitude faster than Monte Carlo. In LITHOS, resist development modules for conventional resists was implemented also allowing for the prediction of 2D resist structures.

SELID: the energy deposition algorithm of LITHOS was later on integrated in SELID software offered by Sigma-C where very accurate simulation algorithms for resist development for both conventional and chemically amplified resists were also implemented allowing for 3D simulation of arbitrary shapes at low CPU times [7, 84].

3.5.2 Proximity Effect Correction

A key application for e-beam lithography simulation is the development and evaluation of proximity correction strategies. This reduces the dependency on the hardware itself and improves process development cycle times.

E-beam Proximity Effect Correction (PEC) is an effective mean of correcting for line-end shortening, critical dimension linearity, and the typical line-edge variations seen between dense and isolated lines. Several software tools have been developed for PEC and are offered for both SEM to lithography conversion kits and high end EBL tools. In the latter case, PEC correction is included in the data preparation software for conversion of GDS files to formats ready to be exposed by the various e-beam tools. In particular, the commercial tools that are offered are:

PROXECCO was a joint development by Fraunhofer Institute for Solid State Technology, Munich, Germany and aiss GmbH, Munich, Germany [85] and was commercialized by aiss. The conventional approach for PEC is the deconvolution method that is very accurate but needs high computing power and data reduction algorithms especially for large layouts. In PROXECCO, the calculation is split to correction related and pattern related steps. This way the grid size, which is the dominant parameter for the CPU time needed, is spatially independent from the pattern and can be coarse. Thus, pattern dimensions are maintained and data reduction is not applied. PROXECCO utilizes an N-Gaussian analytical function or a point-beam spread function and corrects for PEC through dose modulation on systems which support shape-based dose modulation and through either a unique N-pass writing technique or geometric shape manipulation for systems which do not support shape-based dose modulation (MEBES). Nowadays PROXECCO is offered by Synopsys integrated in Data Preparation software CATS [86].

In the same way, in **BEAMER** layout software [87], PEC module has been implemented in order to allow for the preparation of the final layout to be written by the e-beam tool free of proximity effect. In this case, proximity effect could be corrected in all 3 dimensions. In contrast to the 2D solution, the correction for 3D patterning is complex and application dependent. While 3D applications, such as T-gates and bridges, still focus on CD control, others, such as 3D zone plates or holograms require highly accurate resist thickness control. These different applications need to be treated with different PEC algorithms, since the correction targets of "CD control" and "thickness control" are entirely different and contradictory in some cases [88].

3.5.3 Modern Tools

In recent years, the TCAD tools, thanks also to the high computational power available at very low cost by computers, have integrated various modules towards the so-called computational lithography i.e. the accurate prediction of the final profile of any layout with critical dimension in the nanoscale.

In this direction, Sentaurus Lithography, [89], covers a wide range of applications in optical, immersion, extreme ultraviolet (EUV) and electron beam lithography, allowing predictive modeling and thorough analysis of fundamental effects.

The e-beam module of Sentaurus Lithography supports the simulation of both wafer direct-write as well as mask-writing applications. Electron-scattering processes in the resist and wafer or the mask stack determine the corresponding energy deposition functions (EDF's); the implemented models support both low (<10 keV) as well as high electron energies (20–50 keV). From the total energy deposited in the resist film, a physical model determines the resulting resist profiles.

3.6 Examples

Electron beam lithography simulation has been applied for numerous cases. Below, two selected examples related to the patterning of complex layouts on multilayer substrates and to the shot noise effect are discussed

3.6.1 Electron-Beam-Patterning Simulation and Metrology of Complex Layouts on Multilayer Substrates

Complex layout patterning simulation is one of the most difficult cases for simulation, and it becomes more complicated when the substrate consists of multilayers instead of bulk Si. EBL on EUVL (Extreme Ultra Violet Lithography) mask blanks is one of the most demanding cases, since the mask blank consists of a multilayer of 40 double Mo/Si layers on the substrate. For this application, a Monte Carlo algorithm is applied and the obtained simulation results are compared with experimental ones for test patterns with critical dimension down to 100 nm [26]).

For the metrology of both simulated and experimental results, a dedicated pattern matching algorithm is applied for the identification of a test pattern from the layout to the actual fabricated design [71, 83]. The layer stack used consists of 200 nm PMMA spin coated over a film stack of 40 Mo-Si bi-layers (thicknesses: 4.1 nm Si, 2.8 nm Mo) on top of a Si substrate (Fig. 3.25a). Horiguchi et al.'s method [90] was incorporated in order to correctly determine the mean free path of the electron track due to multilayer presence in the Mo-Si structure.

For the modeling of electron scattering in a multi-layer Mo-Si substrate, the Monte Carlo procedure of Salvat and Parellada [91] was applied [92]. The EDF(r, z) was calculated for all resist depths. The simulation used 40,000 electron trajectories. Figure 3.26, shows only 2,000 electron tracks scattering while traversing the resist and Mo-Si multilayer stack.

The EDF(r) at the resist/substrate interface for various substrates and for 100 keV is illustrated in Fig. 3.27. The stacks considered are (a) PMMA/Mo-Si multilayer/Si-substrate, (b) PMMA/Si-substrate, and (c) PMMA/70 nm Cr absorber/70 nm SiO_2 buffer/Mo-Si multilayer/Si-substrate. The last stack corresponds to a typical EUV mask [93].

Due to the high initial electron energy, the backscattering coefficient is mainly determined by the properties of the substrate material (Si), although the high density Mo is present in the Mo-Si multilayer structure. Specifically, the backscattering coefficient (ratio of backscattered to initial number of electrons) is 0.11 in the case of Mo-Si multilayer/Si stack. Approximately, the same value was obtained in the other stacks studied. Its value has been verified also with simulations using the CASINO electron-beam lithography simulator [94]. This is manifested also by the fact that the energy distributions in Fig. 3.27, show small relative differences. Since the experimental data in this work are for Mo-Si multilayers on top of Si bulk, the following discussion is limited to this type of stack (i.e., case a).

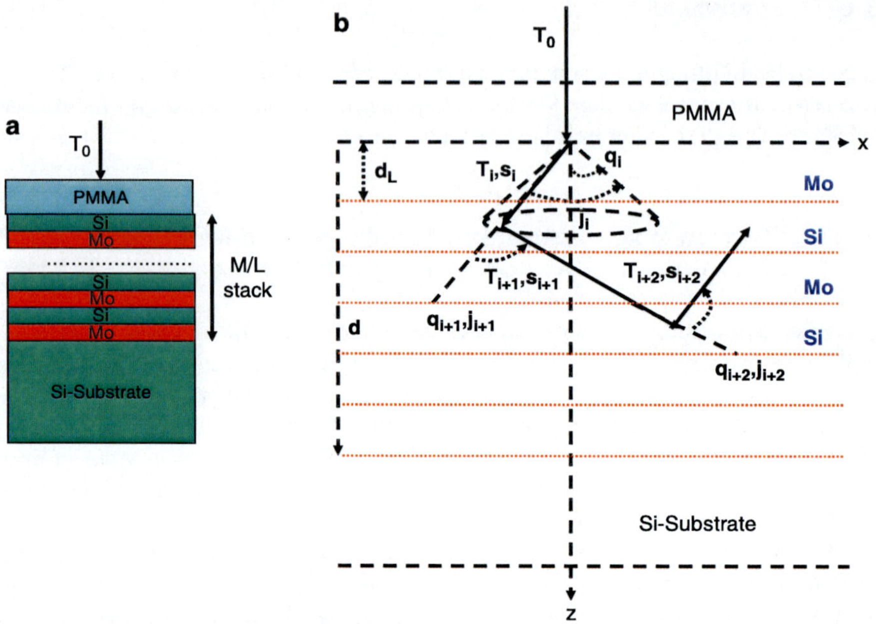

Fig. 3.25 (**a**) Stack model. (**b**) Coordinate system used in the electron track simulation. Between successive scattering events, energy loss is calculated from the continuous slowing down approximation. The electrons are considered as point particles with kinetic energy T scattering elastically with the material atoms. θ and φ are the polar and azimuthal angles and s the free electron path between two successive collisions. Symbol d marks the depth from the resist-substrate interface (d is approximately 280 nm) and dL each layer's thickness (For Si, dL = 4.1 nm, while for Mo, dL = 2.8 nm) [26]

The EDF has been fitted with the typical 3-Gaussian function:

$$f(r) = \eta_1 \exp\left(-\frac{r^2}{a^2}\right) + \eta_2 \exp\left(-\frac{r^2}{b^2}\right) + \eta_3 \exp\left(-\frac{r^2}{c^2}\right) \tag{3.21}$$

with a, b, c width parameters and η_1, η_2, η_3 weight parameters for each Gaussian. The calculated values for the width parameters are a = 28 nm, b = 450 nm, and c = 30 μm, while for the weight parameters the corresponding values are $\eta_1 = 2.2 \times 10^{-7}$, $\eta_2 = 1.1 \times 10^{-9}$, and $\eta_3 = 1.1 \times 10^{-11}$ keV/nm^3/electron.

For the evaluation of the theoretical results, appropriate experimental data were used. Mo-Si multilayer stacks deposited on Si substrates were prepared by ion beam sputtering deposition and the stack period was checked by grazing X-ray reflectance [95]. Exposures on those stacks were performed with a VISTEC EBPG 5HR working at 100 kV. The beam current parameters are 0.4 nA, dose: from 200 up to 1,600 μC/cm^2, in steps of 100 μC/cm^2. The resist used was 250 nm PMMA and the development was performed in IPA: DI water = 7:3 for 1 min.

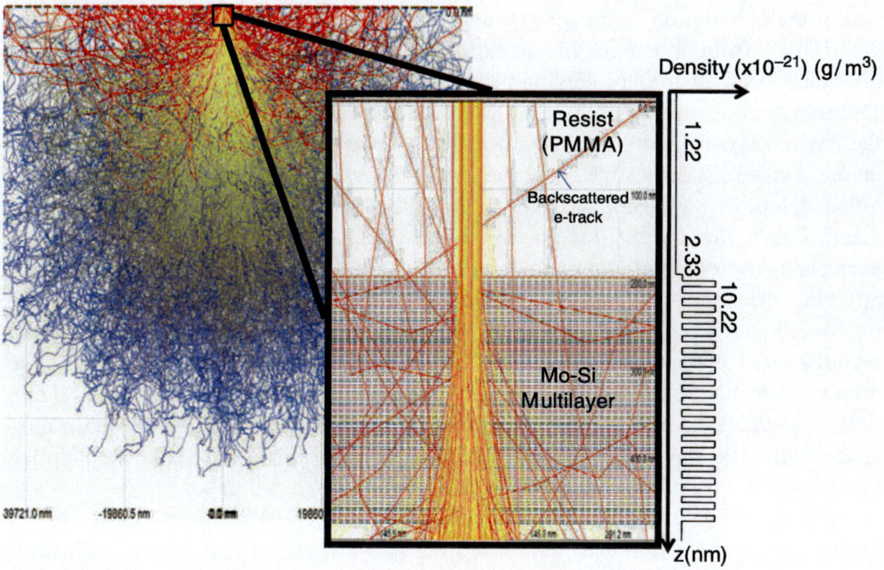

Fig. 3.26 Sample electron tracks for 100 keV electrons impinging on 200 nm PMMA on top of 40 multilayers of Mo-Si [26]

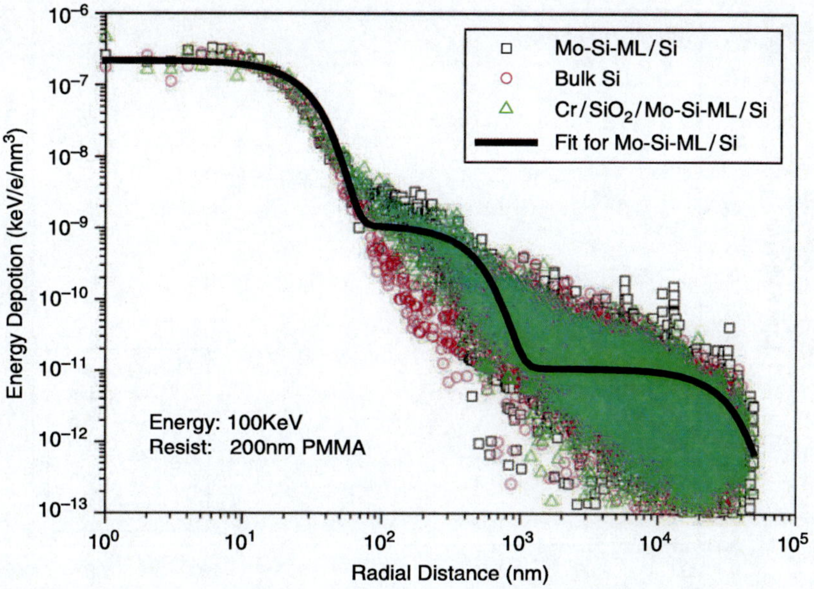

Fig. 3.27 Electron energy deposition function (EDF) in the resist film for the Mo-Si case, the bulk Si-substrate case, and a full EUV stack case. Electron energy is 100 keV. The three-Gaussian fit on the PMMA/Mo-Si multilayer case is also shown [26]

For the calculation of the energy deposition in each cell of the layout's matrix, the EDF(r) from the point beam exposure is convolved with the beam profile (Gaussian beam shape), producing the energy deposition of a single pixel exposure. The energy deposition profile from the single pixel exposure is then convolved with the desired layout pattern, and the total deposited energy in every cell of the 3D layout's matrix is determined. For the simulation of resist development, the threshold development was applied, i.e. the application of an energy threshold value discriminates the dissolved from the undissolved cells. This algorithmic approach even though very simplified can be used for the resist (PMMA) employed [96]. By applying the energy threshold, a 3D resist image is obtained and further image processing can be applied in order to match the simulated results with the initial layout as well as to perform metrology in order to calculate the differences from the nominal dimensions. Several energy thresholds are examined until one is found that best reproduces the experimental metrology for one of the test patterns (dense line-spaces with 100 nm nominal size). Then, this energy threshold value was applied for the simulation of all patterns exposed on PMMA at 100 keV.

In Fig. 3.28, the results from the metrology of the PMMA trench-width of experimental and simulated test structures in the case of the Mo-Si multilayer over bulk Si substrate are shown. The test structures examined are lines/spaces with 100 nm nominal width and 300 nm spaces and 100 nm holes with a pitch of 300 nm in both directions. In the exposure dose range examined, a very good

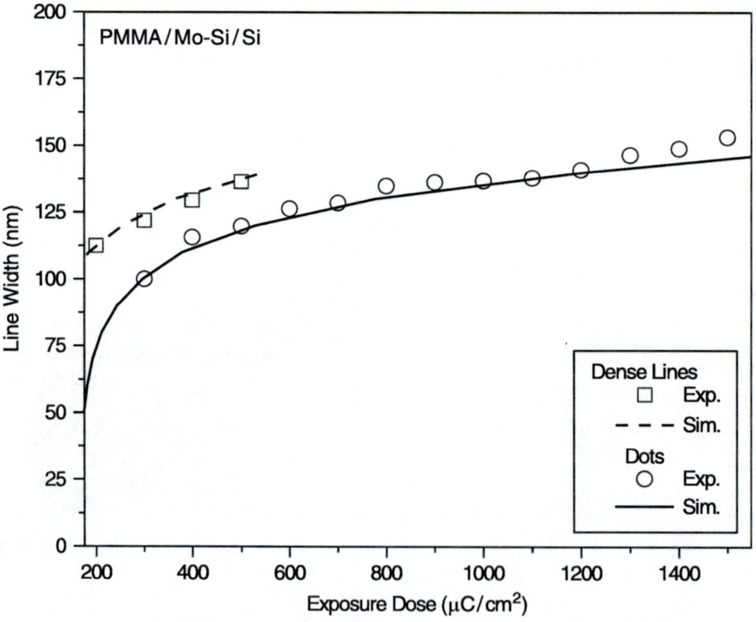

Fig. 3.28 Comparison of experimental and simulated data. *Squares* correspond to 100 nm dense lines/spaces, while *circles* to 100 nm diameter, dots [26]

agreement between the experimental and simulation results is observed. For the simulation of these structures, all exposure conditions applied such as beam diameter, beam step size, exposure dose, etc. were taken into account, while the cell size for the simulation was selected to be 2×2 nm^2. In addition, it should be noticed (not shown here), that the evolution of the trench width and hole diameter in the case of bulk Si substrate is very close to the one when the multilayer exists. This experimental result verifies the limited difference observed in the simulated EDF(r) presented in Fig. 3.27.

For a detailed comparison between the experimental results (top-down SEM images) and the simulation ones certain algorithms for image processing, pattern matching and metrology should be applied in a way suitable for automatic analysis. The pattern matching algorithm enables the identification of the desired layout for metrology on a complex layout containing many printed features. The matching procedure is followed both for the images produced by simulation as well as for the SEM images. These grayscale images are converted to binary, images that have been quantized to two values denoted 0 and 1. For the simulation results, this was accomplished by applying a threshold. In SEM images, histogram equalization is needed before applying a threshold, in order to enhance the contrast. The desired layout is converted to a binary image as well, in the same magnification of the simulated image and the corresponding SEM image. The two images (layout – binary SEM image or layout – binary simulated image) are matched using a cross-correlation method. The matching algorithm returns the coordinates of layout pattern edges in pixels. Having recognized the desired layout in a complex image, the metrology procedure can be automated.

In order to study the differences between the experiment (Fig. 3.29a) and simulation (Fig. 3.29b), automatic metrology is required. Metrology applied in the features reveals a relatively good reproduction of the pattern by the simulator. Thus electron-beam lithography simulations can provide understanding of the fine details of EUV mask making process.

3.6.2 Simulation of Shot Noise Effect on CD and LER of Electron-Beam Lithography in 32 nm Designs

Microelectronic device fabrication is prone to variability issues of the geometrical characteristics of the printed structures due to illumination, discrete resist properties, and processing. The total squared CD variability is the sum of the squares of the CD variability of the previously mentioned sources of variation, if these are considered uncorrelated [97]. In the case of e-beam lithography, shot noise variation originates from the discreteness of the electrons of the beam and needs to be accurately modeled, especially if low dose is used for the realization of very fine features, necessary for the adaptation of e-beam lithography in production. Work has been done in modeling analytically the effects of shot noise, and has been concluded that both shot noise and LER minimization requires increased dose [97] while LER is analogous to inverse square root of dose [98]. Additionally, the

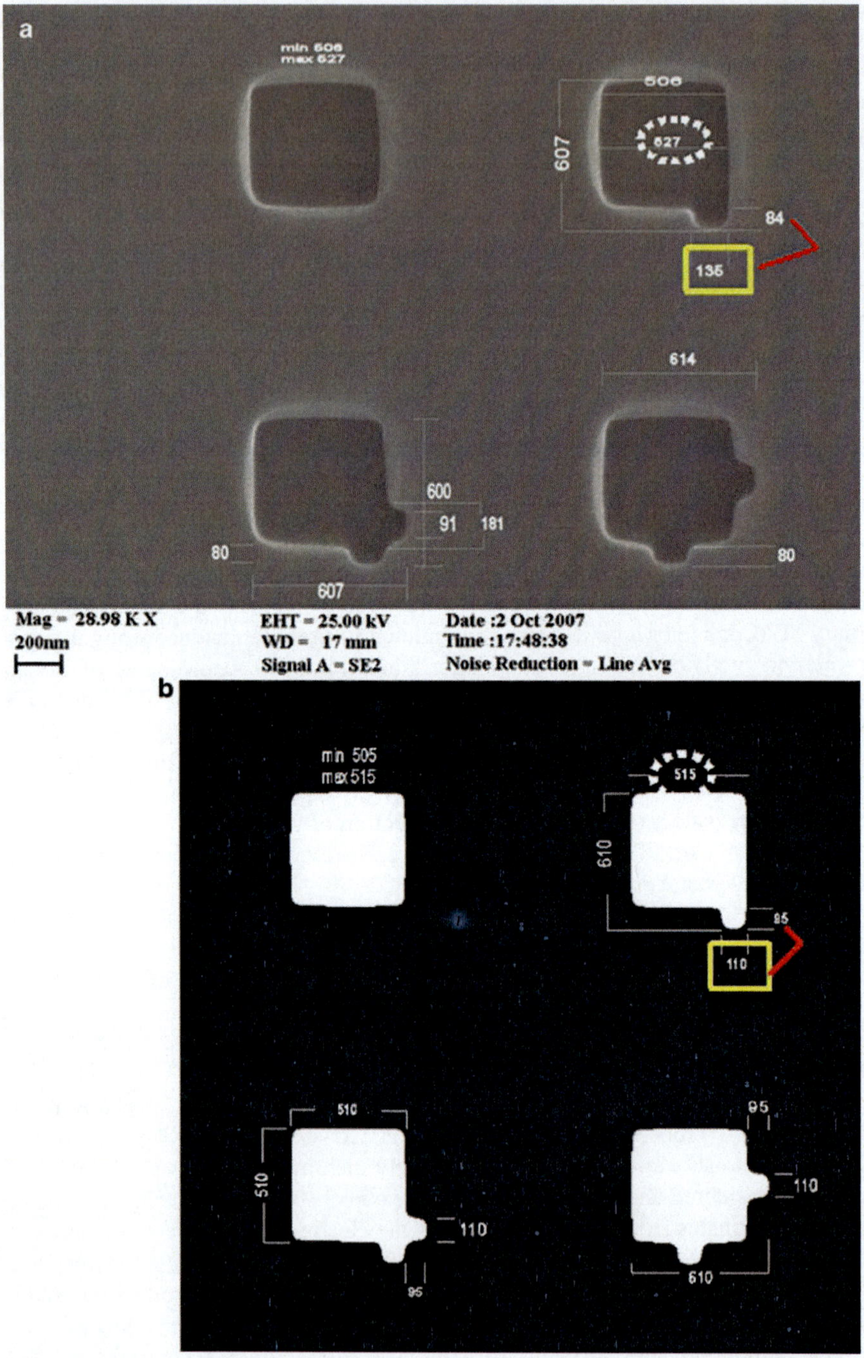

Fig. 3.29 (**a**) Metrology on the experimental SEM image. (**b**) Corresponding metrology on the simulated data [26]

discreteness of resist material is also an important contribution to LER that can be improved at the cost of sensitivity and throughput [99].

In order to print fine features in small pitches (64 nm and less), e-beam exposure variations, resist properties and processing conditions should be simulated in a coupled framework. In this direction, a Monte-Carlo electron beam simulator, [92], was combined with a discrete resist film model, [100], and a critical ionization development algorithm, [101]. An interesting work on the same subject has been published by Neureuther et al. [102] where the exposure, amplification, and deprotection of CA resists are treated as a sequence of statistical events to determine the effect of shot noise. This combination of processes can be physically interpreted as contributing noise from the uncertainty in the number of quanta at each step.

Figure 3.30 shows the simulation flow used. The convolution of EDF(r) with the shape of the actual beam (Gaussian (G) or Blurred Rectangular (BR)) and then once again the convolution with the pattern shape, results in the total deposited energy, LEDF(r). LEDF(r) is combined with the stochastic lithography simulator in order to reproduce the developed pattern. Metrology on the produced edges with specific metrology software [103] results in CD and LER quantification.

Figure 3.31 shows a sample probability density function of the two beam shapes for 32 nm full-width-half-maximum (FWHM). Electron shots are distributed either with Gaussian or blurred rectangular beam profiles. Edge acuity was considered the same (10 μC/nm) for both cases. The difference in profile resulted in different shot noise effects. The layout is written using single-shots of 100 keV (resist thickness of 100 nm on bulk Si) and consists of 10 lines with 10 μm length, 32 nm nominal CD, and 64 nm pitch. The pixel size of the exposure algorithm is 1 nm/pixel. The deposited energy is then used as input in the stochastic model of the resist film.

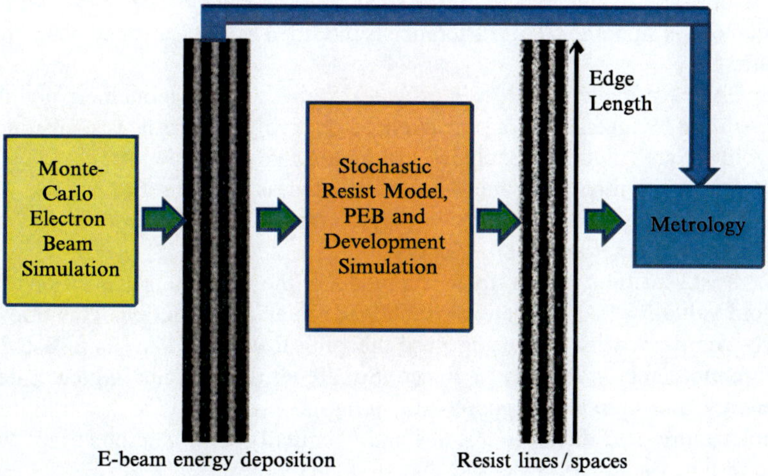

Fig. 3.30 Major steps in the simulation flow. Edge length is marked. LER values are expressed vs. edge length, both after exposure and after development [4]

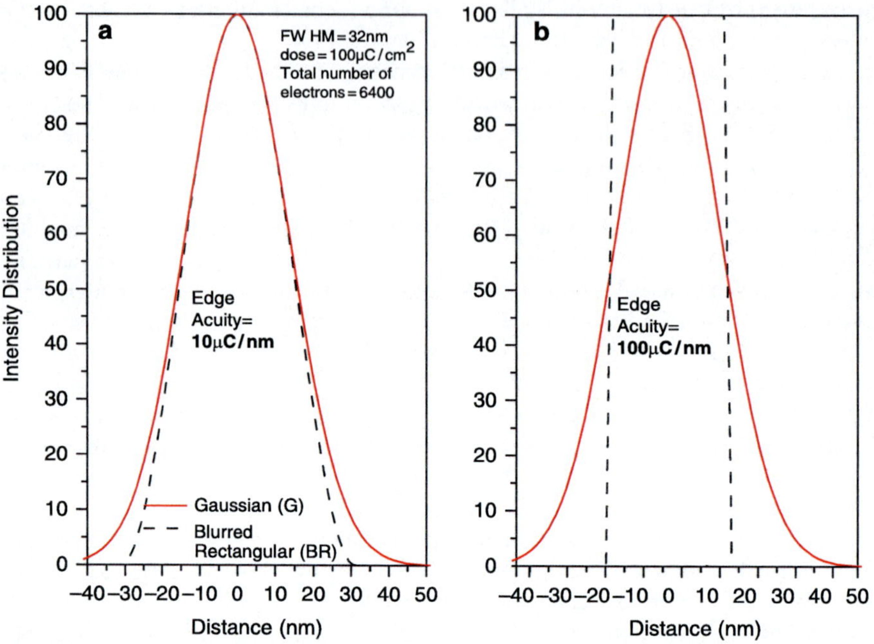

Fig. 3.31 Electron beam profiles used in the current work. In (**a**) edge acuity is 10 μC/nm, while in (**b**) is 100 μC/nm. In the current work, the analysis is performed with the 10 μC/nm acuity [4]

For the determination of each exposure distribution, the point spread function of the impinging electrons is convolved with a Gaussian and the Rectangular profile, respectively. The rectangular profile is steeper at the sides of the pulse, while the Gaussian one is smoother. This difference is modified by the statistical nature of the exposure.

The CA resist model considered polymer chains with 50 monomers per chain, 10% PAG and 5% quencher concentration, respectively. The resist is a collection of self-avoiding and mutually excluded random walks with 5% free volume. The critical ionization model for the dissolution process of the resist is used [102]. Acid diffusion length is selected accordingly (around 5 nm) in order to reproduce approximately the nominal CD = 32 nm.

The resist sensitivity in the stochastic model of the material is just a probability-threshold value for PAG initiation which leads in acid production. This threshold actually decreases with increasing dose meaning that more PAG is initiated and therefore more acid is initially produced. So, higher dose results in greater deposited energy and therefore higher PAG initiation probability or lower initiation threshold, more acid diffusion species and eventually more deprotections, which lead to faster dissolution, wider CD and smoother edges, thus less LER.

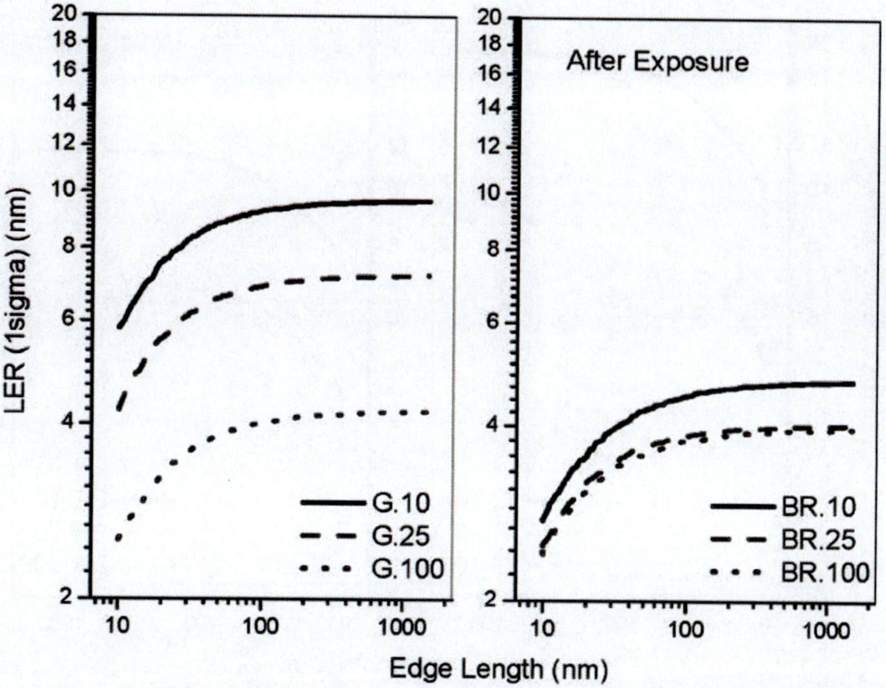

Fig. 3.32 Metrology of LER after exposure (i.e. on energy deposition edges) for the two e-beam profiles and the three exposure doses [4]

Metrology is performed on the edges of the energy deposition top-down profile, and of the corresponding developed resists lines (as is shown qualitatively in Fig. 3.30) for the two e-beam shapes and all the exposure doses.

Figure 3.32 shows the LER vs. edge length for three doses (10, 25, and 100 $\mu C/cm^2$), after exposure. This LER is attributed to shot noise. It is seen that LER decreases with exposure dose, and that BR-LER is less than the G-LER in all doses. Figure 3.33 shows the corresponding data after resist deprotection and development. An increase of LER values compared with LER after exposure of Fig. 3.32, is observed for both beam profiles.

Therefore the coupling of Monte Carlo e-beam simulator and stochastic lithography allows the separate quantification of shot noise LER after exposure, and of the LER due to resist, deprotection and development process. Two exposure profiles were tested: a Gaussian and a blurred rectangular with the same edge acuity. The resist and processing effects on LER were seen to be more important, however, covering the impact of shot noise. This LER contribution is a function of resist polymer radius of gyration and acid diffusion range. However, it is concluded that shot noise LER will pose a lower limit to LER even if resist material and its processing are optimized to result in minimum LER.

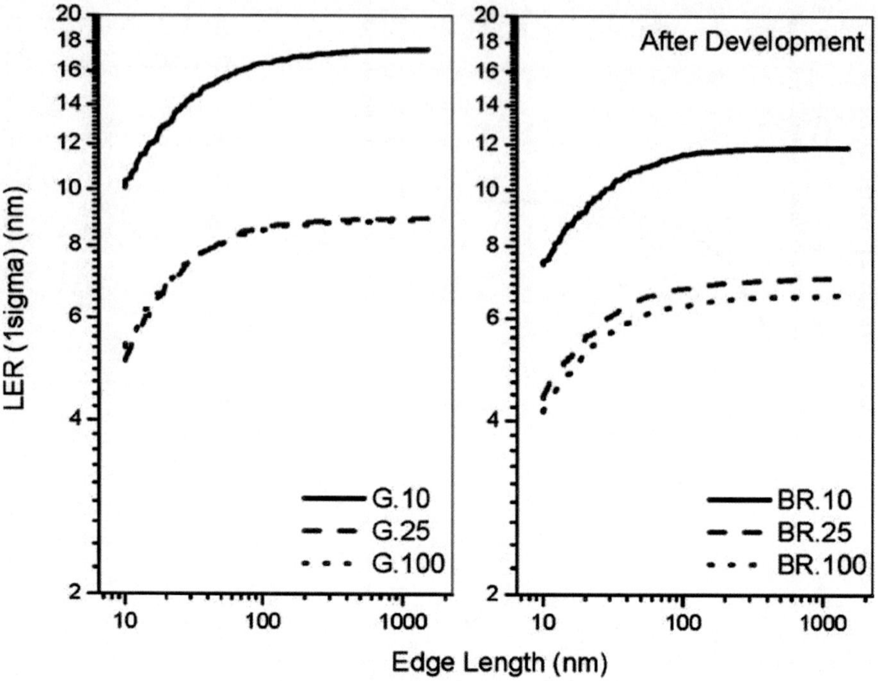

Fig. 3.33 Metrology on energy deposition profile and final resist line edges for the two e-beam profiles and the three exposure doses [4]

Conclusions

Electron beam lithography is a very powerful tool for patterning in the deep nano-scale and is broadly used for the realization of novel structures and devices and for the realization of high-end masks. However, the electron beam lithography process is rather slow compared to optical lithography which is the work-horse for the mass production of IC's. For that reason, the EBL process should be thoroughly optimized prior to the realization of the nanostructures. Furthermore, due to the nature of electron beam patterning, unwanted phenomena such as proximity effects and heating are taking place. For all those reasons, it is necessary that the electron beam lithography process be simulated with high accuracy and reasonable CPU times. The research effort in this direction was actually started in early 70s and even though tremendous progress has been reported and commercial software tools have been realized, is still an open issue mainly due to the need for realization of continuously smaller and denser features at larger areas.

References

1. http://cuervo.eecs.berkeley.edu/Volcano/.
2. Rishton SA, Kern DP. J Vac Sci Technol. 1987;5:135–41.
3. Raptis Jpn I. J Appl Phys I. 1997;36:6562–71.
4. Patsis GP, Tsikrikas N, Drygiannakis D, Raptis I. Microelectron Eng. 2010;87:1575–8.
5. Babin S, Kuzmin IY, Sergeev G. Microelectron Eng. 1998;41–42:191–4.
6. http://www.abeamtech.com/?dir=products/TEMPTATION&pg=about.
7. Raptis I, Nowotny B, Glezos N, Gentili M, Meneghini G. Jpn J Appl Phys I. 2000;39:635–44.
8. Tagawa S, Nagahara S, Iwamoto T, Wakita M, Kozawa T, Yamamoto Y, Werst D, Trifunac AD. Proc SPIE. 2000;3999:204–15.
9. Raptis I, Grella L, Argits P, Gentili M, Glezos N, Petrocco G. Microelectron Eng. 1996;30:295–9.
10. Glezos N, Patsis G, Raptis I, Argitis P, Gentili M, Grella L, Hatzakis M. J Vac Sci Technol B. 1996;14:4252–6.
11. Sha J, Lee J, Kang S, Prabhu VM, Soles CL, Bonnesen PV, Ober CK. Chem Mater. 2010;22:3093–8.
12. Augur RA, Jones GAC, Ahmed H. J Vac Sci Technol. 1984;3:429–33.
13. Phang JCH, Ahmed H. J Vac Sci Technol. 1979;16:1754–8.
14. Adesida JC, Everhart TE, Shimizu R. J Vac Sci Technol. 1979;16:1743–8.
15. Glezos N, Raptis I, Tsoukalas D, Hatzakis M. J Vac Sci Technol B. 1992;10:2606–9.
16. Hawryluk RJ. J Vac Sci Technol. 1981;19:1–17.
17. Gueorguiev YM, Vutova KG, Mladenov GM. Physica C. 1995;249:187–95.
18. Ivin VV, Silakov MV, Vorotnikova NV, Resnick DJ, Nordquist KN, Siragusa L. Microelectron Eng. 2001;57–58:355–60.
19. Mao-xin W, Jian-kun Q, and Yi-Zeng, Microelectron Eng 1985;3:99–102.
20. Fretwell TA, Jones PL. Microelectron Eng. 1994;23:97–9.
21. Kyser DF, Pyle R. IBM J Res Develop. 1980;24:426–37.
22. Stepanova M, Fito T, Szabó Zs, Alti K, Adeyenuwo AP, Koshelev K, Aktary M, Dew SK. J Vac Sci Technol B. 2010;28:C6C48–57.
23. Ivin VV, Silakov MV, Kozlov DS, Nordquist KJ, Lu B, Resnick DJ. Microelectron Eng. 2002;61-62:343–9.
24. Zhou J, Yang X. J Vac Sci Technol B. 2006;24:1202–9.
25. Kyser DF. J Vac Sci Technol. 1983;1:1391–7.
26. Tsikrikas N, Patsis GP, Raptis I, Gerardino A, Quesnel E. Jpn J Appl Phys. 2008;47:4909–12.
27. Williamson Jr W, Duncan GC. Am J Phys. 1986;54:262–7.
28. Joy DC. Inst Phys Conf Ser. 1988;93:23–32.
29. Bethe HA. Selected works of Hans A Bethe with commentary, World scientific series in 20th century physics, vol. 18. Singapore: World Scientific; 1997. p. 80–154. ISBN 9810228767.
30. Dapor M. Phys Rev B. 1992;46:618–25.
31. Raptis I, Glezos N, Hatzakis M. Microelectron Eng. 1993;21:289–92.
32. Raptis I, Glezos N, Hatzakis M. J Vac Sci Technol. 1993;B 11:2754–8.
33. Glezos N, Raptis I. IEEE Trans CAD. 1996;15:92–102.
34. Paul BK. Microelectron Eng. 1999;49:233–44.
35. Mohammad MA, Guthy C, Evoy S, Dew SK, Stepanova M. J Vac Sci Technol B. 2010;28: C6P36–41.
36. Argitis P, Raptis I, Aidinis CJ, Glezos N, Baciocchi M, Everett J, Hatzakis M. J Vac Sci Technol B. 1995;13:3030–4.
37. Chang THP. J Vac Sci Technol. 1975;12:1271–5.
38. Parikh M. J Appl Phys. 1979;50:4371–7.
39. Parikh M. J Appl Phys. 1979;50:4378–82.
40. Parikh M. J Appl Phys. 1979;50:4383–7.
41. Wind SJ, Gerber PD, Rothuizen H. J Vac Sci Technol B. 1998;16:3158–63.

42. Grella L, Di Fabrizio E, Gentili M, Baciocchi M, Maggiora R. Microelectron Eng. 1997;35:495–8.
43. Takahashi K, Osawa M, Sato M, Arimoto H, Ogino K, Hoshino H, Machida Y. J Vac Sci Technol B. 2000;18:3150–7.
44. Lee SY, Anbumony K. Microelectron Eng. 2006;83:336–44.
45. Lee SY, Laddha J. Microelectron Eng. 2000;53:345–8.
46. Wüest R, Strasser P, Jungo M, Robin F, Erni D, Jäckel H. Microelectron Eng. 2003;67–68:182–8.
47. Dill FH. IEEE Trans Electron Dev. 1975;ED-22:440–4.
48. Dill FH, Hornberger WP, Hauge PS, Shaw JM. IEEE Trans Electron Dev. 1975; ED-22:445–52.
49. Meyerhofer D. IEEE Trans Electron Dev. 1980;ED-27:921–7.
50. Kim DJ, Oldham WG, Neureuther AR. IEEE Trans Electron Dev. 1984;ED-31:1730–5.
51. Papanu JS, Soane DS, Bell AT, Hess DW. J Appl Polym Sci. 1989;38:859–85.
52. Hasko DG, Yasin S, Mumtaz A. J Vac Sci Technol B. 2000;18:3441–4.
53. Brainard RL, Trefonas P, Lammers JH, Cutler CA, Mackevich JF, Robertson SA. Proc SPIE. 2004;5374:74–89.
54. Yamaguchi T, Namatsu H, Nagase M, Yamazaki K, Kurihara K. Appl Phys Lett. 1997;71:2388–90.
55. Ferguson RA, Spence CA, Reichmanis E, Thompson LF. Proc SPIE. 1990;1262:412–24.
56. Zuniga M, Wallraff G, Neureuther AR. Proc SPIE. 1995;2438:113–24.
57. Crisalle OD, Keifling SR, Seborg DE, Mellichamp DA. IEEE Trans Semicond Manufact. 1992;5:14–26.
58. Mack CA. J Electrochem Soc. 1987;134:148–52.
59. Ushirogouchi T, Onishi Y, Tada T. J Vac Sci Technol B. 1990;8:1418–22.
60. Namatsu H, Nagase M, Yamaguchi T, Yamazaki K, Kurihara K. J Vac Sci Technol B. 1998;16:3315–21.
61. Ma Y, Shin J, Cerrina F. J Vac Sci Technol B. 2003;21:112–7.
62. Patsis GP, Tserepi A, Raptis I, Glezos N, Gogolides E, Valamontes ES. J Vac Sci Technol B. 2000;18:3292–6.
63. Ocola LE, Orphanos PA, Li WY, Waskeiwicz W, Novembre AE, Sato M. J Vac Sci Technol B. 2000;18:3435–40.
64. Arcus RA. Proc SPIE. 1986;631:124–34.
65. Shih HY, Zhuang H, Reiser A, Teraoka I, Goodman J, Gallagher-Wetmore PM. Macromolecules. 1998;31:1208–13.
66. Tsiartas PC, Flanagin LW, Henderson CL, Hinsberg WD, Sanchez IC, Bonnecaze RT, Willson CG. Macromolecules. 1997;30:4656–64.
67. Flanagin LW, McAdams CL, Hinsberg WD, Sanchez IC, Willson CG. Macromolecules. 1999;32:5337–43.
68. Flanagin LW, Singh VK, Willson CG. J Vac Sci Technol B. 1999;17:1371–9.
69. Burns SD, Schmid GM, Tsiartas PC, Willson CG, Flanagin L. J Vac Sci Technol B. 2002;20:537–43.
70. Schmid GM, Stewart MD, Singh VK, Willson CG. J Vac Sci Technol B. 2002;20:185–90.
71. Tsikrikas N, Drygiannakis D, Patsis GP, Raptis I, Gerardino A, Stavroulakis S, Voyiatzis E. J Vac Sci Technol B. 2007;25:2307–11.
72. Schmid G, Stewart MD, Burns S, Willson CG. J Electrochem Soc. 2004;151:G155–61.
73. Patsis GP, Constantoudis V, Gogolides E. Microelectron Eng. 2004;75:297–308.
74. Patsis GP, Gogolides E. J Vac Sci Technol B. 2005;23:1371–5.
75. Schmid GM, Smith MD, Mack CA, Singh VK, Burns SD, Willson CG. Proc SPIE. 2001;4345:1037–47.
76. Mack CA. J Electrochem Soc. 1987;134:148–52.
77. Yeh TF, Shih HY, Reiser A. Macromolecules. 1992;25:5345–52.

78. Burns D, Schmid GM, Trinque BC, Willson J, Wunderlich J, Tsiartas PC, Taylor JC, Burns RL, Wilson CG. Proc SPIE. 2003;5039:1063–74.
79. http://www.synopsys.com/Tools/Manufacturing/MaskSynthesis/CATS/Pages/ProximityEffectCorrection.aspx#Scelton.
80. DeRose GA, Zhu L, Choi JM, Poon JKS, Yariv A, Scherer A. J Vac Sci Technol B. 2006;24:2926–30.
81. Mack CA. Microelectron Eng. 1999;46:283–6.
82. http://www.kla-tencor.com/lithography-modeling/probeam.html.
83. Glezos N, Raptis I, Hatzakis M. Microelectron Eng. 1994;23:417–20.
84. Rosenbusch A, Cui Z, DiFabrizio E, Gentili M, Glezos N, Meneghini G, Nowotny B, Patsis G, Prewett P, Raptis I. Microelectron Eng. 1999;46:379–82.
85. Eisenmann H, Waas Th, Hartmann H. J Vac Sci Technol B. 1993;11:2741–5.
86. http://www.synopsys.com/Tools/Manufacturing/MaskSynthesis/CATS/Pages/ProximityEffectCorrection.aspx#PROXECCO.
87. Unal N, Mahalu D, Raslin O, Ritter D, Sambale Ch, Hofmann U. Microelectron Eng. 2010;87:940–2.
88. Tsikrikas N, Drygiannakis D, Patsis GP, Raptis I, Stavroulakis S, Voyiatzis E. Jpn J Appl Phys B. 2007;46:6191–7.
89. http://www.synopsys.com/TOOLS/TCAD/PROCESSSIMULATION/Pages/SentaurusLithography.aspx.
90. Horiguchi S, Suzuki M, Kobayashi T, Yoshino H, Sakakibara Y. Appl Phys Lett. 1981;39:512–4.
91. Salvat F, Parellada J. J Phys D. 1984;17:185–202.
92. Patsis GP, Tsikrikas N, Raptis I, Glezos N. Microelectron Eng. 2006;83:1148–51.
93. Rizvi S, Handbook of photomask manufacturing technology. Boca Raton, FL: CRC Press; 2005, Chap. 11.
94. Drouin D, Couture AR, Joly D, Tastet X, Aimez V, Gauvin R. Scanning. 2007;29:92–101.
95. Quesnel E, Hue J, Muffato V, Pellé C, Lamy P. J Vac Sci Technol B. 2004;22:2353–8.
96. Deshmukh PR, Khokle WS. Solid-State Electron. 1989;32:261–8.
97. Kruit P, Steenbrink S. J Vac Sci Technol B. 2005;23:3033–6.
98. Kotera M, Yagura K, Niu H. J Vac Sci Technol B. 2005;23:2775–9.
99. Van Steenwinckel D, Lammers JH, Koehler T, Brainard RL, Trefonas P. J Vac Sci Technol B. 2006;24:316–20.
100. Drygiannakis D, Patsis GP, Raptis I, Niakoula D, Vidali V, Couladouros E, Argitis P, Gogolides E. Microelectron Eng. 2007;84:1062–5.
101. Patsis GP. Polymer. 2005;46:2404–17.
102. Neureuther AR, Pease RFW, Yuan L, Parizi KB, Esfandyarpour H, Poppe WJ, Liddle JA, Anderson EH. J Vac Sci Technol B. 2006;24:1902–8.
103. Constantoudis V, Patsis GP, Leunissen LHA, Gogolides E. J Vac Sci Technol B. 2004;22:1974–81.

Helium Ion Lithography

Principles and Performance

Emile van der Drift and Diederik J. Maas

4

Abstract

Recent developments show that Scanning Helium Ion Beam Lithography (SHIBL) with a sub-nanometer beam diameter is a promising alternative fabrication technique for high-resolution nanostructures at high pattern densities. Key principles and critical conditions of the technique are explained. From existing data, the fundamental factors underlying the sensitivity gain by 1–2 orders of magnitude and the prospects for high resolution at high pattern densities are analysed. State-of-the-art performance of the technique is illustrated with experimental achievements in HSQ and PMMA resists. Exploratory SHIBL work on aluminum oxide resist is presented as a novel approach to overcome potential shot noise effects in pattern definition and to improve masking capabilities in subsequent pattern transfer.

4.1 Introduction

The launch of the helium ion (He^+) microscope by Zeiss/Alis [1] has opened novel perspectives in nanotechnology, both from an imaging and manufacturing point of view. The key point in this development is the sub-nanometer He^+ beam technology. For imaging, the resolution of the novel helium ion microscope (HIM) progressed to about 0.35 nm [2]. In this, it nicely fills the gap between scanning electron and transmission electron microscopy. For manufacturing, thus far only a few studies with the sub-nanometer probe setup were reported, in ion milling [3], ion beam-induced growth [4] and ion beam lithography [5, 6]. Nanostructuring

E. van der Drift (✉)
Kavli NanoLab, Delft University of Technology, Delft, The Netherlands
e-mail: e.w.j.m.vanderdrift@tudelft.nl

D.J. Maas
TNO Science & Industry, Delft, The Netherlands

M. Stepanova and S. Dew (eds.), *Nanofabrication*,
DOI 10.1007/978-3-7091-0424-8_4, © Springer-Verlag/Wien 2012

capabilities in both milling and beam-induced growth will be elaborated on in Chap. 11 on He$^+$-induced processing. In the present chapter, we will focus on the promising potential of scanning He$^+$ ion beam lithography (SHIBL). State-of-the art SHIBL performance is more or less equivalent to the best electron beam lithography (EBL) achievement [7], except for the 1–2 orders of magnitude higher sensitivity as compared to EBL. See Fig. 4.1.

A great beneficial circumstance of SHIBL is the much more directional scattering profile of helium ions in matter with negligible backscattering, as compared to electrons. Figure 4.2 illustrates calculated trajectories of helium ions and electrons both at 30 keV for several depths (1,000, 100 and 20 nm) in silicon [8]. The top panels show that 30 keV electrons have substantial forward, lateral and backscattering over a range of about 6 μm, while most helium ions scatter merely forward in a 30° cone down to a depth of 400 nm. It illustrates two important beneficial features of SHIBL as compared to EBL. Firstly, He$^+$ ions result in an energy deposit over trajectories being about a factor of about 15 shorter and so yield a correspondingly higher resist sensitivity. Secondly, the negligible backscattering accounts for the greatly reduced proximity effect from adjacent pixel exposures. It allows for the break-through in achievable pattern density shown above, whereas for EBL electron backscattering tends to be a show stopper [7]. The middle panels show how the ion and electron scattering have broadened the interaction region by 20–30 nm at a depth of 100 nm, which corresponds to the typical resist layer thickness. The bottom panels zoom in to the upper 20 nm depth, and show a broadening by about 5 and 1 nm for electron and He$^+$ exposure respectively. Apparently, in this layer thickness range the potential for high-resolution exposure with the sub-nanometer probe emerges at full strength.

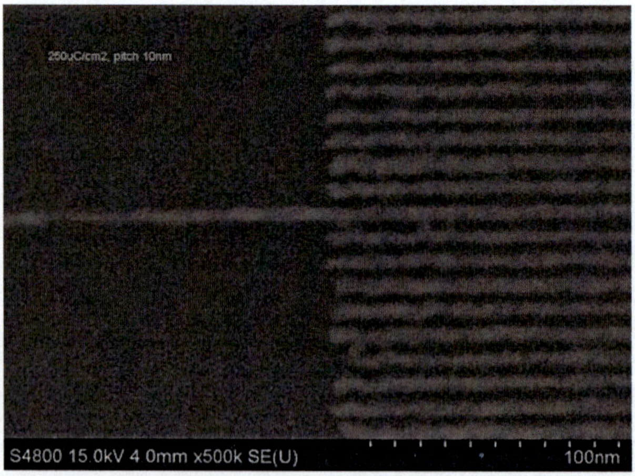

Fig. 4.1 SEM micrographs of 5 nm wide lines at 10 nm pitch written in a 5 nm thick HSQ resist layer by He$^+$ beam lithography. All features are exposed at the same ion dose without any proximity correction. The uniform line width of the isolated line and the *outer* and *inner lines* of the array are indicative of the very low proximity effect

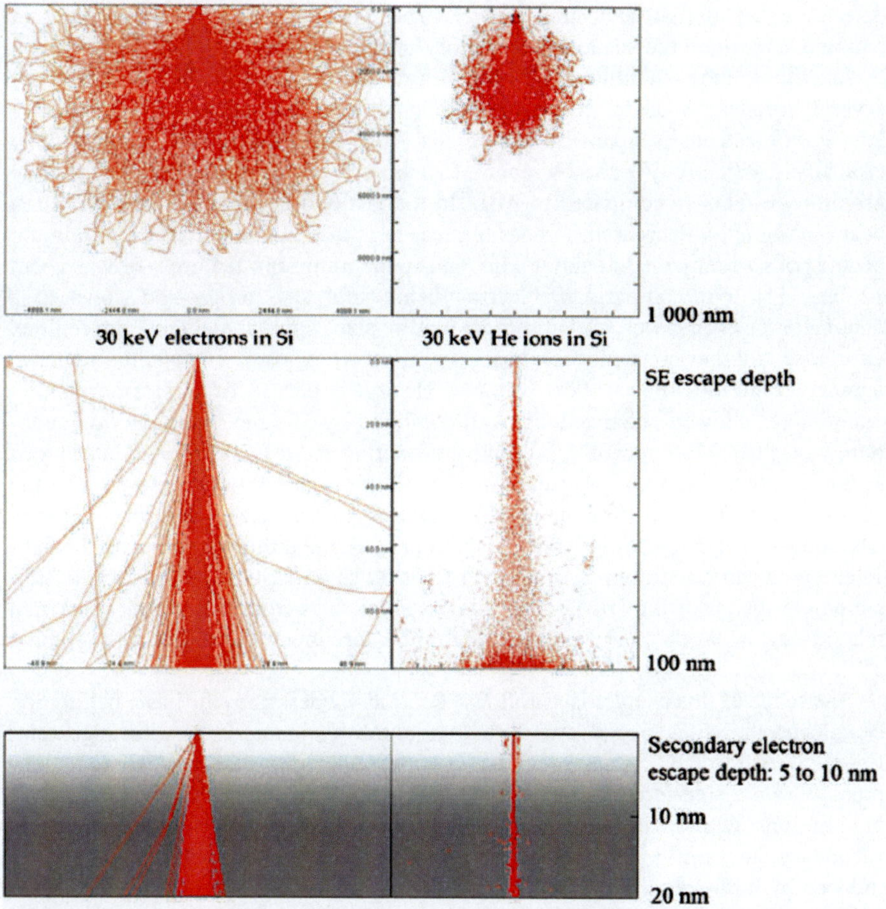

Fig. 4.2 Comparison of scattering of 30 keV helium ions and electrons in silicon showing the more confined interaction volume for helium ion impact (By courtesy of Postek et al. [8])

Another important aspect of particle beam lithography is the generation of secondary electrons (SE,s), which in EBL account for the actual exposure reaction. The bottom panels in Fig. 4.2 indicate that for both electron and He^+ exposure at 30 keV the calculated escape depth of SE's is typically in the range of 5–10 nm. It establishes the potential interaction range of the SE's with the resist molecules. In the literature, quantitative experimental information of SE's is available mainly in terms of escape depth of the SE in the scope of microscopy. These studies show that for He^+ exposure, the number of SE's is higher and the SE energy spectrum is shifted to lower energies as compared to SE's from electron exposure [9]. It is pointed out that in lithography it is the (energy dependent) inelastic collisional cross section of the SE and the bond energy of the chemical bonds involved which determine the actual exposure impact. The SE's establish a lateral blur of nanometers around the scattering trajectory of the primary beam particles.

Ion lithography, including light-ion beams like H^+ and He^+, has existed for several decades. A useful review including all pros and cons when comparing ions with electrons is given by Melngailis [10]. The experimental studies show typically a sensitivity gain by about two orders of magnitude and negligible proximity effects as compared to EBL. In the last two decades the interest in ion beam lithography decreased. A possible reason is that in this period the minimum beam spot size remained at about 8 nm and so the minimum feature size was about 12 nm [11]. For comparison: electron beam spot size progressed down to a nanometer or even lower, and minimum feature sizes approached the 5 nm regime. Of course, another consideration in ion vs. electron exposure is potential material damage from ion impact. For example, He ion exposure from plasma etching experiments showed enhanced sidewall depletion by 250 nm in conducting quantum wires [12]. More recently, proton beam writing at MeV energy was introduced as a useful technique for 3-D fabrication and high-aspect-ratio structures (like the LIGA process). With this approach, features down to 30 nm in hydrogen silsesquioxane (HSQ) resist were obtained [13]. A breakthrough towards the sub-10 nm area came with two recent He^+ litho studies on HSQ [5, 6] where the sub-nanometer beam of the HIM setup was essential. It is from this recent work and related thesis work [14] we will highlight the most important lithographic achievements.

The scope of this chapter is as follows. Section 4.2 describes the basic helium ion beam system, with special attention to the sub-nanometer probe arrangement. Essential instrumental attachments like a pattern generator and provisions to suppress vibronic interference are briefly indicated. In Sect. 4.3, the interaction of helium ions with matter is discussed in more detail. Scattering behavior and secondary electron yields are the important components here, as well as damage induced by ion impact. In Sect. 4.4, we elaborate on the lithographic performance. Three resist systems are considered: the state-of-the-art organic high resolution materials of hydrogen silsesquioxane (HSQ) and poly (methyl methacrylate) (PMMA) and an explorative purely inorganic aluminum oxide resist grown by atomic layer deposition (ALD). Finally, in Sect. 4.5, the progress as yet is summarized as conclusions.

4.2 Helium Ion Beam System

The helium ion microscope (HIM) introduced by Carl Zeiss SMT, Inc in 2006 is innovative for the unique sub-nanometer He^+ ion source. The source is a tungsten wire, the end of which acts as a single-atom emitter of ions, see Fig. 4.3. The end of the wire has a pyramidal shape with an atomically sharp apex and edges. Actually, the apex consists of three atoms and is called a "trimer." When cooled down to below 77 K, the tip acts like a cold finger for the helium gas ambient. The arriving helium ions are ionized by the strong electric field just around the few outer atoms

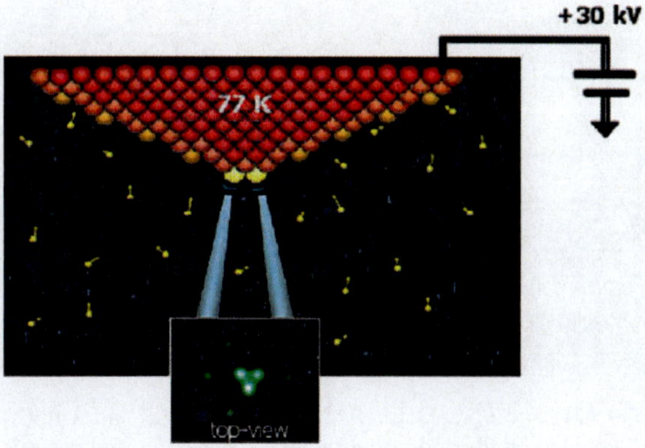

Fig. 4.3 Trimer helium ion source concept (By courtesy of Dr. L. Scipioni/Carl Zeiss SMT, Inc.)

of the apex creating the helium beamlets. See the inset of Fig. 4.3. The emission from only one atom is aligned into the column creating an almost ideal point source with a very narrow energy spread. Measurements of the energy spread [15, 16] indicate that the source has an ionization disc with an approximate thickness of 0.3 Å. The virtual source size has been estimated by calculation to be in the order of 2.5 Å. The beam current can be easily controlled, as it is proportional to the partial helium pressure in the source vacuum. At the moment the helium current from a single atom can be in the order of 10 pA. The acceleration voltage can be varied between 10 and 30 kV. Compared to other ion source concepts, the single atom emitter gives a smaller source, a higher brightness, and a smaller energy spread. It also allows the column to be operated with less demagnification and thus a small beam convergence angle. This leads to a long depth of focus for imaging and a large focus tolerance in nanofabrication.

The ion-optical column is a traditional two-lens electrostatic configuration, which operates in a crossover mode with pre-lens deflection [17]. The microscope is equipped with a secondary electron detector for high-resolution surface imaging.

Figure 4.4a shows the HIM tool at TNO-Delft, upgraded for 3-D nanofabrication. The setup has been equipped with a pattern generator (Raith Elphy Plus) and a gas injection system (Omniprobe, OmniGIS). These extensions enable the system to be used for direct write, lithography and helium ion beam induced etching and deposition [18], see Chap. 11. For nanofabrication of the smallest features it is important not only to have a small beam spot, but also to control accurately the movement of the spot over the surface. Vibrations are one of the major factors, which can make the helium beam dislocate over the sample, creating a blurring effect. The source is placed in the top of the microscope hanging in the gun area and cooled down to a temperature between 70 and 75 K. This configuration makes the source very susceptible to vibronic coupling. For this reason, the source is cooled with a solid N_2 cooler. The alternative, i.e. cooling with boiling liquid N_2, would give too much vibronic interference. As another result of the solid N_2 cooling, the temperature stability of the source is within 1 K during a period of several hours. To

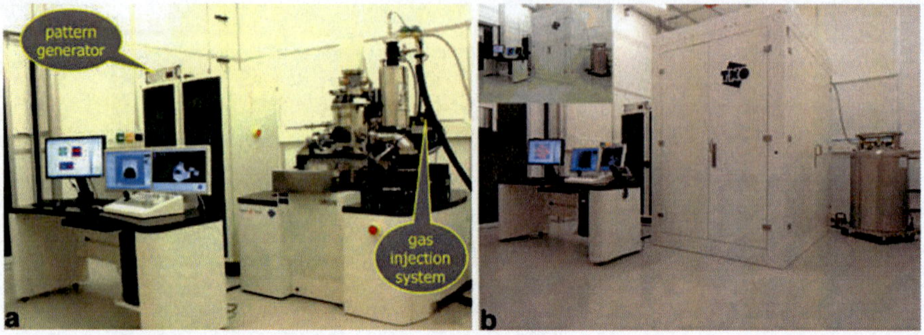

Fig 4.4 The TNO-Delft HIM setup without (**a**) and with (**b**) acoustic enclosure. Nanofabrication is enabled by the addition of a pattern generator and gas injection system to the HIM

isolate the microscope from external acoustic noise, from e.g. the room air conditioning, operators and other sources, Zeiss and TNO have developed an acoustic enclosure for the HIM reducing the sound load at least by 12 dB, as is illustrated by Fig. 4.4b. These recent improvements both allow for higher precision in nanofabrication.

4.3 He^+ Ion: Matter Interactions

Two additional effects contribute to the overall areal exposure of a single beam spot: (1) broadening of the beam exposure by forward scattering, (2) lateral extension of the exposure reaction due to the impact from SEs. Actually, these phenomena are intimately related to the scattering behavior of the primary particle in matter. In this section, we discuss how this works out for SHIBL as compared to EBL. Specific resist molecule aspects like length of molecules, diffusion of fragments, etc. which also may contribute in the resolution, are not taken into account.

4.3.1 Primary Beam Divergence

Firstly, we consider the forward scattering of the helium ions due to elastic collisions in the material. Figure 4.2 clearly indicates that at a depth of 20 nm in silicon the divergence of the helium particles is about five times smaller compared to electrons. SRIM calculations by Sijbrandij et al. [19] on He^+ transmission through titanium give a more quantitative picture of the beam spread. See Fig. 4.5. Mean radial spread from the primary beam axis is about 0.085 nm over the first 5 nm of material. The SRIM study also shows that for carbon material (which closely resembles the composition of organic resists like PMMA) the divergence will not be much different. By extrapolation of the functional behavior to a typical resist thickness of 20 nm, a broadening in the order of 1 nm can be

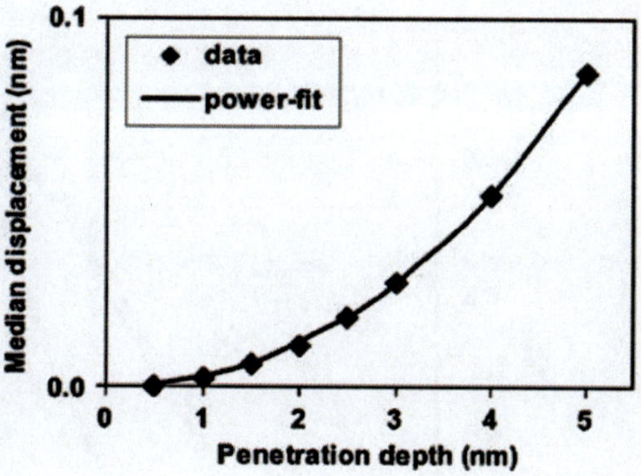

Fig. 4.5 Mean radial displacement (with respect to beam axis) for a 30 keV helium beam passing through a titanium sample as a function of the penetration depth. The *continuous line* is power fit to the data (By courtesy of Sijbrandij et al. [19])

estimated. At this point, one would conclude that resist layer thickness should be limited to sub-10 nm to keep full advantage of the sub-nm beam size. In the following we will learn that other interaction effects are more dominant, relieving somewhat the layer thickness condition.

4.3.2 Secondary Electron (SE) Generation

The SE's are often considered as the most important drivers of the exposure reaction with resist molecules, besides possible contributions from molecular excitations, and other intramolecular electronic processes. SE's are generated by inelastic collisions of the He^+ particles with resist molecules. In the Introduction, we already mentioned the 15 times higher energy deposition per unit of length in He^+ exposure. Two additional factors are also important: (1) the number of SE's generated and (2) the energy distribution of the created SE's. Figure 4.6 shows the 'universal' plot of the calculated normalized SE yield for He^+, H^+ and electron exposure as a function of the normalised particle energy [20]. E_{max} is the ion energy with the maximal SE yield δ_{max} for a given exposed material. The dashed line indicates the position of the helium exposure at 30 keV. Carbon with $E_{max} = 600$ keV is taken as being representative for resist material. With an $E_{max} = 0.4$ keV for electron exposure of carbon [21], typical EBL settings of 30 and 100 keV beam energy are off the X-axis scale ($E/E_{max} = 75$ and 250, respectively). The SE yield from EBL at 30 keV is estimated from Fig. 4.6 by extrapolation according to lnE/E. It indicates a five times lower *normalized* SE yield for electron exposure as compared to He^+ exposure. With δ_{max} values of 4.1 and 1.06 for He^+ [20] and electron [21] exposure, respectively, we find the SE yield for 30 keV He^+ exposure to be about 20 times higher than for electron exposure at the same energy.

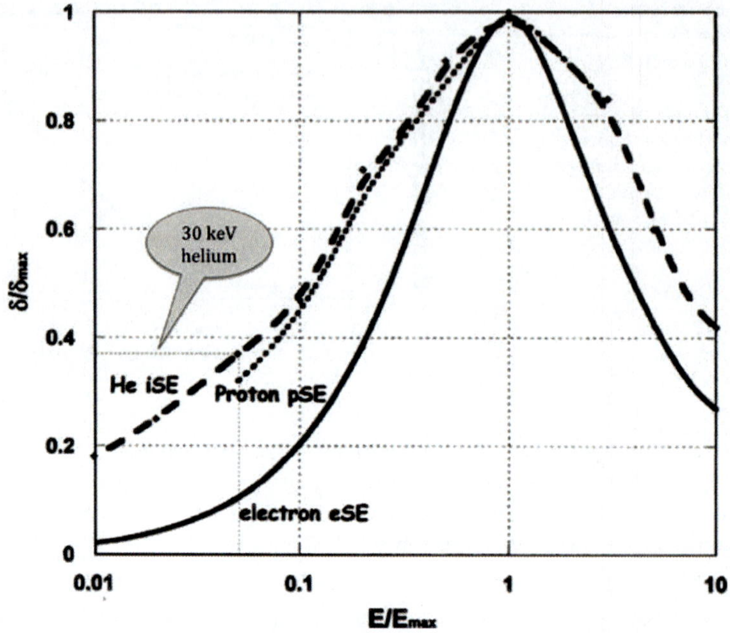

Fig. 4.6 'Universal' curves for normalised secondary electron (SE) yield vs. normalised incident particle energy of helium ions, electrons and protons (By courtesy of Joy [20])

4.3.3 SE Energy Distribution

The involvement of the SE in the exposure process is highly energy dependent. For example, a minimum energy requirement for a SE to participate in the exposure is the energy needed to break a chemical bond. Therefore, we consider measured energy spectra of SE's escaped from Si under He^+ and electron exposure from Vyvenko [22] and Joy [23] respectively. The spectra are from different experiments offering energy distribution information but no comparison in yield. However, the areas under the spectra can be taken as the relative measure of the total yield. Using the estimated 20 times higher yield from He^+ exposure yields the SE energy spectra with comparative intensity information as shown in Fig. 4.7. The SE spectrum of the electron exposure is scaled up by factor 4 for better comparison in the graph. Apparently the SE spectrum for He^+ exposure is at an energy range well below 10 eV with a maximum at about 2 eV. The SE energy range from electron exposure has a maximum at about 7 eV and is broadened up to tens of eV with no clear cut-off.

Several features from Fig. 4.7 need to be addressed. First, the spectra offer the SE characteristics for the exposure of silicon by He^+ and electrons. The comparison can be useful for other materials like resists, albeit only qualitatively. Secondly, the plotted energy is the free electron energy after escape from the material. In the resist the actual SE energy is about 1 eV higher because of the electron affinity in PMMA [24] or SiO_2 [25]. With the energy for bond scission taken as the energy threshold,

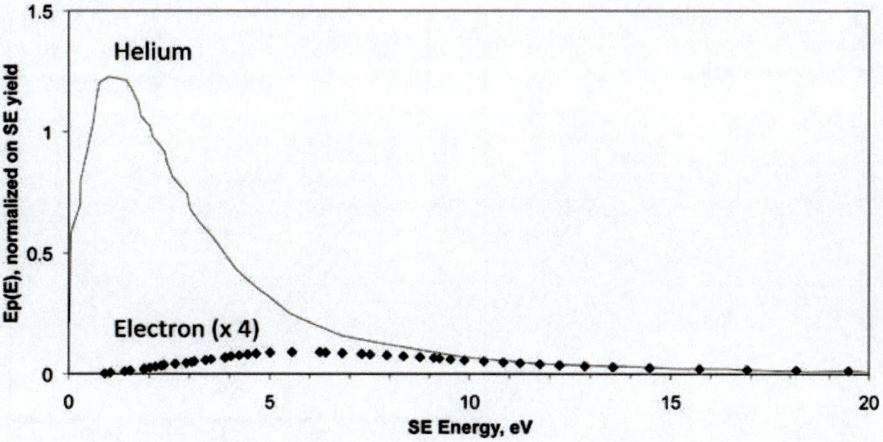

Fig. 4.7 Secondary electron yield vs. energy as generated by electron and helium ion exposure at 30 keV on silicon. The graph is a re-scaling of the plot of Scipioni [9], taking into account the 20 times higher SE yield from He^+ exposure. In lithography, only SE's with sufficient energy to break a bond are effective. Typical resist bond dissociation requires more than four eV. Original data of He^+ and electron exposure by courtesy of Vyvenko [22] and Joy [23] respectively

one can readily obtain the SE fractions, which would potentially contribute to the exposure reaction. Numerical integration of the spectra shows that for a typical threshold of 4 eV about eight times more SE can be effective in bond breaking under He^+ exposure; this factor decreases to about 2 for a threshold of 9 eV. In combination with the larger energy deposition per unit of length by a factor of about 15 (Introduction, Fig.4.2), the overall sensitivity gain becomes apparent. Finally, it is pointed out that measurement of SE energy spectra is sensitive to the surface condition of the sample, influenced by history and preparation. It creates an uncertainty of a few eV in the relative position of the two curves in Fig. 4.7, and so uncertainty when comparing the impact from SE yields from electron and He^+ exposure.

4.3.4 SE Exposure Contribution

The energy dependent role of the SE in the exposure process is expressed by the effective collisional cross section in bond scission reactions. Electron cross-section data of resist for the energy range of interest (0–100 eV) are not available to our knowledge. Electron cross sections are available from electron or ion beam-induced deposition where adsorbed precursor molecules are decomposed by SE impact [26, 27]. An alternative and useful approach is to consider the inelastic scattering length of electrons λ_n in matter. Figure 4.8 shows a compilation of experimental values of λ_n for a variety of elemental materials as a function of the electron energy [28]. The continuous curve is a least squares fit to a general energy dependence

Fig. 4.8 Mean free path of electrons vs. energy for inelastic scattering (By courtesy of Wiley and Sons [26])

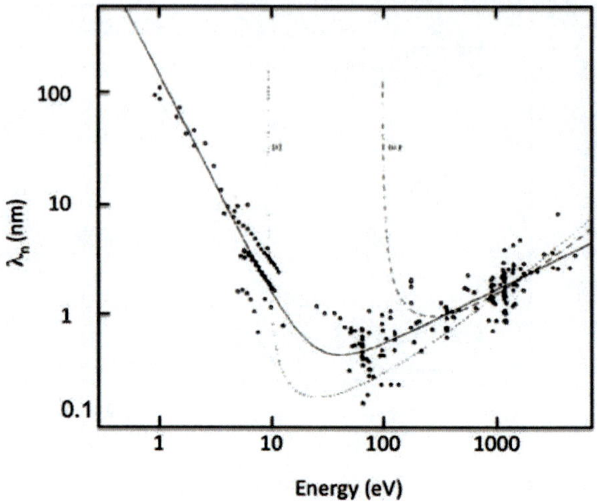

$\lambda_n = a.E^{-2} + b.E^{1/2}$. A recent comprehensive study of low-voltage imaging in scanning electron microscopy (SEM) by Kieft and Bosch [29] confirms this functional behavior. The number of inelastic scatterings of the electrons per unit of path length is inversely proportional to λ_n. Therefore, $1/\lambda_n(E)$ can be taken as a qualitative measure for the energy dependent impact of a SE in the exposure. The $1/\lambda_n(E)$ behavior shows qualitative correspondence with electron dissociation cross-section data of WF_6 [27], which displays a threshold of 6.5 eV and a broad maximum between 25 and 40 eV. Interestingly, the W-F bond strength in WF_6 is about 5 eV [30] which is about 1.5 eV below the threshold in the electron dissociation reaction. Convolution of $\lambda_n{}^{-1}(E)$ with the SE spectra in Fig 4.7 yields the SE impact as an integral over the relevant SE energy range. For a threshold of 4 eV for bond scission, the SE impact from He^+ exposure is about 2.8 times larger. In combination with the factor of 15 for the higher energy deposit per unit of length, it yields an overall sensitivity enhancement of about 42. This factor decreases gradually to 34 for a threshold up to 9 eV. So far, it was implicitly assumed that the energy deposit is uniform over the He^+ or electron trajectory. However, an important difference is that He^+ loses its energy mainly in the first part of its trajectory while electrons lose most energy deeper in the material. See Fig. 4.6 for the different slowing down situations. Especially in thin resist layers, this difference is beneficial for the sensitivity in the exposure.

Figure 4.8 expresses some more useful information, i.e. about the lateral range of the exposure reaction. The relevant SE energy range for He^+ and electron exposure (Fig. 4.7) spans typically the 4–40 eV interval. With the $\lambda_n(E)$ behavior in Fig. 4.8 it points to a calculated inelastic scattering length up to 4 nm. The low-energy SE's with the longer inelastic scattering length are less effective in the exposure. In practice, the minimum resist feature size in HSQ resist is about 5 nm. It points to a radial extension of the exposure reaction by 2–3 nm, both for He^+ and electron exposures. Fine-tuning of the development process (developer strength, time) is the obvious way to trim minimum feature size further down, but the natural lower limit comes in sight.

In summary, the blur in He^+ exposure by several nanometers due to SE's exceeds the nanometer broadening by ion forward scattering in resist layers up to about 20 nm layer thickness. Several factors contribute to the 1–2 orders higher sensitivity in SHIBL as compared to EBL. Firstly, the energy deposition per unit of length is about 15 times larger. A beneficial side-effect is that for He^+ this energy deposit is largest in the first part of the trajectory, while for electrons it is increasingly deeper in the material. Secondly, the overall SE impact from He^+ exposure is typically 2–3 times larger per ion. Herein, the number of SE from He^+ is about 20 times larger, however with reduced cross section for bond scission because of the lower energy range (0–10 eV). Altogether, a sensitivity gain by a factor of more than 40 can be readily explained.

4.3.5 Exposure Damage

An important side effect of ion exposure is potential damage to the irradiated substrate. In lithography with ion beams, such damage can occur in the semiconductor material under the resist layer, e.g. silicon, silicon-germanium or gallium arsenide. The penetration depth of 30 keV He^+ ions is typically several hundreds of nanometers (see Fig. 4.2) while common resist thickness is in the tens of nanometers scale. So, rather than in the resist, most helium ions are stopped in the underlying material and could raise a damage issue there. The important

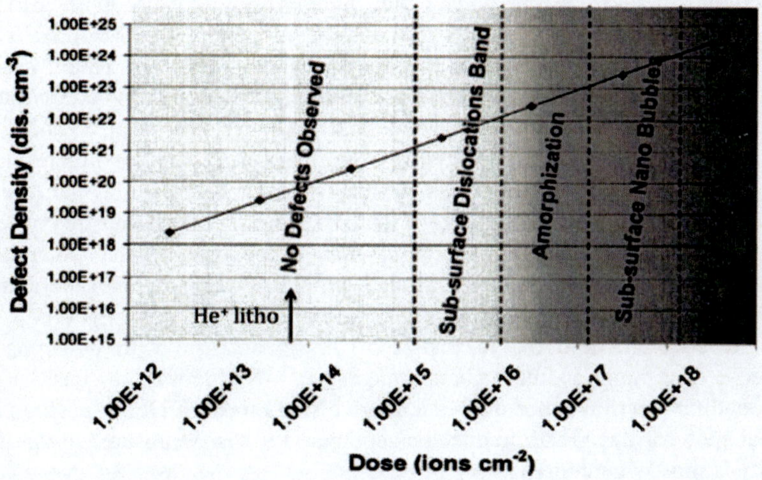

Fig. 4.9 Volumetric defect density from SRIM simulations vs. ion exposure dose. *Arrow* at 7×10^{13} cm^{-2} indicates dose typical for helium ion lithography (By courtesy of Livengood [31])

parameter is the areal dose of the exposure. Figure 4.9 summarizes the physical damage effects involved with ion impact as a function of ion dose [31]. Typical exposure dose for ion lithography in current resists is on the level of 10 μC-cm^{-2} corresponding to the arrow at 7×10^{13} cm^{-2} in Fig. 4.9. So, the damage problem seems limited. However, it is pointed out that electrical damage in terms of depletion and trapping effects may show up already at much lower defect densities, e.g. in gate oxides at the level of 10^{10}–10^{11} cm^{-2} [32].

4.4 Lithography

The comparative lithographic performance study of helium ion and electron beam exposures comprises sensitivity, contrast and resolution. Further, three resist materials are considered here: hydrogen silsesquioxane (HSQ), poly (methyl methacrylate) (PMMA) and aluminum oxide (Al_2O_3). HSQ and PMMA resists are high-resolution standards in state-of-the-art EBL with smallest feature sizes down to about 6 [33] and 10 nm [34] respectively. The e-beam exposure mechanism in HSQ and PMMA is based on local chemical modification. An interesting research question is how the totally different collisional interaction of He^+ ions with these resist molecules will work out in sensitivity, contrast and resolution, the more so now a sub-nm probe size is available.

Exploration of aluminum oxide as resist under He^+ ion exposure has a different motivation. A direct consequence of ever-increasing resolution is reduction of the pixel size. The correspondingly lower number of exposure particles (N) per pixel is a serious threat for the exposure uniformity because of the inherent shot noise (proportional to $\sqrt{N}$). On the other hand, higher resolution also pushes to ever thinner resist layers, which makes the subsequent pattern transfer increasingly more difficult. A natural solution may be inorganic materials, which are less sensitive (so require a higher N) and have a better etch selectivity in pattern transfer by plasma processing. Inorganic resist materials like metal halides [35] and aluminum oxide [36, 37] have been explored by e-beam exposure. Outstanding resolution down to 1–2 nm was ascribed to a mechanism of local material damage and elemental removal with negligible contributions from SE. Hereafter this exposure mechanism is indicated as *damage-controlled* as opposed to the conventional mechanism of local *chemical conversion*. A prohibitive drawback of the damage-controlled mechanism is the extremely high electron dose involved (10^4 C/cm^2 range). In contrast, lithographic studies of this mechanism with Ga^+ ions [38] revealed a dose range in the mC/cm^2 range, i.e. 7–8 orders more sensitive than corresponding electron exposure. An attractive side aspect of He^+ exposure is that physical sputtering is greatly reduced as compared to Ga^+ exposure, so that mask integrity is mostly guaranteed.

In this chapter, we focus on the lithographic results obtained with the sub-nm He^+ probe. For experimental methods and conditions, we refer to the original work of Winston et al. [5] and Sidorkin et al. [6, 14].

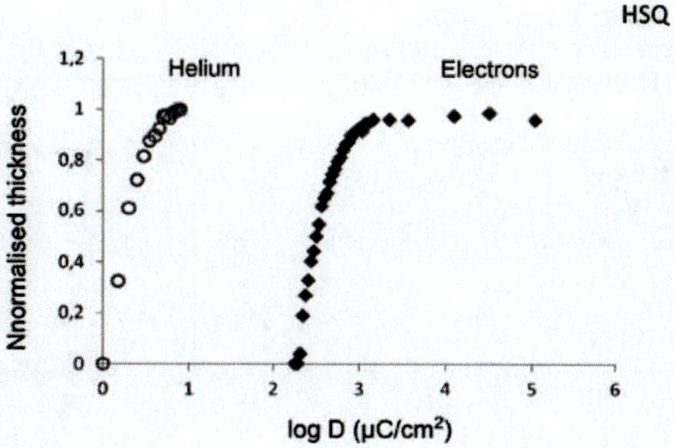

Fig. 4.10 Dose response curves of HSQ resist for He^+ and electron exposure at 30 and 100 keV, respectively

4.4.1 Sensitivity and Contrast

Figure 4.10 shows the normalized thickness of HSQ resist versus exposure dose of 30 keV He^+ ions after development.[1] For comparison, results for exposure with 100 keV electrons are included. To compare He^+ and electron exposures at the same particle energy of 30 keV, the dose of the 100 keV e-beam data is to be scaled down by a factor of 100/30. This scaling factor was established previously from comparative writing experiments at 100 and 30 keV, and proven to be consistent with exposure results at different keV on PMMA [39]. The resulting dose D_{50} at 30 keV, defined as the dose at which the developed structure has a thickness of 50% of the original thickness, is then found to be $1.7 \pm 0.1\ \mu C/cm^2$ for He^+ and $94 \pm 2\ \mu C/cm^2$ for electron exposure respectively. It reveals a sensitivity gain for He^+ exposure by a factor of 55. The contrast, defined from tangent at half-height according to Thompson [40], is in both experiments the same within experimental error, i.e. 2.0 ± 0.5 (electron) and 2.3 ± 0.5 (He^+).

Figure 4.11 shows He^+ and electron 30 keV exposure results for PMMA with average molecular weight of 950,000 (950k). The plotted electron exposure data are from measured 100 keV values scaled down by a factor of 100/30. Both exposure response curves of PMMA 950k show positive tone in the low-dose range and a negative tone in the high-dose range, in agreement with literature [34, 41]. Note the reduction of the layer thickness in negative tone by about a factor of 2 compared to positive tone. Most probably the resist structure is highly densified by the high exposure dose. In the dose range of 100–10,000 $\mu C/cm^2$, a further decrease of the resist height occurs. With initial layer thickness of 70 nm the decrease is about 10 nm, which cannot be accounted for by sputtering, as that would be estimated at less than a nanometer [42]. Apparently, a further densification is going on in this dose range. D_{50} values for He^+ exposure at 30 kV are 2.0 $\mu C/cm^2$ in positive tone and 68 $\mu C/cm^2$ in negative tone. Corresponding D_{50} values for electron exposure are 138 $\mu C/cm^2$ (+ tone) and 7,890 $\mu C/cm^2$ (– tone), respectively. The accuracy in

[1] More extensive results compared to [6, 14], submitted for publication.

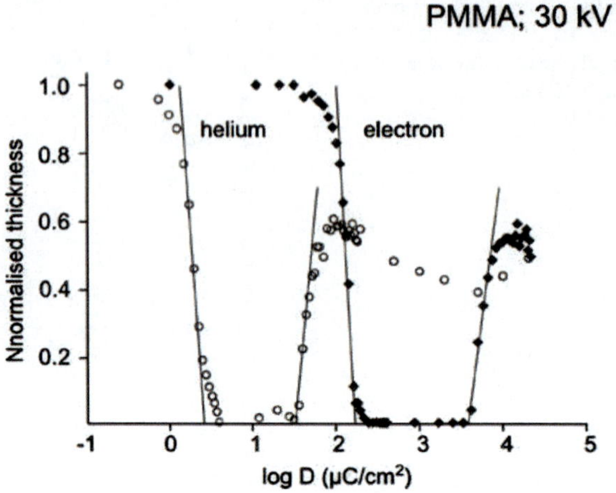

Fig. 4.11 Dose response curves of PMMA for He^+ and electron exposures at 30 keV. The electron dose data are obtained from measured 100 keV data scaled down by a factor of 100/30

Table 4.1 Sensitivity[a] and contrast[a] for HSQ and PMMA 950k for He^+ and electron exposures at 30 keV

Resist	HSQ		PMMA-Pos.		PMMA-Neg.	
Beam	e^-	He^+	e^-	He^+	e^-	He^+
Sensitivity (μC/cm²)	94	1.7	138	2	7,891	68
Contrast	2	2.3	4.2	3.7	3.9	4.7
Enhancement	–	55	–	69	–	116

[a]Accuracies: dose 2%, contrast ±0.5

the observed D_{50} doses is about 2%. Altogether it shows a sensitivity gain for the He^+ exposure by a factor of about 69 (+tone) and 116 (−tone), respectively. Contrast values range between 3.7 ± 0.5 and 4.7 ± 0.5 for both exposures and beam types.

The sensitivity and contrast data of HSQ and PMMA are collected in Table 4.1. The gain observed for PMMA (positive tone) is in agreement with enhancements reported earlier [43, 44]. The gain observed for HSQ is a novel result.[2] The enhancements are ascribed to the higher exposure impact of secondary electrons in combination with a higher energy deposit per unit of length (see Sect. 4.3). The exposure mechanism of both HSQ and PMMA positive tone involves bond scissions. In HSQ, it is the starting point for cross-linking; in PMMA, it yields low-molecular weight products. In the exposure of HSQ, the scission of the Si-H bond [45] is most important and possibly also Si-O bond rupture contributes [46]. For PMMA, the C-C bond scissions are most important. The slightly lower enhancement in HSQ may point to reduced involvement of the scission of the

[2] Enhancement in previous work [6] was based on erroneous ion current measurement.

relatively strong Si-O bonds (~9 eV) under He^+ exposure compared to the weaker bonds of Si–H and C–C (4 eV). It fits to the SE energy spectrum from He^+ exposure, which contains a small energy component beyond the Si–O bond strength (see Fig. 4.7).

The negative tone in PMMA at high exposure dose is ascribed to cross-linking. The concentration of reactive molecular fragments from bond scissions is so high that mutual coupling reactions are favored, and a molecular network is formed. So, the negative tone includes (at least partly) so-called second order reactions, which proceed according to quadratic dependence on the bond scission concentration. This contrasts with molecular fragmentation (positive tone), which goes linear with the concentration of the bond scission product. It explains the substantially higher sensitivity gain for PMMA in the negative tone as compared to the positive tone. The similar contrast values under He^+ and electron exposures in all cases points to more or less identical molecular weight distributions in the exposed area. Only the negative tone PMMA product under He^+ ion exposure seems to be shifted somewhat to a less soluble cross-linked state as manifested by the slightly higher contrast value.

Figure 4.12 shows the sensitivity of PMMA 950k for the 10–30 keV He^+ energy range, both for the positive tone and negative tone (the former scaled up vertically by factor 5 for better comparison in the graph). The continuous lines represent a fit to a functional variation according to S^{-1}, the reciprocal value of the electronic stopping power S of He^+ [47]. The results show unambiguously that the required He^+ exposure dose D_{50} is inversely proportional to the electronic stopping power of He^+, in agreement with the previous section and Ref. [43]. For He^+ exposure in the 10–30 keV range, it implies sensitivity enhancement towards higher beam energy. This is dissimilar to electron exposure where the stopping power as a function of the electron energy in the 10–100 keV energy range shows a descending trend and thus a higher dose is required for higher keV beam energy settings. See also Fig. 4.6 where it is shown by calculation that He^+ and electron exposure at 30 keV are below and beyond the energy of maximal SE yield, respectively. Once the D_{50} and

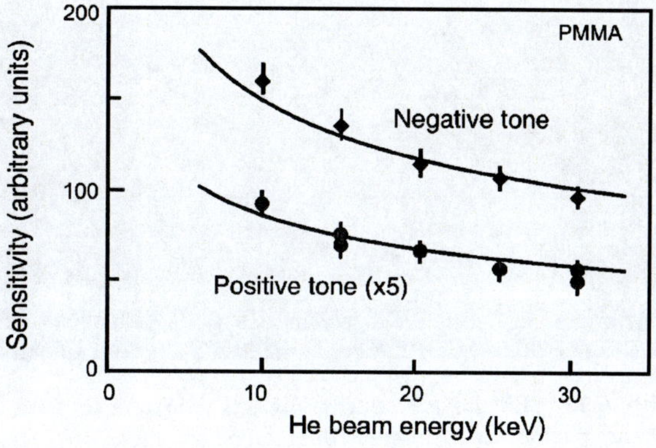

Fig. 4.12 Sensitivity of PMMA under He^+ exposure vs. beam energy. The continuous lines represent functional variation according to S^{-1} with S the stopping power of the He^+ ions. Dose in arbitrary units because of uncalibrated current measurement

contrast values of a resist are known, one can start experiments to determine the resolution under electron and helium ion beam exposure.

4.4.2 Resolution

Resist layer thickness is a crucial parameter in high-resolution lithography. To illustrate its importance, SEM images of He^+ exposed dot arrays at a pitch of 98 nm in 5- and 55-nm-thick HSQ films are shown in Fig. 4.13. The average dot diameter is 6 ± 1 and 14 ± 1 nm, respectively. For both film thicknesses, the exposure dose (100 μs dwell time at 1 pA beam current per pixel) and the development time (5 min) were the same. Nevertheless, the dot size for the thicker layer is about twice as large as for the thinner one. Two effects may contribute: forward scattering of the helium ions (see Sect. 4.3) and SE exposure [48]. Based upon MC calculations, we know that SE exposure at a given location in the thicker layer includes contributions from adjacent areas. For the thinner layer, this proximity exposure is largely reduced. In layers thinner than the escape depth, a fraction of SE's may get lost in the vacuum or in the substrate. A similar minimum dot size of 7.5 nm in 31 nm thick HSQ developed with high-contrast 'salty' developer was reported by Winston [5].

The next step is a test of nanodot and nanoline structures in 5-nm-thick HSQ. The results are shown in Figs. 4.14 (dots) and 4.15 (lines). The pitch in the dot arrays are 48, 24 and 14 nm in Figs. 4.14a, b, and c, respectively. The insets are SEM images at a higher magnification. The dose is for all dot arrays the same, i.e. 1 pA at a dwell time of 100 μs, implying 625 particles per dot. With a shot noise of 25 particles (4%) statistical variations are insignificant given the contrast curve in Fig. 4.10.

The measured dot diameters are collected in Fig. 4.14d, showing a dot diameter 6 ± 1 nm for all pitches. The pitches in the line arrays are 100, 50, 25 and 15 nm.

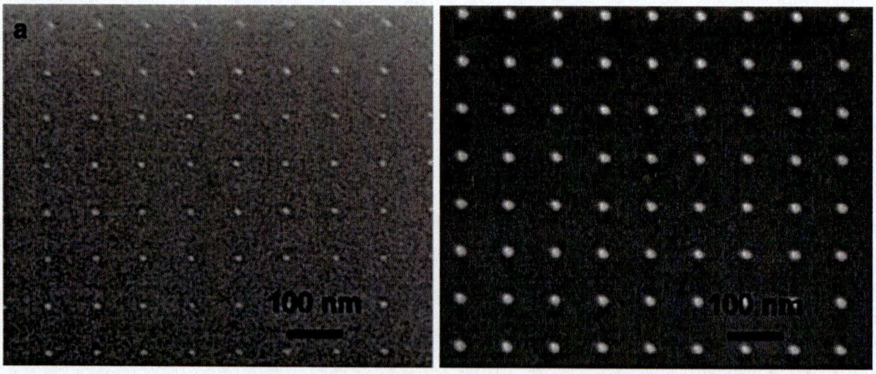

Fig. 4.13 SEM images of arrays of dots written in (**a**) 5 nm and (**b**) 55 nm thick HSQ layers at 98 nm pitch using scanning helium ion beam lithography. Field of view is 900 nm in SE mode at 20 keV. Average dot diameters: (**a**) 6 ± 1 nm and (**b**) 14 ± 1 nm [14]

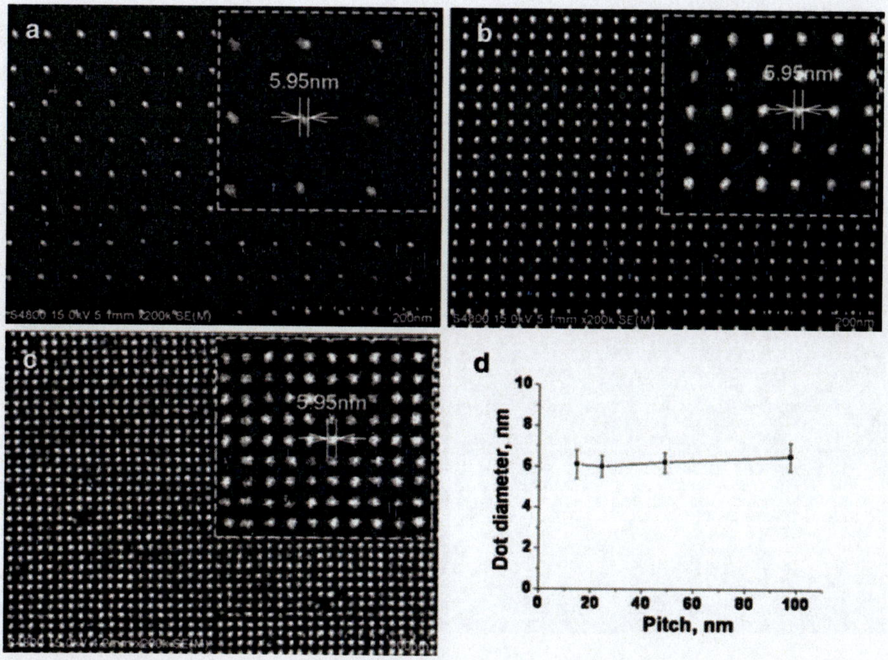

Fig. 4.14 SEM images of arrays of dots written in a 5 nm thick HSQ layer using He^+ exposure at 30 keV at pitches of **(a)** 48 nm, **(b)** 24 nm, and **(c)** 14 nm. Panel **(d)** shows the average dot size versus pitch. The insets are SEM images at a higher magnification. The average dot size for all pitches is 6 ± 1 nm [6]

The area dose for all line features is 500 μC/cm^2. The line width is 6 ± 1 nm. These results are at least competitive with the best e-beam performances. To our knowledge, the best results in e-beam exposure of HSQ are currently 6 nm lines on a pitch of 20 nm, written at 100 keV [49] and 4.5 nm lines on a pitch of 9 nm, written at 30 keV [50]. The latter result was obtained with the high contrast 'salty' developer of NaOH with 4% NaCl. Related work of He^+ exposure of HSQ by Winston [5] shows nested lines of 10 nm width on a pitch of 20 nm. The constant feature size at the same dose for different pattern densities points to a very small proximity effect in He^+ ion beam lithography.

Figure 4.16 shows images of nanostructures written in 20 nm thick PMMA in positive and negative tone using He^+ exposure at 30 keV. The corresponding doses are 250 μC/cm^2 and 2,500 μC/cm^2 respectively. The dose in the high-resolution (HR) experiments differs largely from the large area exposures used in the contrast-sensitivity experiments (Fig. 4.11). The reason is that in the HR experiments the spot size is so small that any overlap from adjacent pixel exposure is excluded.

Throughout the pattern, the feature size is 15 ± 1 nm both in single line and crossed line segments. The constant feature size is an additional indication that proximity effects from He^+ exposures are low indeed.

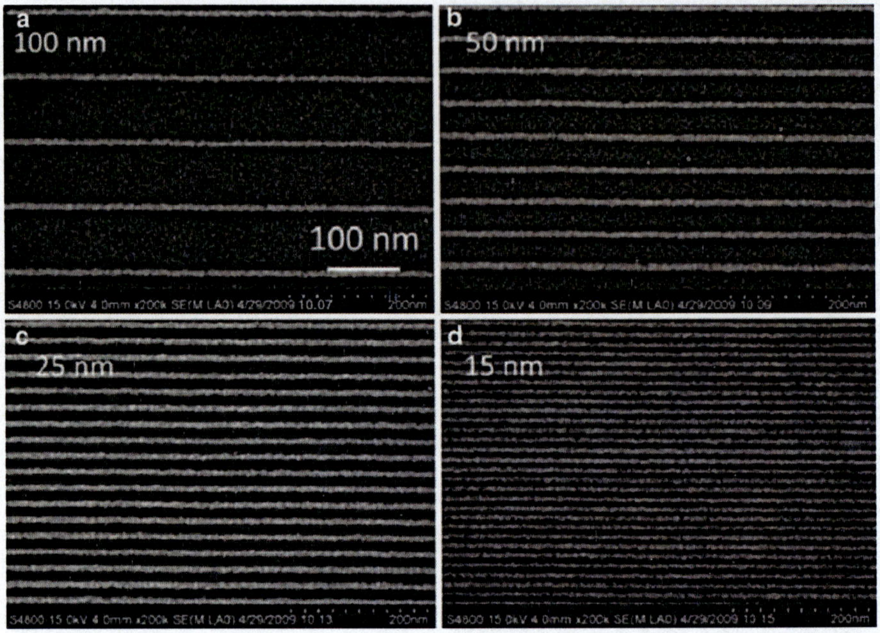

Fig. 4.15 SEM images of arrays of lines written with He^+ exposure at 30 kV in 5-nm-thick HSQ at a pitch of (**a**) 100 nm, (**b**) 50 nm, (**c**) 25 nm and (**d**) 15 nm. Line width for all pitches is 6.5 ± 1 nm [14]

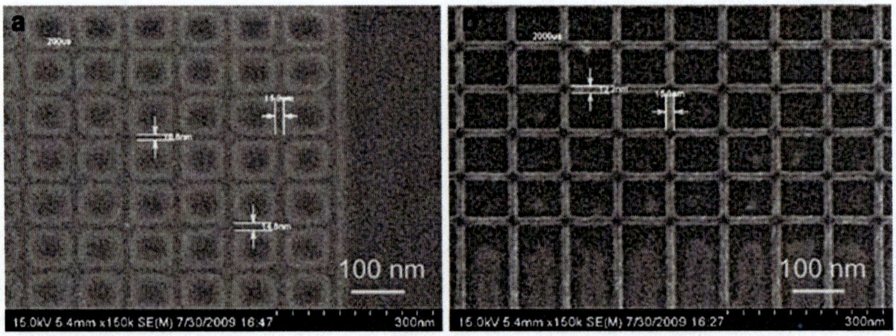

Fig. 4.16 SEM images of grid patterns written in 20-nm-thick PMMA layer with different area doses resulting in (**a**) positive and (**b**) negative exposure modes. Exposure dose: (**a**) 250 $\mu C/cm^2$; (**b**) 2.5 mC/cm^2 [14]

4.4.3 Proximity Effects

The proximity effect, defined as the unintentional exposure of pixels adjacent to the primarily exposed pixel, is a well-known phenomenon in EBL [39]. It is the direct consequence of the scattering of the primary particle back from the substrate to the

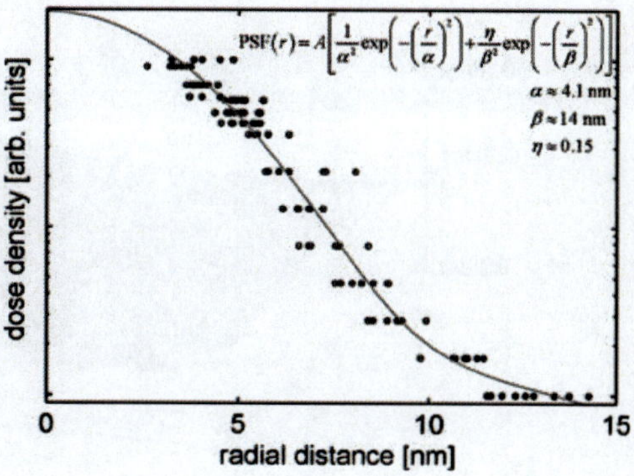

Fig. 4.17 Lithographically estimated point-spread function (PSF) of focused He ion beam exposure of 31 nm thick HSQ resist layer (By courtesy of Winston [5])

resist. The overall exposure $f(r)$ at distance r from the primary exposure can be described well as a superposition of two Gaussians

$$f(r) = A\left[\frac{1}{\alpha^2}\exp\left(\frac{r^2}{\alpha^2}\right) + \frac{\eta}{\beta^2}\exp\left(\frac{r^2}{\beta^2}\right)\right],$$

where the first term describes the primary exposure and the second term refers to the backscattered contribution. The quantities α and β represent the forward and backscattering width respectively, η is the backscatter coefficient which is characteristic for the material involved, and A is a constant. For electrons, the backscatter range β depends on the material and follows an $E^{1.7}$ dependence on the primary electron energy E [51]. Typically, β ranges up to 33 μm in silicon at 100 keV primary beam energy.

Experimental approaches to measure the backscattering include point exposures over a large dose range [52] and so-called doughnut structures [53]. For helium ion exposure, the backscattering has been studied by Winston et al. [5] using the point exposure method. Their experimental result and analysis is shown in Fig. 4.17. The measured forward and backscattering ranges are 4.1 and 14 nm, respectively, while the backscatter coefficient η is found to be 0.15. The numerical outcome fits to the expectations from the scattering behavior (Sect. 4.3). The beam broadening by forward scattering approaches the calculated broadening of 2.5–3 nm at 30 nm depth in silicon (Fig. 4.1, middle panel). The low backscattering is in agreement with Ramachandra et al. [20].

4.4.4 Aluminum Oxide Resist

AFM plots of squares after exposure to He^+ doses of 1 and 20 mC/cm^2 and subsequent development in MF351 for 10 min are shown in Figs. 4.18a, b,

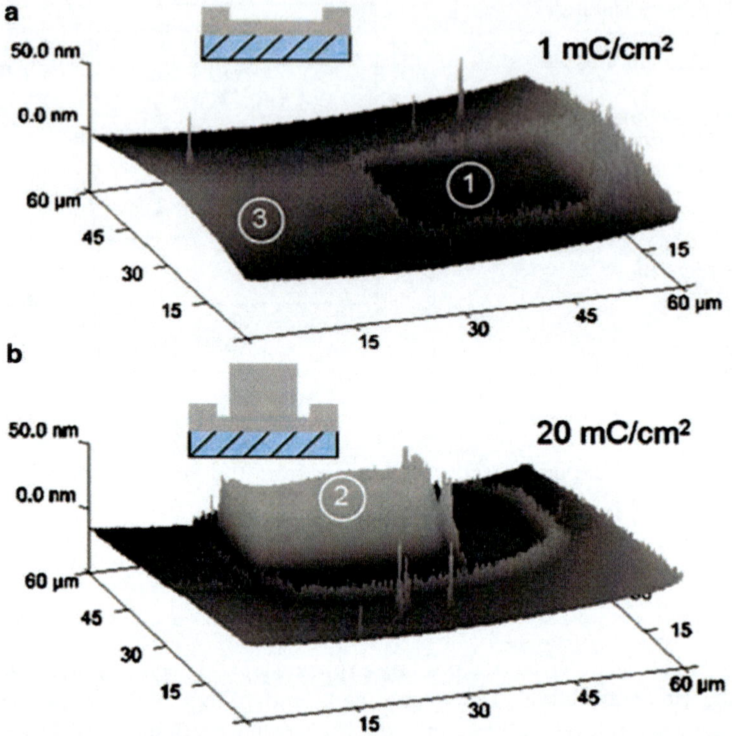

Fig. 4.18 AFM images of 25 × 25 μm He$^+$ exposed squares in 70 nm thick aluminum oxide film after development. Doses per surface area are indicated. The numbers 1, 2 and 3 in the AFM images refer to EDX sampling location

respectively.[3] Insets depict the surface profile. The locations 1, 2 and 3 refer to spots of EDX analysis (not shown) which all point to aluminum oxide material but in largely different intensities from spot to spot. In case of the 1 mC/cm^2 dose the pattern is a square shallow recess of ~10 nm deep. In contrast, there is a square protrusion of approximately ~50 nm surrounded by an asymmetric shallow trench in case of the 20 mC/cm^2 dose. The AFM topography at 1 mC/cm^2 and 20 mC/cm^2 dose is far beyond the estimated 1.2 nm thickness that is removed by sputtering at 20 mC/cm^2 [42]. Thus, at low dose a positive-tone effect shows up and at higher dose a negative tone one. The trench area around the protrusion is attributed to positive tone behavior by exposure from the tail of the defocused beam. A reliable exposure-response curve to determine the contrast and sensitivity of Al_2O_3 could not be made yet because of severe surface irregularities. The origin of the irregularities is still unclear.

[3] Submitted for publication.

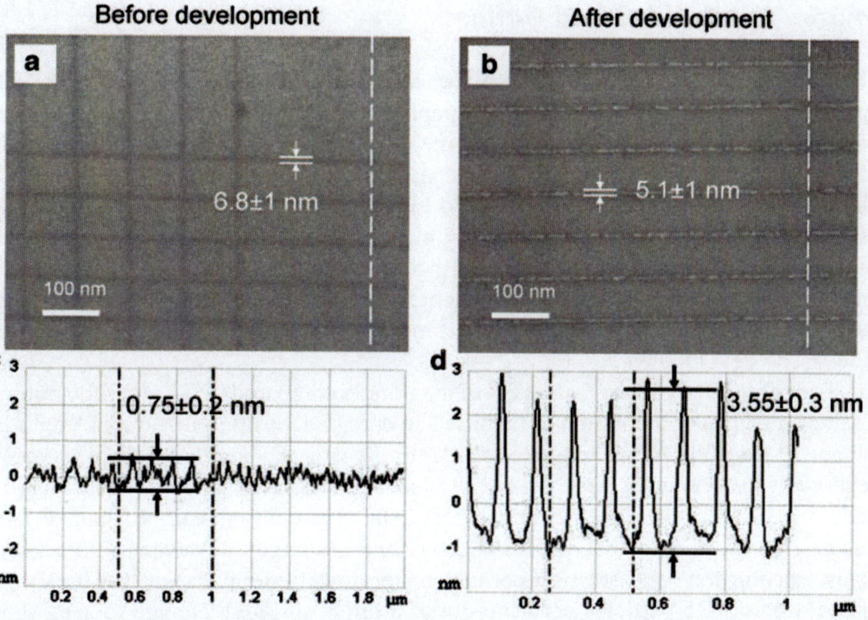

Fig. 4.19 SEM images (**a, b**) and AFM (**c, d**) scans of high-resolution structures obtained with He^+ ion beam exposure: (**a**) before and (**b**) after development, (**c**) and (**d**) are corresponding AFM profiles. Exposure doses are 4 and 14 mC/cm^2 in cases of before and after development. The dashed lines in SEM images are locations of AFM measurement

Figure 4.19 presents SEM images of fine line structures in 5 nm thick ALD grown Al_2O_3 obtained before (a) and after (b) development in MF351 for 5 min. The exposure doses are 4 mC/cm^2 (a) and 14 mC/cm^2 (b). Corresponding AFM profiles are shown in Fig. 4.19c and d. The measured line widths are 6.8 ± 1 nm and 5.1 ± 1 nm before and after development with feature heights of 0.75 ± 0.2 nm and 3.55 ± 0.3 nm respectively. After development, the resulting line height is almost the original film thickness, i.e. negative tone resist behavior, as observed for the larger areas.

In contrast to the positive and negative tone for He^+ exposure in this work, Al_2O_3 behaves merely as a positive tone resist both under Ga^+ ion [38] and electron [36, 37] beam exposure. A hypothesis of the He^+ exposure mechanism could be that He^+ ion impact at low dose creates structural defects, e.g. Frenkel pairs (vacancy – interstitial atom), which results in an increased development speed. At higher irradiation dose the number of defects may increase to the extent that they interact with each other, combine and result in re-ordering of the material towards a less soluble structure with negative tone display. According to Ohta et al. [38], the solubility of RF-sputtered Al_2O_3 can be tuned from soluble to insoluble in 80°C H_3PO_4 by enhancing the ion bombardment and increasing the deposition temperature.

4.5 Conclusions and Outlook

Helium ion beam technology with the novel sub-nm ion probe has excellent perspectives for high-resolution lithography at high densities in the sub-10 nm regime. Single features down to 6 nm and arrays of features with 10 nm pitch in HSQ resist are demonstrated. Key aspects of this approach is the negligible backscattering of helium ions compared to electrons so that proximity effects are small. Resist layer thickness should be limited to a few tens of nanometers to avoid excessive broadening by forward scattering. The lateral extension of the exposure is radially 2–3 nm, which is close to the theoretical limits. Additional lateral trimming to even smaller feature size will require a more advanced development process, e.g. the ‘salty’ development.

The sensitivity gain by 1–2orders of magnitude compared to e-beam lithography is attributed to a combination of factors. The energy deposition per unit of length is about 15 times larger as compared to electron beam exposure. He^+ displays most energy deposition in the first part of its trajectory, which enhances the sensitivity of thin resist even more. The SE impact from He^+ exposure is about a factor of 2–3 higher. Herein, the number of SE’s is about 20 times larger, however, with reduced cross section for bond scission because of the lower energy range (0–10 eV). A pixel dose of 625 particles per nano-dot of 6 nm is still high enough to keep shot noise insignificant. Our explorative study of aluminum oxide as a novel hard-mask resist under helium ion exposure is promising. Its sensitivity for helium ions is in the 10 mC/cm^2 range and features down to 5 nm have been realized.

Notwithstanding the high sensitivity of SHIBL as compared to EBL, the beam current (10 pA) is seriously limiting the process speed in SHIBL. Major improvement in beam current would require source operation at elevated pressure. Another issue is the forward scattering which deteriorates resolution in resist layers beyond 20 nm thickness. It could be greatly suppressed by a larger keV setting. Other light gases like hydrogen (H_2) offer possibly some additional gain in sensitivity

Acknowledgements This research is part of NanoNed, a national research program on nanotechnology, funded by the Dutch Ministry of Economic Affairs of the Netherlands. Our colleagues Ing A.K. van Langen-Suurling, Dr. V. A. Sidorkin, Dr. P.F.A. Alkemade, Prof. H.W.M. Salemink from TU Delft and Dr. E. van Veldhoven from TNO are gratefully acknowledged for their experimental contributions and discussions. The authors acknowledge Prof. D. Joy, Dr. L. Scipioni, Dr. S. Sijbrandij, Dr. Postek, Dr. Livengood, Dr. Vyvenko and Dr. D. Winston for giving permission to use certain figures and data, as indicated in the text.

References

1. Morgan J, Notte J, Hill R, Ward B. Microscopy Today. 2006;14:24–31.
2. Hill R, Faridur Rahman FHM, Nucl Instr Meth A. 2010; in press. doi:10.1016/j.nima.2010.12.123.
3. Scipioni L, Ferranti DC, Smentkowski VS, Potyrailo RA. J Vac Sci Technol B. 2010;28: C6P18–23.

4. Sanford CA, Stern L, Barriss L, Farkas L, DiManna M, Mello R, Maas DJ, Alkemade PFA. J Vac Sci Technol B. 2009;27:2660–7.
5. Winston D, Cord BM, Ming B, Bell DC, DiNatale WF, Stern LA, Vladar AE, Postek MT, Mondol MK, Yang JKW, Berggren KK. J Vac Sci Technol B. 2009;27:2702–6.
6. Sidorkin V, van Veldhoven E, van der Drift E, Alkemade P, Salemink H, Maas D. J Vac Sci Technol B. 2009;27:L18–20.
7. Wang JKW, Cord B, Duan HG, Berggren KK, Klingfus JK, Nam SW, Kim KB, Rooks MJ. J Vac Sci Technol B. 2009;2:2622–7.
8. Postek M, Vladár A, Archie C, Ming B. Meas Sci Technol. 2011;22:1–14.
9. Scipioni L, Sanford C, Notte J, Thompson B, McVey S. J Vac Sci Technol B. 2009;27:3250–5.
10. Melngailis J. Nucl Instr Meth B. 1993;80:1271–80.
11. Atkinson GM, Stratton FP, Kubena RL, Wolfe SC. J Vac Sci Technol B. 1992;10:3104–8.
12. Cheung R, Zijlstra T, van der Drift E, Geerligs LJ, Verbruggen AH, Werner K, Radelaar S. J Vac Sci Technol B. 1993;11:2224–8.
13. van Kan JA, Zhang F, Zhang C, Bettiol AA, Watt F. Nucl Instr Meth B. 2008;266:1676–9.
14. Sidorkin VA, Ph.D. Thesis, Delft University of Technology, Delft, 2010.
15. Ward B, Notte J, Economou NP. J Vac Sci Technol B. 2006;24:2871–4.
16. Hill R, Notte J, Ward B. Phys Proc. 2008;1:135–41.
17. Bell DC. Microsc Microanal. 2009;15:147–53.
18. Maas D, van Veldhoven E, Chen P, Sidorkin V, Salemink H, van der Drift E, Alkemade P. Proc SPIE. 2010;7638:14–23.
19. Sijbrandij S, Notte J, Sanford C, Hill R. J Vac Sci Technol B. 2010;28:C6F6–9.
20. Ramachandra R, Griffin B, Joy D. Ultramicroscopy. 2009;109:748–57.
21. Lin Y, Joy DC. Surf Interface Anal. 2005;27:895–900.
22. Vyvenko O, Petrov YV. Workshop "helium ion microscopy and its application" Forschungszentrum Dresden Rossendorf, 09 Dec 2009.
23. Joy D, Prasad M, Meyer III H. J Microsc. 2004;215:77–85.
24. Sayyah SM, Khaliel AB, Moustafa H. Int J Polym Mat. 2005;54:505–18.
25. Goodman A, O'Neill Jr J. J Appl Phys. 1966;37:3580–3.
26. van Dorp WF, Wnuk JD, Gorham JM, Fairbrother DH, Madey TE, Hagen CW. J Appl Phys. 2009;106:074903.
27. Randolph SJ, Fowlkes JD, Rack PD. Crit Rev Solid State Mat Sci. 2006;31:55–89.
28. Seah MP, Dench WA. Surf Interface Anal. 1979;1:2–11.
29. Kieft E, Bosch E. J Phys D: Appl Phys. 2008;41:215310.
30. Craciun R, Picone D, Long RT, Li S, Dixon DA, Peterson KA, Christe KO. Inorg Chem. 2010;49:1056–70.
31. Livengood R, Tan S, Greenzweig Y, Notte J, McVey S. J Vac Sci Technol B. 2009;27:3244–9.
32. Balk P. The Si–SiO_2 system. Amsterdam: Elsevier; 1988.
33. Grigorescu AE, Hagen CW. Nanotechnology. 2009;20:292001.
34. Duan H, Winston D, Yang JKW, Cord BM, Manfrinato VR, Berggren KK. J Vac Sci Technol B. 2010;28:C6C58–62.
35. Muray A, Scheinfein M, Isaacson M, Adesida I. J Vac Sci Technol B. 1984;3:367–72.
36. Hollenbeck JL, Buchanan RC. J Mat Res. 1990;5:1058–72.
37. Saiffulah MSM, Kurihara K, Humphreys CJ. J Vac Sci Technol B. 2000;18:2737–44.
38. Ohta T, Kanayama T, Tanoue H, Komuro M. J Vac Sci Technol B. 1989;7:89–92.
39. MacCord MA, Rooks MJ. In: Rai-Choudhury P, editor. Microlithography, micromachining and microfabrication, vol. 1. USA: SPIE; 1997. p. 139–250.
40. Thompson LF, Bowden MJ. In: Thompson LF, Wilson CG, Bowden, editors. Introduction to microlithography. Am. Chem. Soc., ACS Symp. Ser. 1983; 219: 161–214.
41. Broers AN, Harper JME, Molzen WW. Appl Phys Lett. 1978;33:392–4.
42. Behrish H, Eckstein W, editors. Topics in applied physics. Sputtering by particle bombardment. Heidelberg: Springer; 2007, vol 110.
43. Ryssel H, Haberger K, Kranz H. J Vac Sci Technol B. 1981;19:1358–62.

44. Hirscher S, Kaesmaier R, Domke W-D, Wolter A, Löschner H, Cekan E, Horner C, Zeininger M, Ochsenhirt J. Microelectron Eng. 2001;57–58:517–24.
45. Namatsu H, Takihashi Y, Yamazaki K, Nagase M, Kurihara K. J Vac Sci Technol B. 1998;16:69–76.
46. Sidorkin V, van der Drift E, Salemink HWM. J Vac Sci Technol. 2008;26:2049–53.
47. Tesmer JR, Nastasi M, editors. Handbook of modern ion beam materials analysis. Pittburgh: MRS; 1995.
48. Sidorkin V, Grigorescu A, Salemink H, van der Drift E. Microelectron Eng. 2009;86:749–51.
49. Grigorescu AE, van der Krogt MC, Hagen CW. J Micro/Nanolith MEMS MOEMS. 2007;6:043006.
50. Yang JKW, Berggren KK. J Vac Sci Technol B. 2007;25:2025–9.
51. Jackel LD, Howard RE, Mankiewich PM, Craighead HG, Epworth RW. Appl Phys Lett. 1984;45:698–700.
52. Rishton SA, Kern DP. J Vac Sci Technol B. 1987;5:135–41.
53. Boere E, van der Drift E, Romijn J, Rousseeuw BAC. Microelectron Eng. 1990;11:351–4.

Nanoimprint Technologies

5

Christophe Peroz, Vincent Reboud, and Clivia M. Sotomayor Torres

Abstract

The advances in nanosciences, micro- and nanotechnology are driving the research and development efforts to fabricate micro- and nano-structures with a high precision in a wide variety of materials using novel lithography methods. These emerging techniques, which include self-assembly, scanning probes, micro-contact printing, and nanoimprint lithography (NIL), are intensively studied to, on the one hand, assess to what degree they meet the demands of ultrahigh precision and high density of nanostructures posed by the semiconductor industry and, on the other hand, to examine them with respect to cost-efficiency to produce components for photonic, data storage, sensing and fluidic or biological applications. This chapter focuses on recent advances in nanoimprint lithography as it is perhaps among the most mature emerging nanofabrication methods. Nanoimprint lithography technology faces some challenges to reach the requirements of the semiconducting integrated circuits manufacturers in terms of overlay accuracy, defectivity and throughput but it meets already some needs of data storage, light extraction, fluidic and biological applications. Significant efforts are currently being made to develop parallel printing on large area and step and repeat techniques. In this chapter, we review the principles of

C. Peroz (✉)
Abeam Technologies, Castro Valley, CA, USA
e-mail: cp@abeamtech.com

V. Reboud
Catalan Institute of Nanotechnology, Barcelona, Spain

CEA-LETI-Minatec Grenoble, Grenoble, France

C.M.S. Torres
Catalan Institute of Nanotechnology, Barcelona, Spain

Catalan Institute for Research and Advanced Studies ICREA, Barcelona, Spain

Department of Physics, Universitat Autonoma de Barcelona, Bellaterra (Barcelona), Spain

M. Stepanova and S. Dew (eds.), *Nanofabrication*,
DOI 10.1007/978-3-7091-0424-8_5, © Springer-Verlag/Wien 2012

nanoimprint lithography and its capability to scale-up the replication of nanostructures by parallel printing, step and stamp and by step and flash, the latter a technique that use UV curable resist. We identify current capabilities of the different variations of nanoimprint lithography and provide examples of the fabrication of three-dimensional structures and nanostructures in inorganic sol-gel materials. Finally, an overview of the wide range of applications realized so far by nanoimprint lithography is given.

5.1 Thermal Nanoimprint

Nanoimprint lithography (NIL) was first reported as a thermoplastic molding technique named thermal NIL [1] and it is sometimes compared to hot embossing lithography [2]. It was then demonstrated that NIL is an adequate, flexible and cost-effective technique to structure thin polymer films showing resolution down to 5 nm [3]. NIL was included into the semiconductor industry (ITRS) roadmap in 2003 as a possible lithography technology for the 32 nm node then expected in 2010 and for the 22 nm node expected in 2013. In the 2004 update of the ITRS roadmap, NIL appears as a possible lithography technology for the 16 nm node [4]. The main advantages of these imprint-based techniques, as will be discussed below, are their potential to reach cost-efficient, high throughput and high resolution production of nanostructures. Despite the enormous development of imprint-based techniques, several challenges have yet to be addressed in order to meet industrial requirements. Therefore, issues like throughput, multilevel capabilities, alignment, instrumentation and polymer stability matters have to be further explored.

5.1.1 General Principles

Thermal NIL is based on the deformation of a thin polymer film with a rigid stamp (Fig. 5.1a). Thermal NIL uses a hard template (called mold/stamp) such as silicon, to imprint a thermoplastic polymer which is heated to a temperature above the polymer glass transition temperature (T_g) while applying a relatively high pressure. After a specific time, which depends on the templates design and topography, the polymer is cooled to a temperature below its T_g. Then stamp and substrate can be separated. To minimize the adhesion between the imprinted resist and the mold a fluorinated based material is deposited on the mold surface [5] to act as an anti-adhesive agent. The thermal NIL process cycle is shown in Fig. 5.1b. The polymer or resist material is heated to make it less viscous and then shaped by the applied pressure so that it flows into the cavities of the stamp. When the cavities of the mold are filled, the thermoplastic is cooled down below its T_g while the pressure is sustained. After relieving the pressure, the stamp is carefully separated from the imprinted substrate and can be reused for the next NIL cycle. The inherent residual

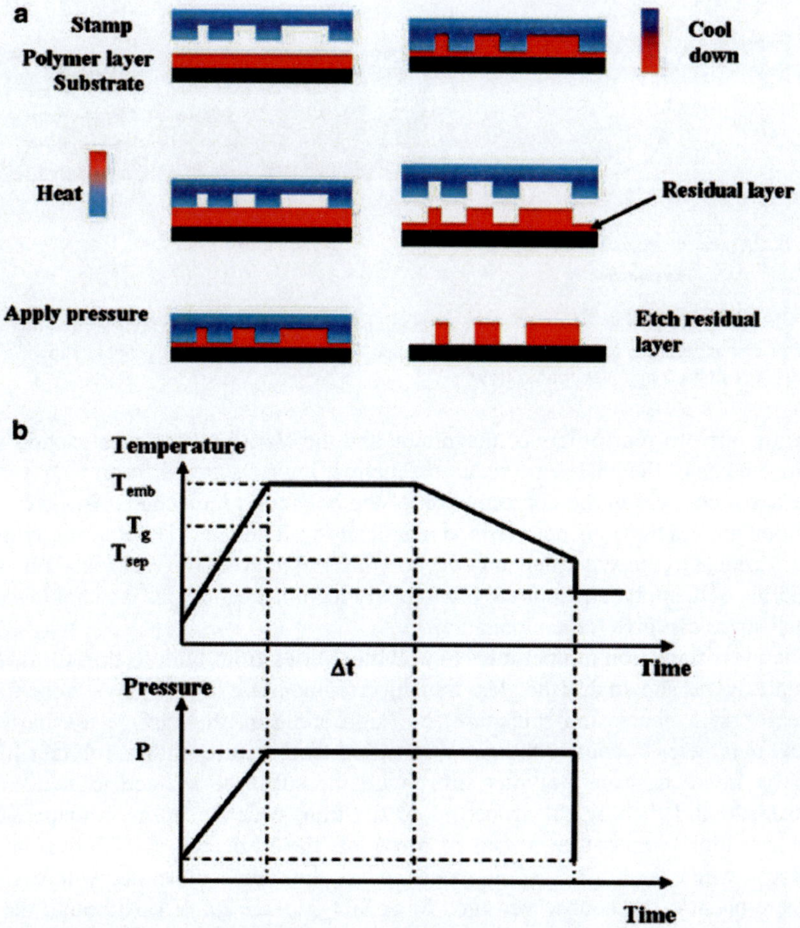

Fig. 5.1 (**a**) Schematics of the NIL process, (**b**) schematics of a typical process sequence during thermal NIL showing the applied pressure, the heating and cooling lines versus time. Typical values of pressure lie in the range of 10–100 bar, the imprint temperature is usually around 30–70°C above T_g, while the embossing time ranges from minutes to 30 min

resist layer can be removed by an anisotropic etching process allowing the use of the imprinted thermoplastic as a masking layer for further processing steps.

5.1.2 Squeeze Flow of Thin Films and Problems of the NIL Process

The dependence of Young's modulus with the temperature of polymers governs the NIL process. For example, the Young's Modulus of an amorphous polymer is shown in Fig. 5.2a as a function of temperatures. When the glass transition temperature is reached, large macromolecular segments are able to move individually, the modulus

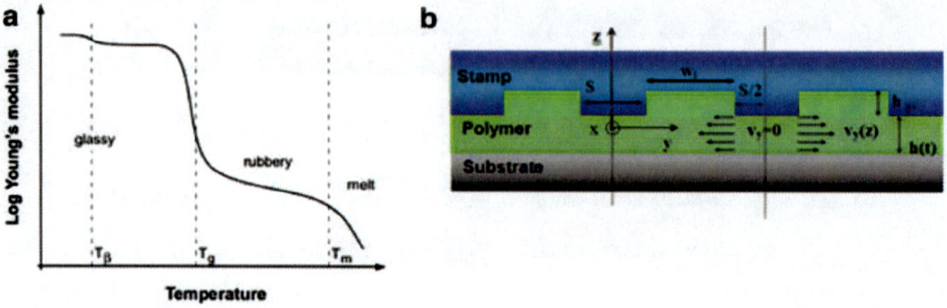

Fig. 5.2 (**a**) Thermal transitions of an amorphous polymer showing the Young's modulus as a function of temperature for an amorphous polymer, (**b**) schematic of flow process showing the velocity distribution in a Newtonian flow

drops by three to four orders of magnitude and the Van der Waals interaction and entanglements of the chains are reduced. Thermal imprints are performed above the T_g to allow changes in the conformation of the polymeric molecules. However, the transition around the T_g is not thermodynamically well defined. The imprint temperature is critical to shape permanently the polymer and to avoid internal relaxation and reordering of the polymer chains. In the rubbery region, the modulus is rather low and a small stress can give large elongation.

The glass transition temperature of polymer varies from bulk to thin films. For example, it was shown that the glass transition temperature of thin polystyrene films decreases with decreasing thickness [6, 7]. In addition, the surface tension is a critical parameter to control the glass transition temperature of thin polymer films [8]. The interactions of polymer films with the substrate surface influence the viscoelastic and rheological properties of the film, which in turns influences the glass transition temperature. It was reported that the T_g decreases with decreasing thickness when the film is spin-coated on an aluminum substrate, whereas the opposite phenomena is observed on a Si or SiO_2 surface [9, 10]. Although the T_g of polymers is sometimes not well defined, a simple model for the squeezed polymer flow underneath the stamp protrusion can be used to predict the final height of the imprinted polymer and the necessary embossing time. The model applies for rigid stamps with constant filling factor. The stamp consists of N parallel and periodic straight line protrusions of length L, widths s_i, and distances w_i (Fig. 5.2b).

Before embossing the initial spin-coated polymer film has a thickness of h_o and the depth of the stamps protrusions is h_{pr}. The initial height, h_o, should be chosen so that in the final thickness, h_f, of the residual layer is that required for window opening. From the above equation we derive h_o:

$h_o = h_f + \frac{h_{pr}}{\Lambda}\sum_{i=1}^{N} w_i$, where Λ is the pitch of the grating defined by $\Lambda \equiv \sum_{i=1}^{N}(s_i + w_i)$.

By solving the Navier–Stokes equation with non-slip boundary condition at the stamp and the substrate surfaces, and by applying the continuity equation with the assumption that the polymer melt is an incompressible Newtonian fluid, the film thickness under the stamp protrusion, *h(t)*, for a fully inserted stamp can be determined when a constant imprint force F is applied [11]. The following expression is known as Stefan equation:

$$\frac{1}{h^2(t)} = \frac{1}{{h_o}^2} + \frac{2F}{\eta_o L s^3} t.$$

Inserting the final thickness $h_f = h(t_f)$ in the previous equation, the imprinting time can be estimated:

$t_f = \frac{\eta_o s^2}{2P}\left(\frac{1}{h_f^2} - \frac{1}{h_o^2}\right)$, where P is the pressure under each stamp protrusion $\mathrm{P} = \frac{\mathrm{F}}{\mathrm{sL}}$.

As a direct consequence of the Stefan equation, small protrusions will sink faster than the larger one. In addition, the process times to imprint polymer thin films can be reduced to a few seconds by increasing the initial layer thickness leading to a thick residual layer which may be not suitable for some applications. A high imprinting temperature will decrease dramatically the polymer viscosities (Fig. 5.2a), and as consequence the imprinting time will be reduced as the stamp cavities will be filled faster. A similar behaviour can be reached by optimizing the viscosity of the resists used.

5.1.3 Stamp Bending and Homogeneity

Stamp bending has to be taken into account, resulting in an inhomogeneous residual layer. Bending is always observed at the border of a grating with a large unstructured area. The prediction of the final h_f residual layer thickness (RLT) over the whole imprinted area is essential to allow a high fidelity pattern transfer into the substrate. Based on the assumption that the resist is an incompressible viscous fluid, the imprinting process can be generally described by the non-stationary Navier–Stokes equations that in the primitive variables velocity–pressure $(\mathbf{V}, P)$ are written as follows:

$$Re\left[\frac{\partial V}{\partial t} + (V \cdot \nabla)V\right] = -\nabla P + \Delta V, \quad \nabla \cdot V = 0, \quad V = (v^x, v^y, v^z),$$

where $Re = (\rho V_0 L_0)/\eta$ is the Reynolds number, $V_0 = V(stamp)$ is the stamp velocity, L_0 is the characteristic lateral size of the stamp, ρ and η are the density and the dynamic viscosity of the resist, respectively. The equations above are given for non-dimensional variables: lengths and velocity components are scaled by the characteristic factors L_0 and V_0, respectively, and the pressure is scaled by $P_0 = (\eta V_0)/L_0$.

By a series of approximations, a coarse-grained method [12] and appropriate boundary conditions, a method is derived for predicting velocity and pressure relations suitable for large-scale (100 mm) wafer simulations. The approach makes

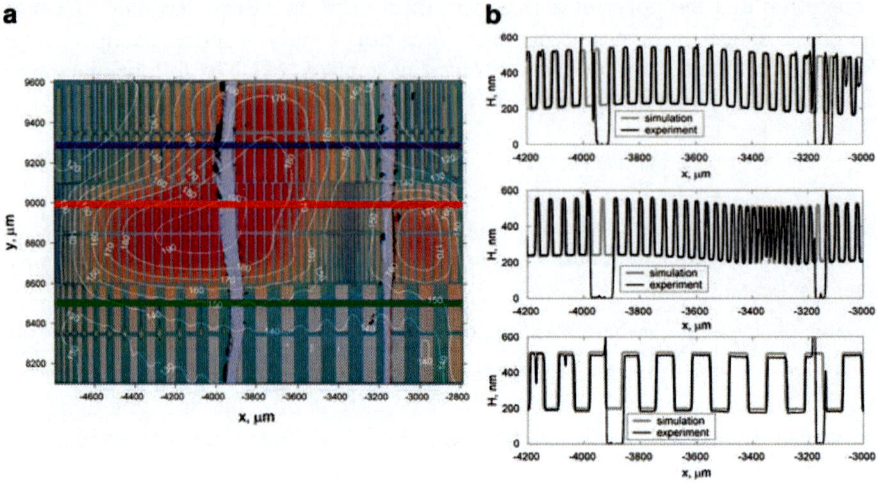

Fig. 5.3 (**a**) Optical microscopy images of the test structure imprinted in the resist at 180°C [103]. *Horizontal lines* (*blue, red, green*) indicate zones of profilometer measurements of resist thickness, (**b**) comparison of measured and simulated distributions of resist thickness for the test structure on the zone defined by the *blue*, *red*, *green lines*

it possible to analyze real 10 cm^2 samples and to consider in a qualitative manner elastic/inelastic stamp and substrate deformation. The test structure measures 2 × 2 mm^2 and contains protrusions of varying width and coverage. The coarse cell has a size in range of 1–50 μm. Figure 5.3a shows an image of the structure through a microscope objective, acquired using a CDD and simulated isolines of bending for a chirped grating imprinted into the mr-I 8030 resist from *Microresist Technology* at a temperature of 190°C. Imprinted structures show inhomogeneities of the RLT related to the non-uniform deformation of stamp. Vertical scratches observed in the images have been used for the determination of RLT.

White isolines specify the calculated distribution of the stamp deformation where the numbers denote the elastic displacement in nanometers. The depth of the stamp cavities is 300 nm. The initial resist thickness is 340 nm. For the modelling, the resist dynamic viscosity is taken to be 10^4 Pa s. The stamp velocity is supposed to be 1 nm/s. Comparisons of the experimental profile of three cross-sections taken with a profilometer and the bending profile calculated by the coarse grained method are shown in Fig. 5.3b. These results indicate that the experimental and simulated values of the RLT agree very well with a maximum difference smaller than 10%. It shows that the coarse-grained simulation software can be used for quantitative predictions of the residual resist thickness showing the potential of the software as an efficient tool for the optimization of stamp geometry.

5.1.4 Anti-adhesion Treatments of Stamps and Automatic Demolding by Step-and-Stamp Process

In NIL, the stamp has to be coated with a self-assembled antisticking monolayer to prevent the resin from adhering to the stamp during the demolding step. As the majority of stamps used in NIL are silicon, surface treatments containing alkyltrichlorosilanes or perfluoro-alkyltrichlorosilanes are mostly used. Functionalization of silicon surfaces using silane chemistry is a well-known process. A commonly used molecule is tridecafluoro-(1,1,2,2)-tetrahydrooctyl-trichlorosilane (F_{13}-TCS). Chlorosilanes are known to react spontaneously with hydroxylated silicon or silicon dioxide surfaces under elimination of hydrochloric acid. Since water reacts with chlorosilanes and polymerizes them, it is important to perform the reaction in a water-free environment [13]. It was shown that a surface composed of only –CF3 groups would have the lowest surface free energy of any system at 6.7 mJ/m^2 [14, 15]. It has been shown that the surface energies of the self-assembled layer of trichloro(3,3,3 trifluoropropyl) silane (FPTS) and trichloro(1H, 1H, 2H, 2H perfluorooctyl) silane (FOTS) decrease with increasing annealing temperature and immersion time for surface energies below 10 mJ/m^2 [16].

Two main techniques are commonly used to pattern large areas: parallel imprinting [17] as in Roll to Roll [18] and Step & Stamp lithography [19]. One key issue for NIL is the demolding force when the stamp is separated vertically from the embossed polymer. Damages on the structures can occur during this critical step. The anti-adhesion treatment plays a key role for a good demolding. In addition, the lifetime of the antisticking layer is crucial for a high throughput of NIL. The NPS300 Step & Stamp apparatus from SET Inc. (Smart Equipment Technology) can be used to optimize thermal imprint processes and to show the capability of scaling up the number of imprints. Molecular bonding was used to attach a 1×1 cm^2 quartz stamp on a 5×5 cm^2 quartz glass to guarantee a good planarity on the whole surface of the stamp. Figure 5.4a presents a schematic of the imprinting process, for the stamp shown in cross-section in Fig. 5.4b.

During a conventional imprinting process, the polymer-coated substrate is heated above the glass transition temperature of the polymer. When the glass transition temperature is reached, force is applied to the stamp and it is pressed onto the polymer surface. The stamp is kept in contact with the resist during a period that allows the resist to flow under the stamp. The stamp is kept under pressure during the cooling cycle until a temperature 10–40°C below the glass transition temperature of the resist is reached. The stamp is released and stepped to the next site to be patterned. The demolding temperature is a key parameter for the success of the imprint and to extend the lifetime of the antisticking layer. To optimize the demolding temperature, 300 nm of mr-I-PMMA was spin-coated onto the silicon substrate. The T_g of mr-I-PMMA is about 105°C. An imprint temperature of around 170°C is required for conventional NIL apparatus where the substrate holder and the piston are at the same temperature. In the case of the NPS 300, the substrate holder has to be at a temperature below the T_g of the polymer to avoid melting the imprinted structures after molding. A higher temperature of the arm holding the stamp is then needed; a temperature of 25°C was chosen for the stamp and the substrate was kept at 95°C. The stamp is heated with IR-radiation from a halogen lamp mounted in the bonding arm, and the substrate is heated using a halogen lamp in the substrate holder. The quality of the printed pattern depends on the levelling accuracy of the arm with respect to the substrate holder. The levelling resolution of the bonding arm

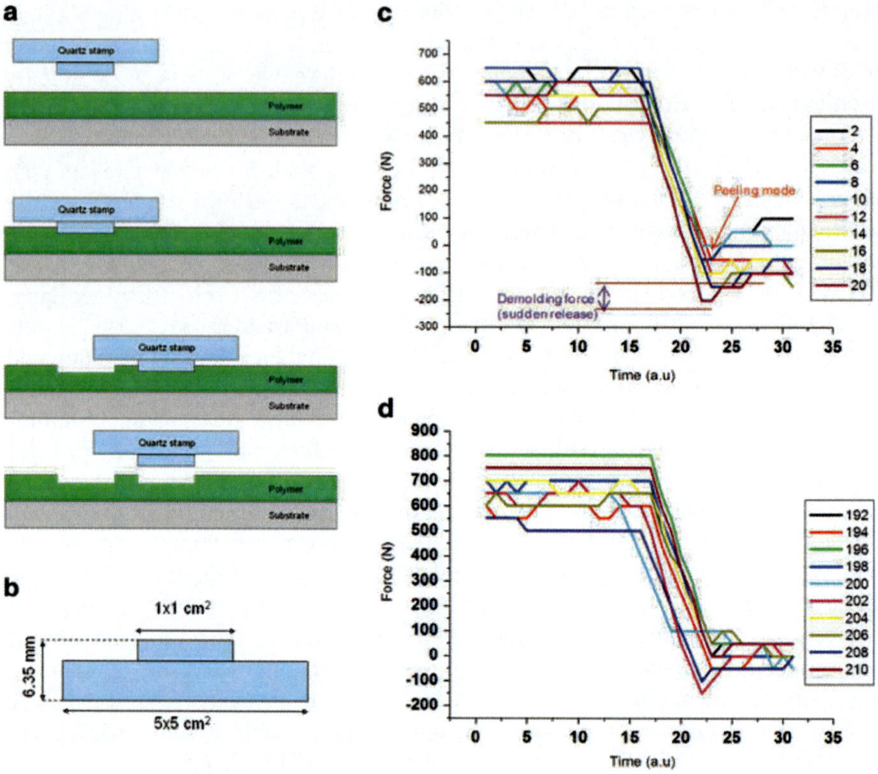

Fig. 5.4 (**a**) Schematic of Step & Stamp imprint process, (**b**) cross-section of the stamp, details of the force curve (**c**) for the first 20 imprints and (**d**) the last 20 imprints on the 210 imprints with an unstructured stamp (demolding temperature: 80°C)

is 20 mrad which allows satisfactory levelling accuracy with the stamps used. Figure 5.4c and d presents the force history at the moment of the demolding for the first 20 imprints and the last 20 imprints on the 210 performed imprints. A force of 35 kN was applied during 180s when the stamp reached the temperature of 250°C. The separation speed of the stamp with the substrate was taken equal to 20 μm/s with a sampling rate of 100 samples/s. The demolding temperature was varied from 70°C and 90°C. The demolding force can be deduced from discontinuities of the recorded force (Fig. 5.4c).

These discontinuities are due to a sudden release of the stamp from the polymer and can be identified as bumps in the force curve. The absence of a bump means that the demolding force is too weak to be measured, and it is an indication that demolding happens in a peeling motion (non-parallel separation) by bending the substrate upwards and subsequently demolding from the borders to the centre slowly. When an instantaneous parallel demolding process happens, one bump is observed, this is the ideal case. The demolding force is calculating for 210 imprints for four demolding temperatures (70, 80, 90, 100°C).

The demolding force as a function of four demolding temperatures is shown on Fig. 5.5. The reported values are the average values on the 210 imprints. An

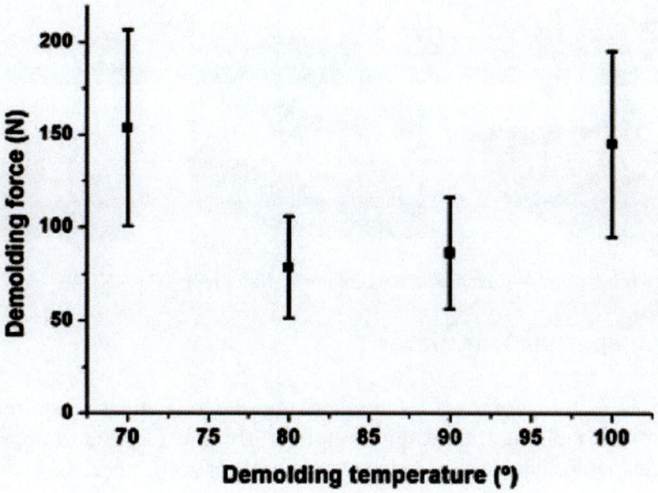

Fig. 5.5 Demolding force measured at different demolding temperatures (with the associated force variance)

optimum imprint temperature is found at around 85°C. The demolding temperature rises significantly above and below this optimum value. This phenomenon could be explained by an increased friction due to the mismatch of thermal expansion between polymer and silicon, the increased adhesion and friction when the polymer become softer.

5.2 Ultra-Violet Assisted Nanoimprint Lithography

The other class of nanoimprint techniques consists of imprinting a liquid photo-curable resist materials at room temperature and low pressure. The process is commonly called Ultra-Violet assisted Nanoimprint Lithography (UV-NIL). UV-NIL was first proposed by J. Haisma in 1996 [20] and its principle is illustrated in Fig. 5.6. During the contact between structured mold and resist film, an external light source is used to cross-link the resist. The filling of template cavities during the imprinting step is mainly driven by capillarity forces and the imprint pressure P_{imp} is typically below 100N. Over the last 10 years, the UV-NIL approach gained several important advantages in terms of throughput and overlay [21, 22] and is in the pre-production phase for various "conventional applications" such as hard disk media [23]. The UV-NIL process is performed at ambient temperature and eliminates the inherent thermal mismatch between mold and imprinted resist in thermal NIL. UV-NIL is also less sensitive to pattern density variation. Key issues limiting the emergence of UV-NIL for several applications are mainly quality template fabrication with high-resolution features, the control of the RLT underneath the imprinted structures and defect counts related to the imprint process.

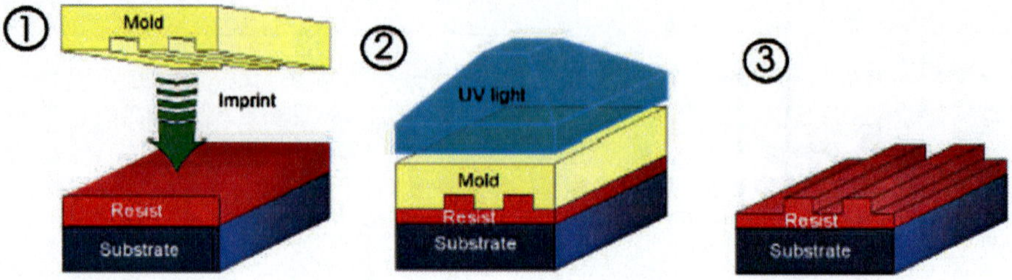

Fig. 5.6 Principle of ultra-violet assisted nanoimprint lithography (UV-NIL)

5.2.1 Transparent Templates

The use of UV light source for cross-linking the NIL resist requires either a transparent substrate or a transparent template. The most common approach is to fabricate transparent nanostructured molds in hard or soft materials.

5.2.1.1 Hard Molds

Selection of hard-mold materials is driven by their compatibility with conventional micro/nanofabrication processes, hardness, transparency and durability. The most common choice of mold substrates is quartz (fused-silica). These substrates exhibit a high transparency to UV wavelengths and their sub-100 nm patterning is well established for conventional photomasks. As nanoimprint is a 1:1 replication technique, it requires the fabrication of high-resolution molds based on hard materials like quartz. The common and most mature route consists of patterning quartz molds by electron beam lithography (EBL) and plasma etching. Numerous efforts are underway to develop high-resolution etching of nanostructures into inorganic films such as silica. Due to the poor etching selectivity of organic resist materials vs. quartz, metallic masks of chromium or nickel are commonly used [24]. These metal films also reduce charging effects during EBL exposure and increase pattern resolution and fidelity. Molds with features sizes down to 20 nm have been successfully reported and replicated by UV-NIL (see Fig. 5.7a) [25]. As an alternative to EBL, the self-assembly of block copolymers [26, 27] has recently been used to replace the lithography step in the fabrication of master molds (Fig. 5.7b) [27]. The main challenges for this approach are to reduce defects numbers and increase the pattern transfer to hard materials.

Another type of template consists of patterning directly a hard inorganic resist material onto transparent substrates. Various EBL resists based on inorganic sol-gel materials are currently in development and will be tested as hard masks for NIL by resist manufacturers and laboratories [28]. Advantages of this approach include simplification of the mold fabrication process and suppression of any etching steps associated with the deterioration of pattern resolution and fidelity. Nanoimprint templates with sub-10 nm sized features have been fabricated onto hydrogen silsesquioxane (HSQ) resist and have been used for imprinting small areas [29] (see Fig. 5.8). High-resolution molds need very thin EBL resist films, which

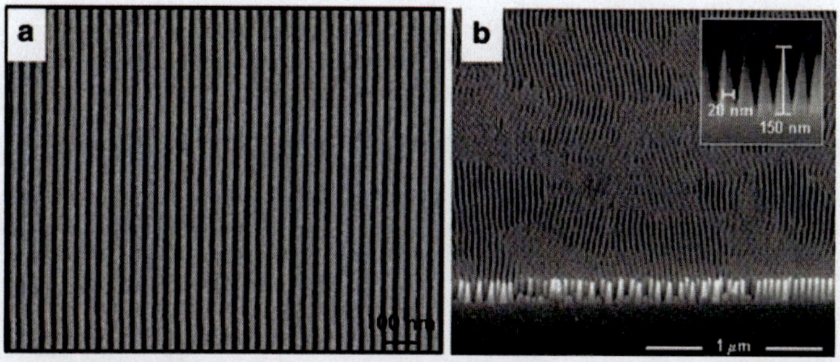

Fig. 5.7 (**a**) SEM picture of pattern into quartz template for SR-NIL, minimum features sizes are around 20 nm from Ref. [25] (Courtesy of Molecular Imprints Inc.). (**b**) SEM cross-section view of SiO_2 mold fabricated by self-assembly of copolymers and dry etching, reproduced with permission from Ref. [27]

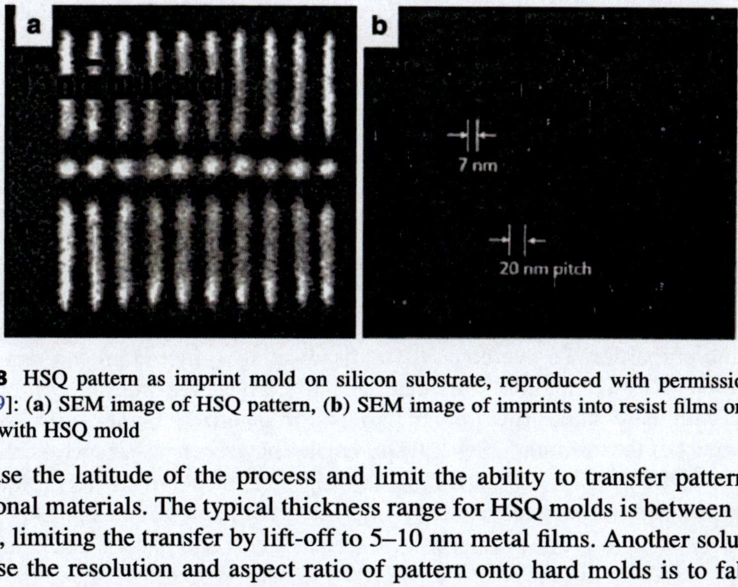

Fig. 5.8 HSQ pattern as imprint mold on silicon substrate, reproduced with permission from Ref. [29]: (**a**) SEM image of HSQ pattern, (**b**) SEM image of imprints into resist films on quartz wafers with HSQ mold

decrease the latitude of the process and limit the ability to transfer patterns into functional materials. The typical thickness range for HSQ molds is between 20 and 50 nm, limiting the transfer by lift-off to 5–10 nm metal films. Another solution to increase the resolution and aspect ratio of pattern onto hard molds is to fabricate molds into well-developed materials like silicon and imprint onto transparent substrate, in this case the UV illumination is inverse and goes through the substrate.

5.2.1.2 Soft Templates

Hard molds have some drawbacks, such as high cost, poor conformal contact to substrate and brittleness. One attractive alternative is to use elastomeric templates replicated from hard master-molds. The inherent soft stamp material properties largely reduce the risk of mechanical damage to the master template and thus enhance its lifetime. Soft templates are replicated from hard master-molds by a simple method (Fig. 5.9): a polymeric liquid solution is poured on the top of nanostructured hard molds and after cross-linking of the polymer by annealing,

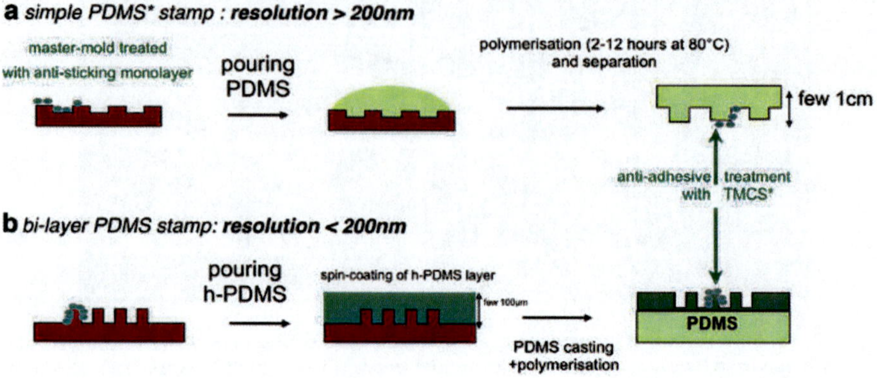

Fig. 5.9 Process flow for fabrication of soft templates replicated from a master mold: (**a**) soft PDMS template, (**b**) bilayer hard/soft PDMS stamp

the template is peeled off. Polydimethylsiloxane (PDMS) is one of the most common elastomers used as a soft template [30]. Multiple advantages of this elastomer are its low Young modulus (300–900 kPa), allowing a conformal contact over large areas, and its low surface energy (15–25 mJ/m^2) for easy demolding after imprinting. Soft templates are particularly suitable for full wafer imprint. The resolution is limited by the softness of the template and aspect ratio of nanostructures which can collapse if the material is too soft. Tool manufacturers are currently developing tools and specific template materials for soft NIL, and some applications like micro-lenses for CMOS image sensors are currently in production. However, soft templates have generally poor imprint durability and demolding properties. To overcome these disadvantages, hybrid bilayer templates are developed by combining soft materials, for large area imprint conformity, covered with hard nanostructured materials for localized higher stiffness and improvement of the resolution. Several approaches have been tested such as bilayers of soft/hard PDMS [31], PDMS/UV-cured resist [32] or polymer/SiO_2 [33]. Sub-15 nm features have been successfully imprinted with soft templates [34]. These solutions seem to be very advantageous in terms of cost and pattern resolution.

5.2.1.3 UV-NIL Resist Materials

Development of UV-curable photoresists which are compatible with imprinting processes is one of the keys for success of NIL technology. NIL tool makers and resist manufacturers are currently developing new resist materials according to the targeted applications. The properties of UV-NIL resist are as follows:

- *Low viscosity* for homogenous and fast filling of template's cavities
- *Small shrinkage* after imprinting to ensure high pattern fidelity
- *Fast UV-curing* for high throughput
- *High plasma etching selectivity* for easy pattern transfer
- *Small polymer molecules sizes* for high-resolution patterning
- *Low volatility* in the atmosphere for easy alignment

The requirements for NIL resists are often conflicting, and UV-curable resist are usually created with balanced properties. Resist materials can be seen as a multi-component polymer containing a photoacid generator (photoinitiator), various monomers and some anti-adhesive materials. The photoinitiator is used for free-radical polymerization [35] of the resist and determines curing energy and time during the NIL process. The required curing energy is in the range of a few hundreds of mJ/cm^2 and curing time is dependent on the power of the UV-system. The photoinitiator usually represents a few percents of the resist solution [36]. Acrylate and vinyl-ether monomers are the main components of the UV-NIL resist [37, 38]. The radical photocuring systems based on acrylate monomers with silicon components are used for their fast radical polymerization and high etching resistance to oxygen plasma [39]. Vinyl-ether derivative monomers are also associated with a decrease in the viscosity of the resist [40] and reducing the shrinkage caused by the photo-polymerization of the acrylic monomers. A balance between low and high molecular weight polymers limits the shrinkage of imprinted patterns to values as low as 3% [41]. The viscosity of the resist is usually kept below 50 mPa s to maintain good imprint uniformity and obtain an ultra-thin residual layer underneath the pattern. The polymer molecule's size for resists is commonly in the range of 1–2 nm and does not pose a problem for NIL resolution with templates with sub-10 nm patterns. Fluorine-based additive components or other antisticking molecules can be added to the resist to decrease the adhesion force during demolding [42]. The anti-adhesion additives migrate to the surface of the resist, i.e., to the interface with the mold, to decrease the surface energy of the resist and enhance the easiness of demolding.

5.2.2 Full-Wafer Scale Imprinting Process

The simplest technique for UV-NIL is to imprint a full wafer at low pressure in one single step. Soft stamps are preferred because they can conform to large area substrates and non-flat surfaces. The volume and displacement of resist are important and limit the resolution and the number of defects of the imprint process. This approach is suitable for small area patterning and applications, which have high tolerance to defects like most of the photonic applications [43]. One of the main constrains is the control of the quality of the imprint and the homogeneity of the RLT over large areas. The fabrication of high-resolution large-area templates is also problematic and expensive. To overcome these problems, templates are fabricated by multiple imprinting of smaller master-molds by Step and Repeat nanoimprint. This solution has attracted increasing interest over recent years.

5.2.3 Step and Repeat Nanoimprint Lithography

By analogy with an optical lithography stepper, successive imprints can be repeated in order to pattern a large area: this is the step and repeat imprint (SR-NIL) process.

The general principle of SR-NIL is depicted in Fig. 5.10. SR-NIL offers a solution to imprinting large areas with high resolution patterning by limiting the resist displacement during the imprint step and thereby reducing the number of defects. The fabrication of molds is also reduced to smaller sizes, which simplifies the process and reduces the cost. There are two main types of SR-NIL process depending on the dispensing method of the NIL resist.

The most popular solution is to dispense droplets of very low viscosity resist, typically $\eta < 5$ mPa s, on a predetermined die and to imprint the resist at low pressure and at room temperature. The method was developed first in the group of Willson [44] and is commonly called "step and flash imprint lithography" (S-FIL). In this technique, an array of droplets of photopolymerizable organosilicon resist (Fig. 5.11a) is imprinted with a small quartz mold by capillary action [45]. NIL resist materials are dispensed as droplets onto a transfer layer film, which plays the

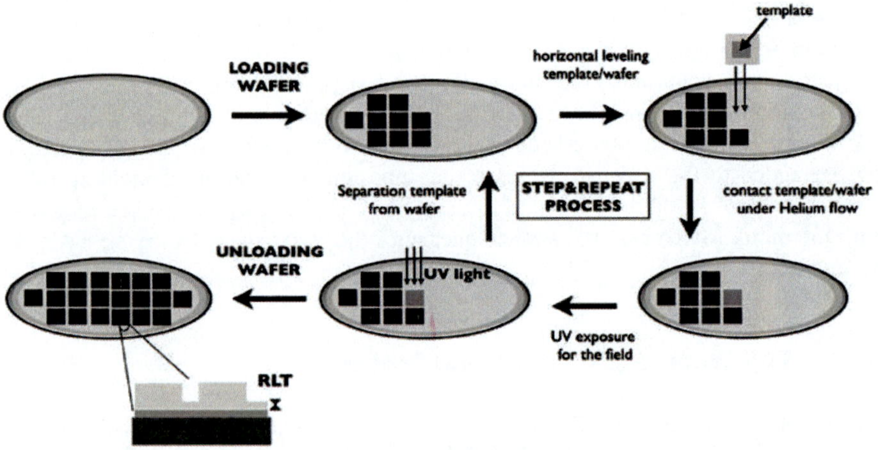

Fig. 5.10 Schematics of step and repeat nanoimprint lithography

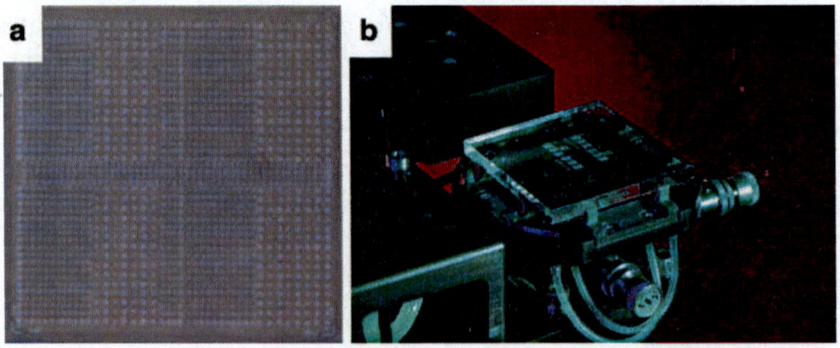

Fig. 5.11 (**a**) Optical images of a resist droplet patterns before imprinting. (**b**) Quartz template with nanostructures on the stage of an S-FIL tool (Pictures are by courtesy of Molecular Imprints Inc.)

Fig. 5.12 Two hundred millimeter wafer imprinted by SR-NIL with EVG 770 Gen II NIL Stepper, courtesy of the EVG Group. Imprints are performed with soft templates (By courtesy of G. Kreindl, EV-Group)

role of adhesion layer to improve the adhesion of the NIL resist and, at the same time, is used as a hard etching mask for further transfer. Numerous companies have developed S-FIL tools for up to 300 mm wafer processing (Figs. 5.11b, 5.12). The S-FIL process is well established now for silicon wafers and it mainly meets the requirements for the 22 nm node in the semiconductor industry, especially in terms of resolution [46]. The number of defects and the throughput, which are closely related, are primary obstacles for the emergence of SFIL in the semiconductor industry. The current lowest defectivity is larger than 100 defects/cm^2 for 32 nm half-pitch devices, whereas the requirement is 0.1 defects/cm^2 for conventional lithography [47]. Another issue is the pattern transfer from the resist into a functional layer. The S-FIL tool developed by Molecular Imprints has demonstrated resist patterning of isolated lines down to 11 nm over an 8 in. wafer, but the pattern transfer from resist to another material is limited to 28 nm half-pitch gratings at the time of writing [48]. The gap between the minimum feature sizes which can be imprinted and what can be really transferred into functional materials is governed by the value and homogeneity of the RLT.

The second strategy consists of spin-coating the NIL resist on the full wafer before loading and imprinting [22, 49–52]. The main difference with S-FIL is the requirements of the resist properties. A specific trade-off between resist, mold, tool and imprint conditions needs to be found. The resist should have a high enough viscosity to stay stable on the substrate after spin-coating and before imprinting, but it has to be low enough to fill up efficiently and quickly the nanometer cavities of the mold. UV-NIL resist is typically required with dynamic viscosity around 30 mPa s. Stability and wettability of the resist before the imprinting is also crucial. The thickness of the spin coated resist film is pre-determined to minimize the RLT value, and the film plays the role of a cushion layer. Recently, a study has shown

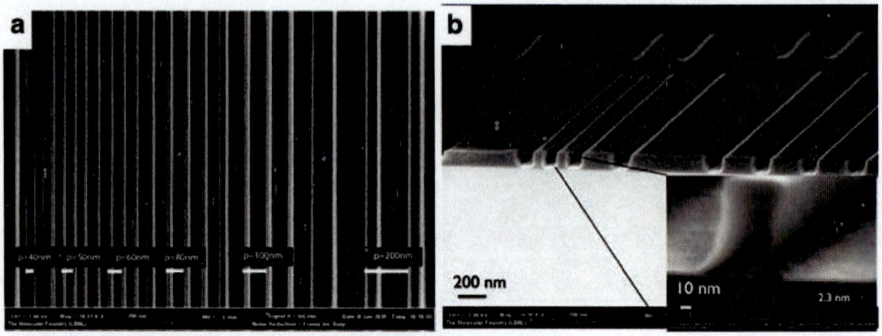

Fig. 5.13 Imprinted patterns by step and repeat on pre-spin coated resist film [43]. (**a**) SEM top view of variable pitch gratings from 40 to 200 nm, (**b**) SEM cross-section image of a grating with sub 5 nm RLT

some advantages of this approach for imprinting with ultra-thin RLT (Fig. 5.13) which can be suitable for the fabrication of optical nano-devices [52].

5.3 Alternative Ways

A multitude of variations of thermal and UV imprint process are developed. Two alternative ways are briefly reported below.

5.3.1 NIL by Reverse Printing

Three-dimensional micro/nanostructures are of particular interest to fabricate, amongst others, photonic devices [53], biosensors [54], nanofluidic devices [55] or nano-electromechanical systems (NEMS) [56]. In the microelectronic industry, direct 3D patterning can reduce the number of alignment steps during the fabrication of, for example, T-gate transistors [57]. Several intrinsic three-dimensional lithography techniques exist, each of them having their own advantages and disadvantages. Lithography techniques which could be used to fabricate 3D structures can be classified in two categories, which we name conventional and non-conventional techniques. High resolution electron beam lithography [58] and focused ion beam lithography [59, 60] have been used for sub 100 nm patterning of 3D structures. The main drawback of these processes is the low throughput, low exposure depth and the limited exposure area. On the other hand, x-ray lithography [61, 62] has been used to imprint 3D structures on a larger surface with a high aspect ratio and high throughput but this process is quite complex and expensive. High aspect ratio and arbitrary shape features [63, 64] have been achieved with two-photon polymerization techniques, but the process complexity and low throughput are the main disadvantages. To fabricate complex structures with a higher throughput, other methods that involve a combination of several techniques have been recently investigated. The combination of two-photon lithography and a phase mask has been used to realize 3D photonic crystal structures [65]. Simultaneously, the combination of nanoimprint lithography with x-ray lithography or wet etching processes [66, 67] produces complex shape structures with particular applications, for example structures suitable for diffractive optics. The RUVNIL technique [68] is based on the combination of reversal nanoimprint, which by nature has the ability to construct 3D nanostructures [69]. This new technique has three main advantages. First, the stamp does not have to be treated with an anti-adhesive layer. Second, no residual layer remains after imprinting and third, 3D device-like structures can be obtained using the same polymer for each layer by repeating the procedure. Potentially, this method could be used to build up structures with several layers suitable, for example, to fabricate 3D photonic crystals with deterministic defects, diffractive optical elements as well as embedded nanochannels devices for bio applications.

There are at least two possible modes of transferring polymer layers on pre-patterned or flat surfaces as shown in Fig. 5.14a and b. These modes are named selective and non-selective pattern transfer modes, depending on the existence or

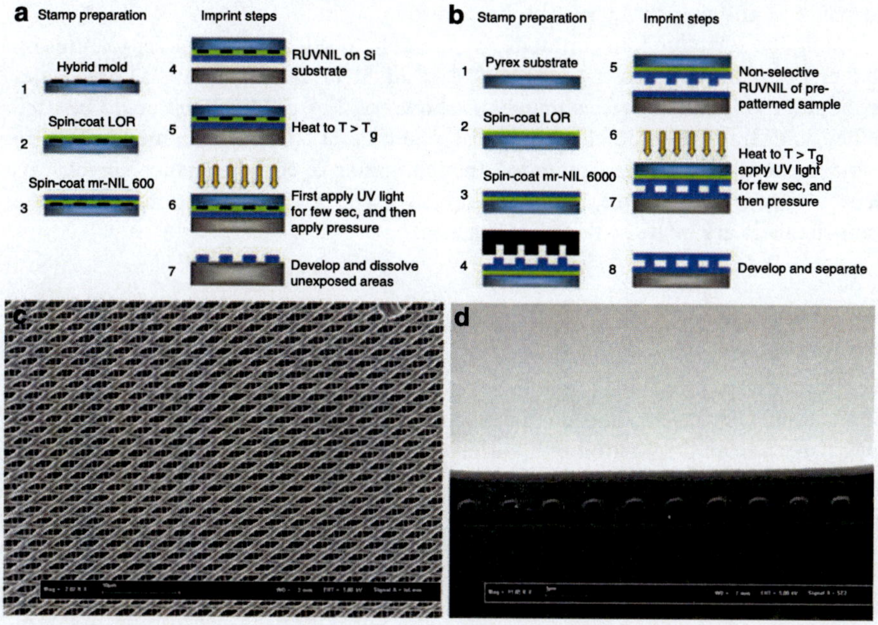

Fig. 5.14 (**a**) Schematics of selective RUVNIL transfer, the stamp preparation steps (1–3) and Imprint steps of RUVNIL technique (4–7), (**b**) SEM images of a stack of three layers (the imprint parameters are 90°C imprinting temperature, 3 s UV light exposure time, and a pressure of 20 bars), (**c**) schematic of the non-selective RUVNIL transfer, (**d**) cross-section SEM image of embedded nano channels surrounded by the same UV crosslinkable polymer mr-NIL 6000

not of a residual layer in the final imprinted structures [70]. On Fig. 5.14a, a polymer layer is spin-coated on a hybrid stamp with incorporated metal protrusions (selective pattern transfer mode), while on Fig. 5.14b, the hybrid stamp, with no metal protrusions, includes a pre-patterned polymer configuration (non-selective pattern transfer mode).

The RUVNIL lithography illustrated in Fig. 5.14a has several steps. First, a thin film of resist, such as a Lift Of Resist (LOR) is spin coated on the hybrid stamp (2). This sacrificial polymer layer is used as an adherence promoter, a planarization layer and to protect the stamp from contamination by the photocuring resist. Then, a film of a UV crosslinkable polymer (mr-NIL 6000 from micro resist technology) is spin coated on the first layer (3). The polymer bilayer is reverse imprinted onto a flat or pre-patterned surface (4). Stamp and substrate are then heated to a temperature above the Tg of mr-NIL 6000 (5) and exposed to UV light. Stamp and substrate are separated just after a post-baking step (6) ensuring a very good adhesion between the polymer and the underlying substrate. Finally, unexposed polymer areas as well as the sacrificial layer are removed leaving behind the negative features of the original stamp (7). The oxygen plasma etching step, usually necessary in standard NIL, is redundant in this process. By repeating the same procedure over prepatterned polymer surfaces, the successive layers adhere to each other by applying pressure and temperature. In this way, three dimensional bridged (or suspended) polymer structures can be realized.

Using the selective transfer mode, we are able to selectively transfer features on flat or pre-patterned substrate without a residual layer. Therefore, a three layer woodpile-like structure was obtained as shown on Fig. 5.14c which could be used either as diffractive optic elements or as photonic crystal configurations [71]. By using the non-selective transfer mode, the fabrication of enclosed nanochannels was demonstrated (Fig. 5.14d), which could be used in bio-applications where the nanochannels are made of the same materials.

5.3.2 Imprinting of Inorganic Sol-Gel Materials

The direct imprinting of inorganic sol-gel films is very promising for numerous applications where the imprinted layer will be used as an active film. Sol-gel chemistry allows preparation of a variety of materials with easy tunability of their properties. The imprinting of sol-gel materials uses thermal activation. A liquid sol solution is imprinted and heated in contact with the NIL mold. The thermal energy is used to condense (cross-link) the sol, which becomes a gel and then a solid film depending on the total thermal energy. The demolding is usually performed at ambient temperature [72]. Under appropriate annealing these materials can provide a completely inorganic and highly resistant coating. Annealing conditions also allow tuning mechanical, optical, electrical properties of imprinted films. Few preliminary works treating this topic reported the process to be a time consuming and complex procedure, requiring high pressures and temperatures and plasma treatment [73–78]. For example Marzolin et al. reported in Ref. [74] on 60 h

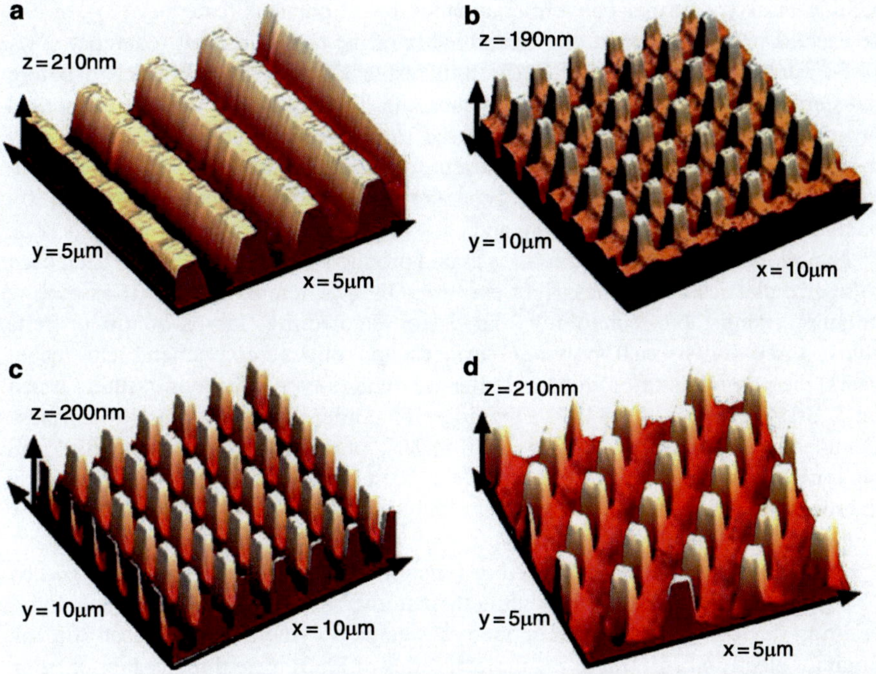

Fig. 5.15 AFM images of nanostructures obtained by double imprint on silica sol-gel film [79], with different imprint time t_1 at a constant temperature of $T_1 = 110°C$: the relative condensation τ_{SiOH} at the end of the first imprint is (**a**) 0.30, (**b**) 0.32, (**c**) 0.33 and (**d**) 0.37. The stamp is a line grating. It has been rotated by 90° between step 1 and step 2

pre-drying of the coatings and 12 h printing time at 20°C while Rizzo et al. [78] used 2 days imprinting at 80°C. Recently, it has been demonstrated that sol-gel films present original thermo-rheological properties which can be used for fast and low cost imprinting of inorganic sol-gel films [72–79]. Some sol-gels films present a material-related condensation threshold. Below the threshold, a fluid state of the material is obtained upon heating, which allows low pressure imprinting. By adjusting the condensation during and/or at the end of the imprint step, it is possible to perform multiple imprinting and even to tune the mechanical properties of the materials. This is illustrated by the variety of patterns that can be obtained from simple gratings, ranging from amplitude-modulated lines to dot patterns (Fig. 5.15). This approach opens a route for creating 3-dimensional nanostructures.

5.4 Applications

Currently, NIL offers as many applications as other lithography techniques thanks to its potential capability of cost-efficient large-scale patterning, high resolution and high throughput to realize nanostructures. The imprinted polymers can act as an etching mask for further conventional nanofabrication steps (category (1)), or can be used as-printed thanks to the functionality of the thin film itself (category (2)).

For standard NIL techniques, the thin residual layer has to be etched before transferring the nanostructures onto the substrate. This alternative fabrication method of nanoimprinted mask transfer, compared to next-generation optical lithography technologies as deep ultraviolet or extreme ultraviolet lithography, has been used to fabricate surface-acoustic-wave generators and filters for mobile phones [80], patterned media for hard-disk [81], and sub-wavelength polarizers for displays [82].

Nanoimprinted polymer structures as end products by direct NIL have been used to realize plastics electronics, light emitting OLEDS, templates for self-assembly, polymer stamps, bio-technology like tissue engineering, lab-on-a-chip or cells study, three-dimensional polymer surfaces and optical components. In recent years, the potentials to produce polymer waveguide-type wavelength filters based on a Bragg grating [83, 84], waveguides [85], microring resonator [86], Mach-Zehnder interferometers [87], lasers [88, 89], plasmonic components [90], and photonic crystals [91, 92] all by NIL, has been demonstrated. NIL is limited by the roughness of the fabricated stamps which can be reduced by a controlled reflow of the polymer [93].

Printable polymers can be doped or functionalized and be directly patterned by NIL [94]. Active polymers can be directly patterned by NIL to realize organic light-emitting devices [95, 96] and conductive organic polymers towards cost-efficient organics electronics [97]. For example, a series of PhCs were directly fabricated in a printable polymer loaded with dye-emitting molecules and showed lasing oscillation at different photonic band-edge frequencies with a minimum threshold of 3 $\mu J/mm^2$ [98]. Semiconducting nanocrystals can also be incorporated in the polymer matrix and be imprinted with photonics crystals geometries with no degradation of the photoluminescence intensity of NC's after patterning by NIL. In fact, an enhancement factor equal to 2.4 at room temperature for the lattice constant of 580 nm [99] has been achieved. This enhancement shows the potential of using NIL to realize high-efficiency LED's with polymers. To further enhance light extraction, the spontaneous recombination rate of the emitters can be increased by coupling the exciton to surface plasmons [100, 101]. This kind of coupling can be achieved by mixing dye and gold nanoparticles in printable polymers [102].

Conclusions

Nanoimprint technologies are collectively one of the most promising alternatives to conventional lithography techniques for fabricating nanopatterns in various applications. NIL has decisive advantages in terms of resolution and versatility to pattern functional materials like conductive polymers or inorganic sol-gel films. NIL techniques are already used for producing commercial micro-lenses and are being intensively developed for the next generation of hard-disks or for flexible electronics, but the demonstration of their commercial cost-efficiency is still missing. NIL opens also a unique route for 3-dimensional nanopatterning and developing new applications.

Acknowledgments The authors thank S. Dhuey and E. Wood from The Molecular Foundry at Lawrence Berkeley National Laboratory, D. Resnick from Molecular Imprints and G. Kreindl from EVG Group. The support of the EC-funded project NaPaNIL (Contract no. NMP 214249) is gratefully acknowledged. The content of this work is the sole responsibility of the authors.

References

1. Chou SY, Krauss PR, Renstrom PJ. Appl Phys Lett. 1995;67:3114.
2. Chou SY, Krauss PR, Renstrom PJ. J Vac Sci Technol B. 1996;14:4129.
3. Austin MD, Ge H, Wu W, Li M, Yu Z, Wasserman D, Lyon SA, Chou SY. Appl Phys Lett. 2004;84:5299.
4. International Technology Roadmap for Semiconductors 2004 Update, www.itrs.net/Common/2004Update/2004_07_Lithography.pdf
5. Schift H, Saxer S, Park S, Padeste C, Pieles U, Gobrecht J. Nanotechnology. 2005;16:171.
6. van Zanten JH, Wallace WE, Wu WL. Phys Rev E. 1995;52:R3329–R3332.
7. Dalnoki-Veress K, Forrest JA, Murray C, Gigault C, Dutcher JR. Phys Rev E. 2001;63:031801.
8. Kim EJ, Tomaszewski JE, de Pablo JJ, Nealey PF, White CC, Fryer DS, Peters RD, Wu WL. Macromolecules. 2001;34:5627.
9. Sharp JS, Forrest JA. Phys Rev E. 2003;67:031805.
10. Murray CA, Wubbenhorst M, Dutcher JR. Eur Phys J. 2003;12:S109.
11. Heyderman LJ, Schift H, David C, Gobrecht J, Schweizer T. Microelectron Eng. 2000;54:229.
12. Sirotkin V, Svintsov A, Zaitsev S, Schift H. Microelectron Eng. 2006;83:880.
13. Beck M, Graczyk M, Maximov I, Sarwe E-L, Ling TGI, Keil M, Montelius L. Microelectron Eng. 2002;61–62:441.
14. Nishino T, Meguro M, Nakamae K, Matsushita M, Ueda Y. Langmuir. 1999;15:4321.
15. Shafrin EG, Zisman WA. J Phys Chem. 1960;64:519.
16. Chen JK, Ko FH, Hsieh KF, Chou CT, Chang FC. J Vac Sci Technol B. 2004;22:3233.
17. Heidari B, Maximov I, Sarwe E-L, Montelius L. J Vac Sci Technol B. 1996;14:2961.
18. Mäkelä T, Haatainen T, Majander P, Ahopelto J, Lambertini V. Jpn J Appl Phys. 2008;47:5142.
19. Haatainen T, Majander P, Makela T, Ahopelto J, Kawaguchi Y. Jpn J Appl Phys. 2008;47:5164.
20. Haisma J, Verheijen M, Heuvel KD, Berg JD. J Vac Sci Technol B. 1996;14:4124.
21. Stewart MD, Johnson SC, Sreenivasan SV. J Microlith Microfab Microsyst. 2005;4:011002.
22. Bender M, Fuchs A, Plachetka U, Kurz H. Microelectron Eng. 2006;83:827.
23. Yang XM, Xu Y, Lee K, Xiao S, Kuo D, Weller D. IEEE Trans Magn. 2009;45:833.
24. Bailey TC, Resnick DJ, Mancini D, Nordquist KJ, Dauksher WJ, Ainley E, Talin A, Gehoski K, Baker JH, Choi BJ, Johnson S, Colburn M, Meissl M, Sreenivasan SV, Ekerdt JG, Willson CG. Microelectron Eng. 2002;61:461.
25. www.molecularimprints.com
26. Ruiz R, Kang H, Detcheverry F, Dobisz E, Kercher D, Albrecht T, de Pablo J, Neley P. Science. 2008;321:936.
27. Park HJ, Kang MG, Guo LJ. ACS NANO. 2009;3:2601.
28. Mancini DP, Gehoski KA, Ainley E, Nordquist KJ, Resnick DJ, Bailey TC, Sreenivasan SV, Ekerdt JG, Willson CG. J Vac Sci Technol B. 2002;20:2896.
29. Morecroft D, Yang JKW, Schuster S, Berggren KK, Xia Q, Wu W, Williams RS. J Vac Sci Technol B. 2009;27:2837.
30. Michel B, Bernard A, Bietsch A, Delamarche E, Geissler M, Winkel D, Stutz R, Wolf H. Chimia. 2002;56:527.
31. Schmid H, Michel B. Macromolecules. 2000;33:3042.

32. Li Z, Gu Y, Wang L, Ge H, Wu W, Xia Q, Yuan C, Chen Y, Cui B, Williams RS. Nano Lett. 2009;9:2306.
33. Lee D, Cho EH, Kim HS, Lee BK, Lee MB, Sohn JS, Lee CH, Suh SJ. J Vac Sci Technol B. 2008;26:514.
34. Muehlberger M, Boehm M, Bergmair I, Chouiki M, Schoeftner R, Kreindl G, Kast M, Treiblmayr D, Glinsner T, Miller R, Platzgummer E, Loeschner H, Joechl P, Eder-Kapl S, Narzt T, Lausecker E, Fromherz T. "Nanoimprint lithography from CHARPAN Tool exposed master stamps with 12.5 nm hp" Microelectron Eng. 2011;88:2070–2073.
35. Studer K, Decker C, Beck E, Schwalm R. Prog Org Coat. 2003;48:92.
36. Vogler M, Wiedenberg S, Mühlberger M, Bergmair I, Glinsner T, Schmidt H, Kley E-B, Grützner G. Microelectron Eng. 2007;84:984.
37. Kim EK, Stacey NA, Smith BJ, Dickey MD, Johnson SC, Trinque BC, Willson CG. J Vac Sci Technol B. 2004;22:131.
38. Kim EK, Stewart MD, Wu K, Palmieri FL, Dickey MD, Ekerdt JG, Willson CG. J Vac Sci Technol B. 2005;23:2967.
39. Song S, Kim S-M, Choi B-Y, Jung G-Y, Lee H. J Vac Sci Technol B. 2009;27:1984.
40. Decker C, Decker D. J Macromol Sci Pure Appl Chem A. 1997;34:605.
41. Wu C-C, Hsu SL-C, Liao W-C. Microelectron Eng. 2009;86:325.
42. Kim JY, Choi D-G, Jeong J-H, Lee E-S. Appl Surf Sci. 2008;254:4793.
43. Byeon K-J, Hong EJ, Park H, Cho J-Y, Lee S-H, Jhin J, Baek JH, Lee H. Thin Films. 2011;519:2241.
44. Michaelson T, Sreenivasan SV, Ekerdt J, Willson CG. Proc SPIE. 1999;3676:379.
45. Bailey TC, Johnson SC, Sreenivasan SV, Ekerdt JG, Willson CG, Resnick DJ. J Photopolym Sci Technol. 2002;15:481.
46. Malloy M, Litt LC. J Photopolym Sci Technol. 2010;23:749.
47. Higashiki T. Challenges to next generation lithography. MNE 2009 Conference, Ghent – Belgium, 29 Sep. 2009.
48. Resnick D. Oral presentation in EIPBN 2010 conference. Anchorage, 2010.
49. Peroz C, Dhuey S, Volger M, Wu Y, Olynick D, Cabrini S. Nanotechnology. 2010;21:445301.
50. Otto M, Bender M, Richter F, Hadam B, Kliem T, Jede R, Spangenberg B, Kurz H. Microelectron Eng. 2004;73:152.
51. Otto M, Bender M, Zhang J, Fuchs A, Wahlbrink T, Bolten J, Spangenberg B, Kurz H. Microelectron Eng. 2007;84:980.
52. Peroz C, Dhuey S, Goltsov A, Volger M, Harteneck B, Ivonin I, Bugrov A, Cabrini S, Babin S, Yankov V. Microelectron Eng. 2011;88:2092.
53. Vlasov YA, Bo XZ, Sturm JC, Norris DJ. Nature. 2001;414:289.
54. Khandurina J, Guttman A. J Chromatogr A. 2002;943:159.
55. Wu H, Odom TW, Chiu DT, Whitesides GM. J Am Chem Soc. 2003;125:554.
56. Yang Z, Yu Y, Li X, Bao H. Microelectron Reliab. 2006;46:805.
57. Li M, Chen L, Chou SY. Appl Phys Lett. 2001;78:3322.
58. Yamazaki K, Namatsu H. Microelectron Eng. 2004;73–74:85.
59. Munnik F, Benninger F, Mikhailov S, Bertsch A, Renaud P, Lorenz H, Gmur M. Microelectron Eng. 2003;67–68:96.
60. Freeman D, Madden S, Davies BL. Opt Express. 2005;13:3079.
61. Romanato F, Businaro L, Vaccari L, Cabrini S, Candeloro P, De Vittorio M, Passaseo A, Todaro MT, Cingolani R, Cattaruzza E, Galli M, Andreani C, Di Fabrizio E. Microelectron Eng. 2003;67–68:479.
62. Romanato F, Tormen M, Businaro L, Vaccari L, Stomeo T, Passaseo A, Di Fabrizio E. Microelectron Eng. 2004;73–74:870.
63. Sun HB, Kawakami T, Xu Y, Ye JY, Matuso S, Misawa H, Miwa M, Kaneko R. Opt Lett. 2000;25:1110.

64. Tormen M, Businaro L, Altissimo M, Romanato F, Cabrini S, Perennes F, Proietti R, Sun H-B, Kawata S, Di Fabrizio E. Microelectron Eng. 2004;73–74:535.
65. Jeon S, Malyarchuk V, Rogers JA, Wiederrecht GP. Opt Express. 2006;14:2300.
66. Tormen M, Businaro L, Altissimo M, Romanato F, Cabrini S, Perennes F, Proietti R, Sun H-B, Satoshi K, Di Fabrizio E. Microelectron Eng. 2004;73–74:535.
67. Tormen M, Carpentiero A, Vaccari L, Altissimo M, Ferrari E, Cojoc D, Di Fabrizio E. J Vac Sci Technol B. 2005;23:2920.
68. Kehagias N, Reboud V, Chansin G, Zelsmann M, Jeppesen C, Reuther F, Schuster C, Kubenz M, Gruetzner G, Sotomayor Torres CM. J Vac Sci Technol B. 2006;24:3002.
69. Cheng X, Guo LJ. Microelectron Eng. 2004;71:277.
70. Kehagias N, Chansin G, Reboud V, Zelsmann M, Schuster C, Kubenz M, Reuther F, Gruetzner G, Torres CMS. Microelectron Eng. 2007;84:921.
71. Kehagias N, Reboud V, Chansin G, Zelsmann M, Jeppesen C, Schuster C, Kubenz M, Reuther F, Gruetzner G, Torres CMS. Nanotechnology. 2007;18:175303.
72. Peroz C, Heitz C, Barthel E, Søndergård E, Goletto V. J Vac Sci Technol B. 2007;25:L27.
73. Brendel R, Gier A, Menning M, Schmidt H, Werner JH. J Non-Cryst Solids. 1997;218:391.
74. Marzolin C, Smith SP, Prentiss M, Whitesides GM. Adv Mater. 1998;10:571.
75. Li M, Tan H, Chen L, Wang J, Chou SY. J Vac Sci Technol B. 2003;21:660.
76. Harnagea C, Alexe M, Schilling J, Choi J, Wehrspohn RB, Hesse D, Gosele U. Appl Phys Lett. 2003;83:1827.
77. Okinaka M, Tsukagoshi K, Aoyagi Y. J Vac Sci Technol B. 2006;24:1402.
78. Rizzo G, Barila P, Galvagno S, Neri G, Arena A, Patane S, Saitta G. J Sol Gel Sci Technol. 2003;26:1017.
79. Peroz C, Chauveau V, Barthel E, Søndergård E. Adv Mater. 2009;21:555.
80. Cardinale GF, Skinner JL, Talin AA, Brocato RW, Palmer DW, Mancini DP, Dauksher WJ, Gehoski K, Le N, Nordquist KJ, Resnick DJ. J Vac Sci Technol B. 2004;22:3265.
81. McClelland GM, Hart MW, Rettner CT, Best ME, Carter KR, Terri BD. Appl Phys Lett. 2002;81:1483.
82. Ahn SW, Lee K-D, Kim JS, Kim SH, Lee SH, Park JD, Yoon PW. Microelectron Eng. 2005;78:314.
83. Seekamp J, Zankovych S, Helfer AH, Maury P, Sotomayor Torres CM, Bottger G, Liguda C, Eich M, Heidari B, Montelius L, Ahopelto J. Nanotechnology. 2002;13:581.
84. Ahn SW, Lee KD, Kim JS, Kim SH, Park JD, Lee SH, Yoon PW. Nanotechnology. 2005;16:1874.
85. Kehagias N, Zankovych S, Goldschmidt A, Kian R, Zelsmann M, Sotomayor Torres CM, Pfeiffer K, Ahrens G, Gruetzner G. Superlattice Microst. 2004;36:201.
86. Chao CY, Guo LJ. J Vac Sci Technol B. 2002;20:2862.
87. Paloczi GT, Huang Y, Yariv A, Luo J, Jen AKY. Appl Phys Lett. 2004;85:1662.
88. Peroz C, Galas JC, Shi J, Le Gratiet L, Chen Y. Appl Phys Lett. 2006;89:243109.
89. Arango F, Christiansen MB, Gersborg-Hansen M, Kristensen A. Appl Phys Lett. 2007;91:223503.
90. Reboud V, Kehagias N, Zelsmann M, Fink M, Reuther F, Gruetzner G, Torres CMS. Opt Express. 2007;15:7190.
91. Belotti M, Torres J, Roy E, Pepin A, Chen Y, Gerace D, Andreani LC, Galli M. Microelectron Eng. 2006;83:1773.
92. Tamborra M, Striccoli M, Curri ML, Alducin JA, Mecerreyes D, Pomposo JA, Kehagias N, Reboud V, Torres CMS, Agostian A. Small. 2007;3:822.
93. Chao CY, Guo LJ. IEEE Photon Technol Lett. 2004;16:1498.
94. Sotomayor Torres CM, Zankovych S, Seekamp J, Kam AP, Cedeno CC, Hoffmann T, Ahopelto J, Reuther F, Pfeiffer K, Bleidiessel G, Gruetzner G, Maximov MV, Heidari B. Mater Sci Eng C. 2003;23:23.
95. Wang J, Sun X, Chen L, Chou SY. Appl Phys Lett. 1999;75:2767.
96. Cheng X, Hong Y, Kanicki J, Guo LJ. J Vac Sci Technol B. 2002;20:2877.

97. Cedeno CC, Seekamp J, Kam AP, Hoffmann T, Zankovych S, Torres CMS, Menozzi C, Cavallini M, Murgia M, Ruani G, Biscarini F, Behl M, Zentel R, Ahopelto J. Microelectron Eng. 2002;61:25.
98. Reboud V, Lovera P, Kehagias N, Zelsmann M, Reuther F, Gruetzner G, Redmond G, Torres CMS. Appl Phys Lett. 2007;91:151101.
99. Reboud V, Kehagias N, Zelsmann M, Striccoli M, Tamborra M, Curri ML, Agostiano A, Fink M, Reuther F, Gruetzner G, Sotomayor Torres CM. Appl Phys Lett. 2007;90:011114.
100. Köck A, Gornik E, Hauser M, Beinstingl W. Appl Phys Lett. 1990;57:2327.
101. Barnes WL. J Lightwave Technol. 1999;17:2170.
102. Reboud V, Kehagias N, Striccoli M, Placido T, Panniello A, Curri ML, Zelsmann M, Reuther F, Gruetzner G, Sotomayor Torres CM. J Vac Sci Technol B. 2007;25:2642.
103. Kehagias N, Reboud V, Sotomayor Torres CM, Sirotkin V, Svintsov A, Zaitsev S. Microelectron Eng. 2008;85:846.

Part III

Deposition at the Nanoscale

Atomic Layer Deposition for Nanotechnology

6

A. Foroughi-Abari and K. Cadien

Abstract

Atomic layer deposition (ALD) is a thin film chemical vapor deposition technology that is uniquely able to deliver extremely conformal, pin hole free, nanometer thick films. It is finding a large number of applications in nanotechnology such as display technology, integrated circuit (IC) fabrication, solar cells and catalysis. In this chapter we will discuss the background behind ALD, its fundamentals, and some of the nanotechnology applications, deposition in nanotubes, deposition of nanoparticles, and deposition of nanometer thick films.

6.1 Background

ALD is a deposition technique in which the reactant gases (precursors) of two sequential reactions are pulsed into the reaction chamber alternately. After each cycle, one atomic layer is formed on the surface of the substrate. ALD cycles can be repeated as many times as desired to grow a given thickness of the film material. ALD is considered to be an ideal deposition technique for production of ultra-thin films especially when the geometry is complex.

There is a controversy over the origin of ALD. Some have given the credit to Russian scientists led by Professor Aleskovskii working on a process to grow TiO_2 and GeO_2 and other oxides during 1960s [1]. The concept of the ALD process was first proposed by Professor Aleskovskii in his PhD thesis published in 1952. Initial experiments were done on high surface area silica, but later single crystal substrates were also used. The results were first published by Kol'tsov and co-workers

A. Foroughi-Abari • K. Cadien (✉)
Department of Chemical and Materials Engineering, University of Alberta, Edmonton, AB, Canada
e-mail: kcadien@ualberta.ca

M. Stepanova and S. Dew (eds.), *Nanofabrication*,
DOI 10.1007/978-3-7091-0424-8_6,

between 1965 and 1967 under the name of "Molecular Layering". The experiments were conducted under supervision of Professor Aleskovskii.

The more commonly referenced origin of ALD is the work done in Finland by Tuomo Suntola and co-workers in 1970s. They named their technique "Atomic Layer Epitaxy" or ALE, and filed a patent in which they demonstrated the employment of alternating element reactants Zn and S to form layers of ZnS. The same technique was used for deposition of SnO_2 from Sn/O and GaP from Ga/P. Suntola and co-workers later showed that the same process could use compounds as reactants instead of elements: a $TaCl_5/H_2O$ process to form Ta_2O_5 and a $AlCl_3/H_2O$ process to form Al_2O_3. The first commercial application of ALD was in production of thin film electroluminescent displays which was first used in the large-area display that operated in the Helsinki airport from 1983 to 1998 [2]. The name ALE was later changed to ALD in the early 1990s, and it is currently the most common name for the technique [1].

There are numerous application areas for ALD such as nanoelectronics, micro electro mechanical systems (MEMS), optical coatings and catalysts. Nanoelectronics is one of the major current application areas for ALD. Further shrinkage of modern transistors requires replacement of silicon dioxide (or nitrided silicon dioxide) by high-k dielectric materials such as hafnium oxide or aluminum oxide due to electron tunneling through the gate oxide layer. As the thickness of the oxide layer decreases, current leakage increases which leads to lower device performance and high power consumption. Deposition of these high-k dielectric materials is possible using ALD [3, 4].

6.1.1 ALD Fundamentals

The basic steps in an ALD cycle are illustrated in Fig. 6.1. The first precursor is introduced into the chamber and is chemically adsorbed on the surface of the substrate until the surface is saturated (Fig. 6.1a). Once the surface becomes saturated, no further precursor molecule is adsorbed. The carrier gas (typically N_2 or Ar) sweeps away the excess gases out of the chamber (Fig. 6.1b). The second precursor is pulsed into the chamber and reacts with the adsorbed species on the surface until the surface becomes saturated (Fig. 6.1c). Excess precursors and by-products are again removed by pumping and carrier gas purging (Fig. 6.1d). At the end of these four steps, one monolayer of the desired material is deposited on the substrate. By repeating step a to d numerous times, a film with a specified thickness can be deposited [5].

Each cycle may take times from 0.5 to a few seconds and a film with thickness of 0.01–0.3 nm may be deposited [6]. The growth rate depends on the size of the precursor molecules and also the number of available adsorption sites on the surface. In the case of large precursor molecules, due to steric hindrance between molecules, a lower growth rate has been observed [6].

Deposition of aluminum oxide using trimethylaluminum (TMA) and H_2O is a classic ALD process, in which the following half-reactions occur:

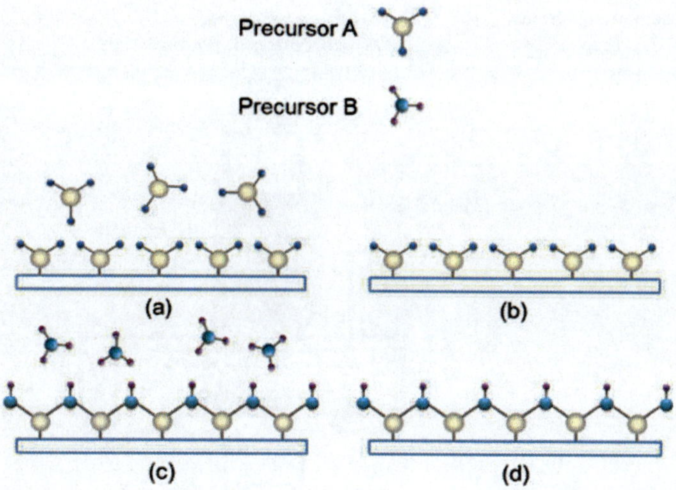

Fig. 6.1 Schematic of ALD cycle: **(a)** the precursor A is introduced until the surface is saturated; **(b)** unadsorbed reactants are purged; **(c)** the precursor B is introduced and reacts with adsorbed molecules from the precursor A; **(d)** excess reactants and byproducts are purged

$$Al - OH^* + Al(CH_3)_3 \rightarrow Al - O - Al - CH_3{}^* + CH_4 \quad (6.1)$$

$$Al - CH_3{}^* + H_2O \rightarrow Al - OH^* + CH_4 \quad (6.2)$$

where the asterisks denote the surface species [7, 8].

ALD is a self-limiting process which means that once a monolayer is formed on the surface, no more atoms can adsorb on the surface. In other words, as available sites for precursor adsorption are saturated, the process limits itself and does not proceed until the next precursor is introduced. This unique feature makes ALD an ideal method for depositing uniform and conformal films on complex structures.

6.1.1.1 Plasma ALD

Plasma ALD was developed to enable the deposition of metals and the low temperature deposition of oxides and nitrides. In plasma ALD, the metal molecule pulse is thermal ALD while a remote plasma is used for the other precursor with gases such as argon, oxygen, hydrogen, or nitrogen. For example, for plasma ALD of alumina, TMA is used for the metal precursor and plasma oxygen is used for atomic oxygen. A schematic diagram of an ALD plasma reactor is shown in Fig. 6.2.

ALD reactions, such as TMA with water, have been modeled and shown to have intermediate metastable states separated by activation barriers. These barriers must be overcome for the reaction to proceed. In thermal ALD, the required thermal energy comes from the heating of the substrate. In plasma ALD, the energy is provided by both the thermal energy as well as remote plasma energy. The plasma

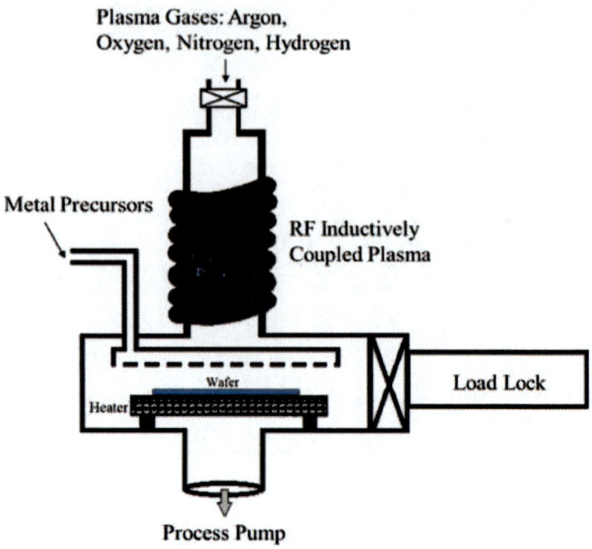

Fig. 6.2 Schematic diagram of a plasma ALD reactor

is used for gas dissociation to produce highly reactive atomic radicals. These radicals, once they reach the surface, accelerate the reactions and allow them to take place at much lower temperatures. When plasma ALD is used, the reaction rates at the surface increase, fragmentation of precursor molecules increases, and the product molecules are more easily removed by bombardment. The deposition rate depends on the separation between the plasma source and the substrate, the plasma pulse duration, pressure, and plasma power [9].

A well-studied example of usage of plasma is for the deposition of aluminum oxide using TMA and plasma oxygen. The difference with the thermal ALD is that instead of water, oxygen plasma is used as the source of oxygen. With the help of plasma, room-temperature deposition on thermally fragile, heat sensitive polymeric materials has been successfully carried out without destroying them [10, 11]. Another benefit of plasma ALD compared to thermal is that cycles times are much shorter during plasma ALD which results in faster depositions and higher throughput. Figure 6.3 shows the difference in growth rate between the thermal and plasma ALD of aluminum oxide. The plasma ALD of aluminum oxide is clearly faster than the thermal ALD. One of the factors that significantly limits the speed of the thermal ALD process is the time required to completely purge the water after each pulse due to its polar nature. This poses special challenges at low deposition temperatures. If the water remains at the surface due to improper purging, the films may contain high amounts of hydrogen and possibly other impurities. Therefore, the thermal ALD of aluminum oxide at deposition temperatures of 373 K and less is impractical [11]. This is where the plasma presents its advantage by enabling us to lower the process temperature down to room temperature and yet maintain acceptable quality.

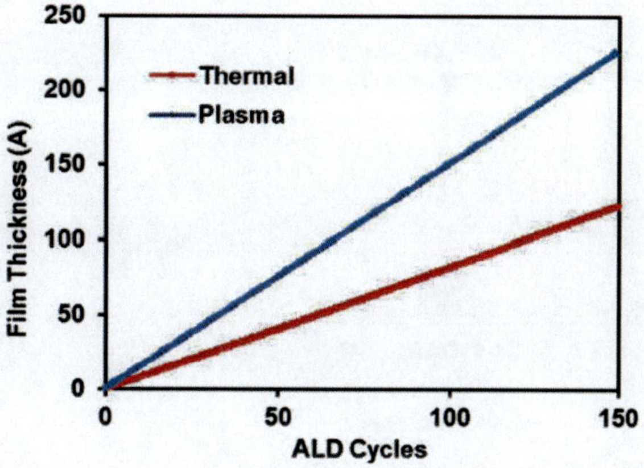

Fig. 6.3 Comparison between growth per cycle for thermal and plasma ALD of aluminum oxide at 373 K. Data from in-situ ellipsometry during film growth

6.1.1.2 ALD Precursors

The choice of proper precursor is an important consideration in the design of an ALD process. The precursors must have sufficient thermal stability to remain in the gas phase before reaching the substrate and yet be reactive enough to chemisorb on the substrate. Decomposition of the precursor before reaching the substrate is detrimental to the self-limiting growth mechanism. The precursor must also have enough vapor pressure when heated to be transferred into the reaction chamber either as a pure gas or as a vapor mixed with a carrier gas. Byproducts produced in the reaction chamber, should not be reactive to prevent process defect generation and excessive film accumulation [5]. Precursors may be solid, liquid or gases and this affects the type of delivery system used.

Trimethylaluminum (TMA) and water are examples of ideal ALD precursors. Once TMA molecules are exposed to hydroxyl groups (OH) on the surface (water absorption on a surface leads to OH formation), methane is formed and aluminum atoms with one or two methyl groups are bonded to the surface. After removing the excess TMA, water vapor is added to the system and the methyl groups, attached to the aluminum atoms, are replaced by hydroxyl groups. These OH groups make the surface reactive toward TMA and therefore this cycle can be repeated [12], as shown in Fig. 6.4. In panel 1, the water attaches to the substrate surface creating an OH group and this is followed by a purge step in which water vapor is removed. TMA then is reacted with the OH group and methane forms as a byproduct. Excess TMA and the byproducts are then purged. The process is then repeated in panels 3 and 4 to build up the alumina layer.

Precursors of hafnium are very important since ALD hafnium dioxide is used by the semiconductor industry as a high κ gate oxide. Precursors for hafnium are typically hafnium amides, tetrakis(dimethylamino)hafnium ($Hf(NMe_2)_4$) and tetrakis(ethylmethylamino)hafnium ($Hf(NEtMe)_4$) [13, 14], and precursors based on much larger molecules such as, bis(methylcyclopentadienyl)dimethylhafnium-(IV)

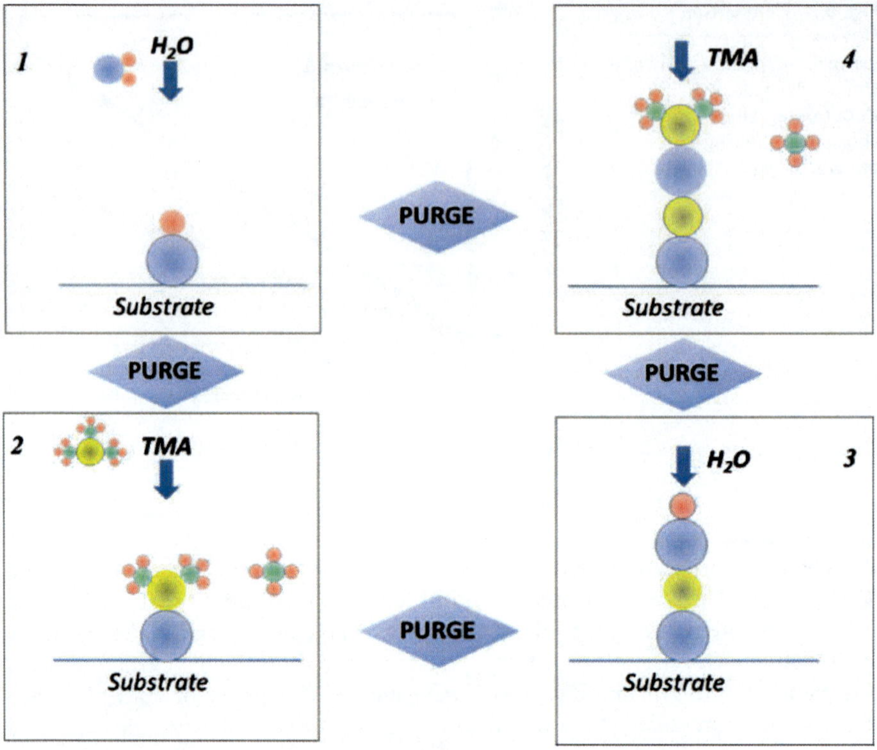

Fig. 6.4 Schematic diagram of the ALD of trimethylaluminum and water

(HfD-02) and bis(methylcyclopentadienyl)methoxymethylhafnium (IV) (HfD-04) [15]. This suite of four precursors have a wide range of vapor pressures as shown in Fig. 6.5. If a precursor pressure of 0.25 Torr is required, the ($Hf(NMe_2)_4$) source would be set at 310.1 K, the $Hf(NEtMe)_4$ at 341.6 K, the HfD-02 at 357.9 K, and the HfD-04 at 361.4 K. This wide range of vapor pressure permits a wide range of ALD deposition temperatures and rates as shown in Fig. 6.6. The optimum ALD deposition temperature range for the four precursors is summarized in Table 6.1.

Precursors are sold by weight, typically in 25 g quantities. Since the precursors are typically organometallic compounds with large molecules, the actual metal content can be 50% or less. Some expensive precursors, such as tetrakis (dimethylamino)hafnium, are sold in 10 g ampoules. This precursor contains 50.3% Hf, so a 10 g ampoule has 5 g of hafnium. Precursors also have a shelf life, even at room temperature, and a much shorter life when kept at the ALD temperature. For example, for the four hafnium precursors discussed earlier, thermogravimetric analysis (TGA), shown in Fig. 6.7, indicates that ($Hf(NMe_2)_4$) and ($Hf(NEtMe)_4$) start to breakdown at 336 and 374 K, within the range of typical precursor temperatures.

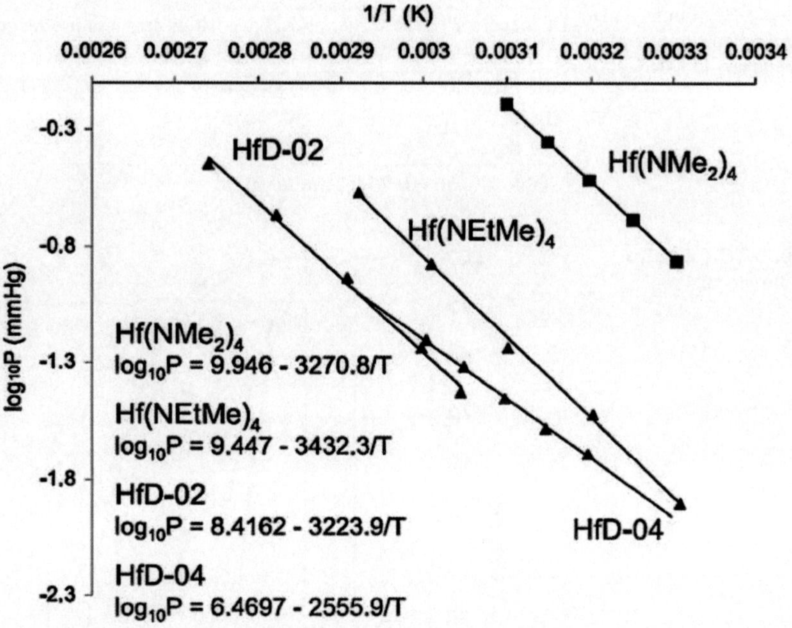

Fig. 6.5 Vapor pressure data for Hf sources from SAFC Hitech [15]

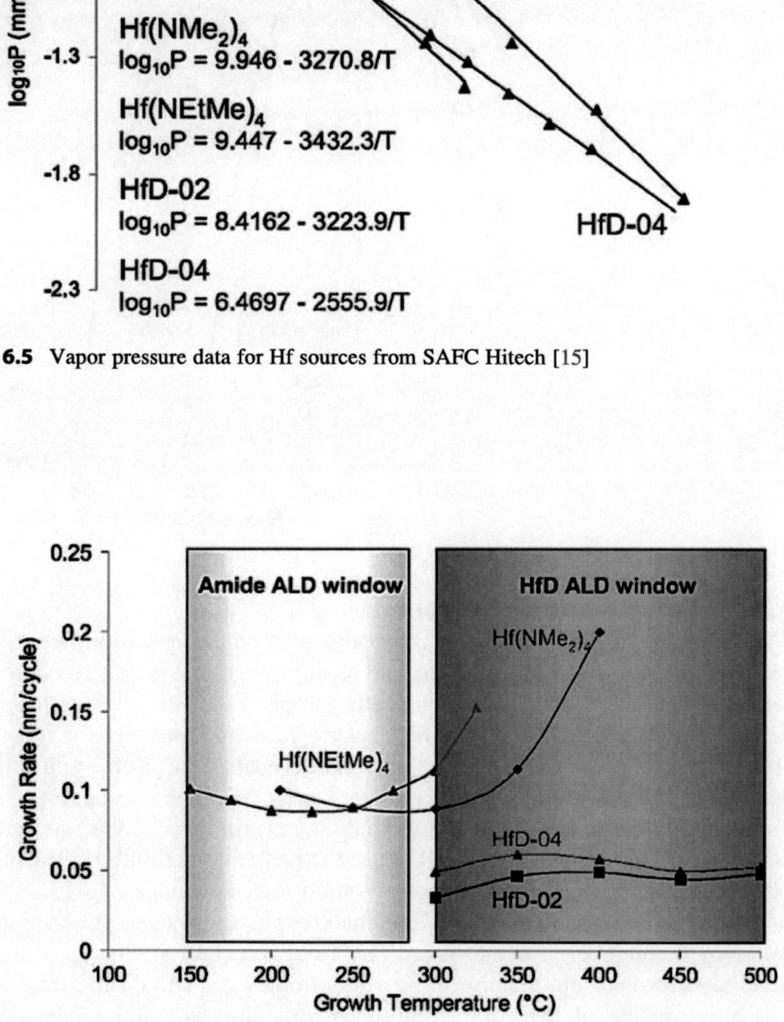

Fig. 6.6 Growth rate as a function of temperature for the four Hf precursors from SAFC Hitech [15]

Table 6.1 Optimum ALD deposition temperatures for Hf precursors [13–15]

Hf precursor	ALD deposition temperature range, K[a]
$(Hf(NMe_2)_4)$	~473–613
$Hf(NEtMe)_4$	~323–548
HfD-02	~583–773
HfD-04	~573–773

[a]Stable growth rate (±0.01 nm/cycle)

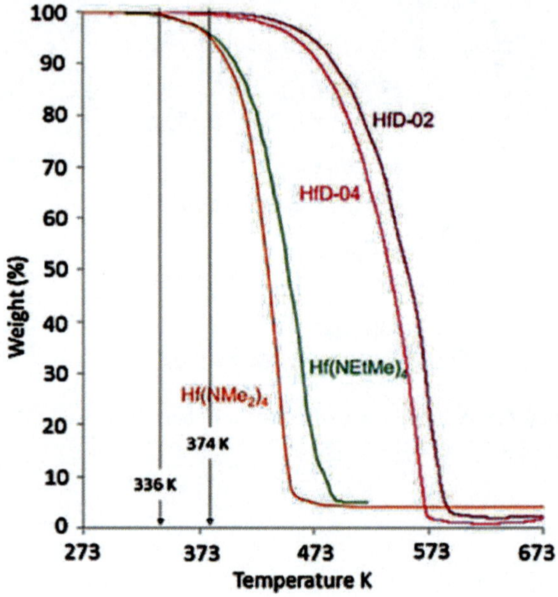

Fig. 6.7 TGA data for hafnium precursors [15]

6.1.1.3 Advantages and Limitations

ALD has some important advantages compared with other deposition techniques especially in the case of ultra-thin films and where the geometry is also complex. Conformality is perhaps the most advantageous property of ALD which allows for deposition of thin films in very complex topographic structures and high aspect-ratio features [16, 17]. The self-limiting growth also enables large area uniformity, reproducibility and excellent adhesion. Unlike CVD, the process temperature can be relatively low (less than 400°C) and since there is no gas phase nucleation, defects are much lower. This makes ALD very attractive to the high tech industry [18]. Accurate thickness control is another interesting advantage of ALD which makes it superior to other techniques. The thickness is simply and yet accurately controlled by the number of cycles used. This feature is essential in deposition of ultra-thin layers where the material properties strongly depend on the thickness. ALD is also capable of depositing multilayer structures with relatively abrupt transitions since films are formed layer by layer. This unique feature is essential in production of nano-laminates.

Although many materials have been deposited successfully using ALD, still there is an increasing demand for deposition of new materials and therefore new suitable precursors have to be developed. Slow deposition rate is also considered to be a limitation for ALD [19]. Selectivity is also an issue in ALD since precursors once entered into the chamber, tend to cover almost every existing surface.

6.2 Nanotechnology Applications

6.2.1 ALD Deposition in High Aspect Ratio Features

As mentioned earlier, ALD is exceptionally conformal, and an indicator of this is the ability of ALD to coat the inside of nanopores with aspect ratios (pore length divided by pore diameter) of more than 1000. Several researchers have explored the fabrication of nanoporous anodized alumina [20, see also Chap. 14] and there are several papers on the use of ALD to line the interior of the pores [21–25]. This allows the creation of a nanoporous structure from any material that can be deposited by ALD. However, research has shown that anodic aluminum oxide membranes are contaminated with dehydrated aluminum oxalate (from the anodization process) [26]. The oxalate can only be removed by a high temperature anneal, or it can be encapsulated with a 15 nm alumina ALD layer as shown in Fig. 6.8a, b.

A detailed study of pore coating of aspect ratios of up 5000:1 was published in 2003 [22]. In this study, the thermal ALD of ZnO, from dimethylzinc(DMZ) and water, was used to coat different size pores in anodized alumina. The ALD half reactions for ZnO are given below:

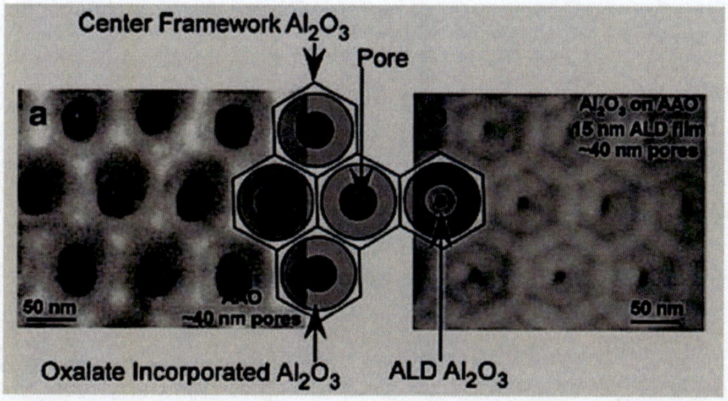

Fig. 6.8 SEM micrographs of (**a**) a porous anodized alumina membrane showing the discoloration due to the oxalate around the pore openings, (**b**) a 15 nm ALD alumina coated membrane showing 10 nm pores. The schematic diagram inset in the figure explains the features in micrographs (From Ref. [26])

$$\mathrm{ZnOH^* + Zn(CH_2CH_3)_2 \rightarrow ZnOZn(CH_2CH_3)^* + C_2H_6} \quad (6.3)$$

$$\mathrm{Zn(CH_2CH_3)^* + H_2O \rightarrow ZnOH^* + C_2H_6} \quad (6.4)$$

ALD ZnO was deposited in high aspect ratio anodized alumina nanopores with pore length, *L*, of 120 μm and pore diameters, *d*, of 19, 46, and 65 nm. The researchers found that the coating uniformity in a nanotube depended on exposure time and pore diameter, as shown in Fig. 6.9a, b, respectively. They showed that the uniformity of the ZnO coating in the nanopores depended on the square root of the exposure time (pulse width) and linearly with pore diameter. From Monte Carlo simulations, they were able to show that for uniform pore coverage, the exposure time to coat the pore, *t* seconds, is given by:

$$t = 2.3 \times 10^{-7} P^{-1} \sqrt{m}\,\Gamma \left(\frac{L}{d}\right)^2$$

where *P* is the reactant pressure in Torr, Γ is the density of ALD reactive sites in $10^{19}\ \mathrm{m}^{-2}$ and *m* is the reactant molecular mass in amu. They found that for DMZ, Γ is 0.84, *m* is 123, and for $P = 5$ Torr, and $L/d = 5000$, the predicted exposure time is 11 s. One of the conclusions from this work is that uniform pore coating can consume a large amount of precursors and experimental strategies must be used to reduce precursor consumption.

Recent research has shown that the interior of porous alumina nanopores can also be coated with ALD Pt and Ir [25]. Figure 6.10 shows the top down SEM view showing the variation of the tubes diameter with the number of ALD cycles. This structure was used to amperometrically sense glucose. There has also been interest in using coated nanotubes as catalytic membranes for the oxidative

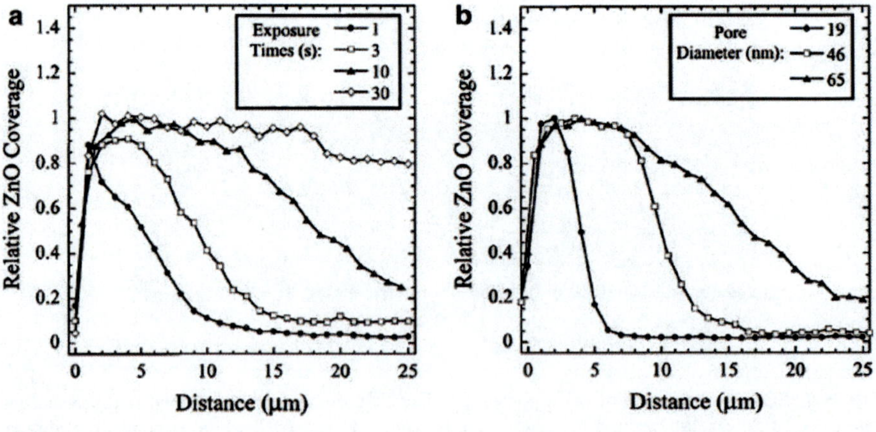

Fig. 6.9 The ZnO coverage dependency for ALD 64 cycles on (**a**) exposure time for $d = 65$ nm and $L = 50$ μm, and (**b**) pore diameter for a 5 s exposure time (From Ref. [22])

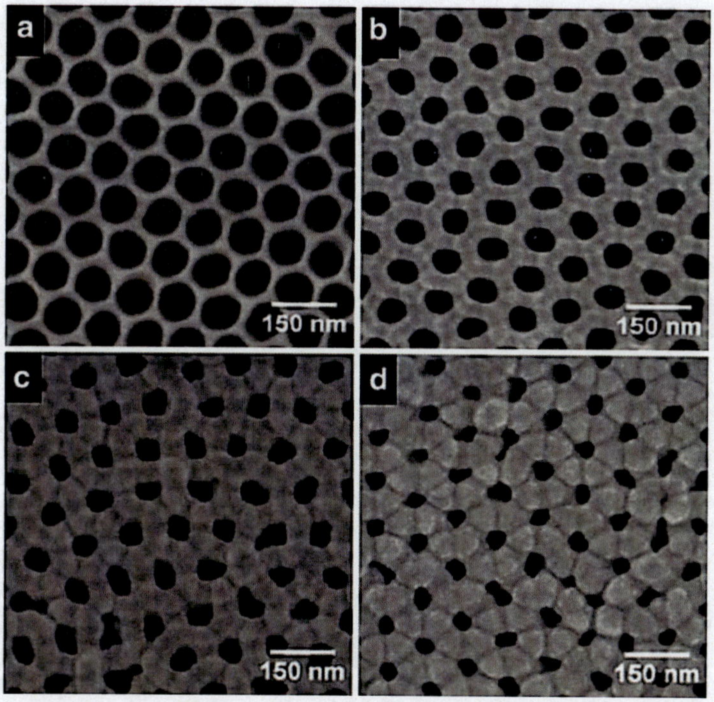

Fig. 6.10 Top down SEM images of porous anodized alumina after (**a**) 0, (**b**) 200, (**c**) 300, and (**d**) 400 ALD cycles of (trimethyl)methylcyclopentadienylplatinum(IV) (Pt(MeCp)Me_3) and O_2 (From Ref. [25])

dehydrogenation of cyclohexane [23], in dye synthesized solar cells [24], and energy storage [27].

6.2.2 ALD Films for Nano-Electronics

One of the largest applications of ALD is for nanoelectronics where ALD is used for deposited gate oxides and to deposit diffusion barriers for copper interconnects. This section will focus on the application of ALD to gate oxides.

The MOSFET (metal oxide semiconductor field effect transistor) is the building block of the CMOS (complementary metal oxide semiconductor) devices that are ubiquitous in integrated circuits. In a MOSFET, voltage is placed across the gate oxide to control the flow of carriers (electrons or holes) in the channel as shown in Fig. 6.11a. The gate oxide that has been used for the past 40 years is thermally grown silicon dioxide (SiO_2). Due to scaling, the oxide thickness had gone from 100 nm thickness in 1969 to ~2 nm by 2005 while the leakage current increased exponentially (Fig. 6.11b). It is important to understand why this happened in order to understand the solution to the problem. IC's have become ubiquitous because as

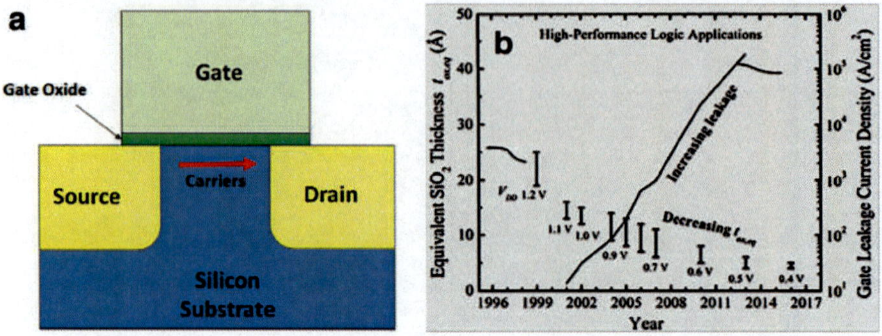

Fig. 6.11 (**a**) Schematic diagram of a MOSFET, (**b**) leakage current versus time for SiO_2 gate oxide (From Ref. [28])

devices shrink, more devices can be put on the same area of silicon for about the same price. This relationship underlies Moore's Law [29], named after the founder of Intel Corporation. One way to understand this is to consider the fact that the first microprocessor, the Intel 4004, had about 2,000 transistors while today, a modern Intel Core i7® CPU has close to a billion transistors, depending on the amount of cache memory. This was enabled by the shrinkage of the minimum feature size from 10 μm (1969) to 45 nm in 2007.

An important metric for transistor performance is the speed at which the transistor switches. This speed is related to the voltage swing of interest (*V*), the drive current (*I*), and the capacitance (*C*) of the device by the relationship, *CV*/*I* [30]. The voltage swing and the drive current are device parameters, while the capacitance is related to dimensions (area (*A*) and thickness (*t*)) and material properties of the gate oxide as given by the equation below

$$C = \frac{A\varepsilon_0\varepsilon_r}{t} \qquad (6.5)$$

where ε_0 is the dielectric constant of vacuum and ε_r (also referred to as "*k*") is the relative dielectric constant of the oxide. As dimensions shrink, *A* gets smaller and *t* needs to shrink in order to keep *C* from getting smaller and the device from slowing down. However, as the gate oxide thickness approach 2 nm, direct tunneling of carriers through the oxide starts to become significant, and leakage current increases rapidly (Fig. 6.11b). Tunneling current depends on the electric field, *E*, ($E = V/t$) which depends inversely on the gate oxide thickness. Thus, the only way to reduce the leakage current is to increase the gate oxide thickness. The only way to keep capacitance high if *t* increases, is to use a material with a higher ε_r. A plot of dielectric constant and band gap of various oxides is given in Fig. 6.12. Band gap is an important property for dielectrics since the break down voltage is directly related to band gap and gate oxides are subjected to very large electric fields of the order of ~5×10^6 V cm^{-1}.

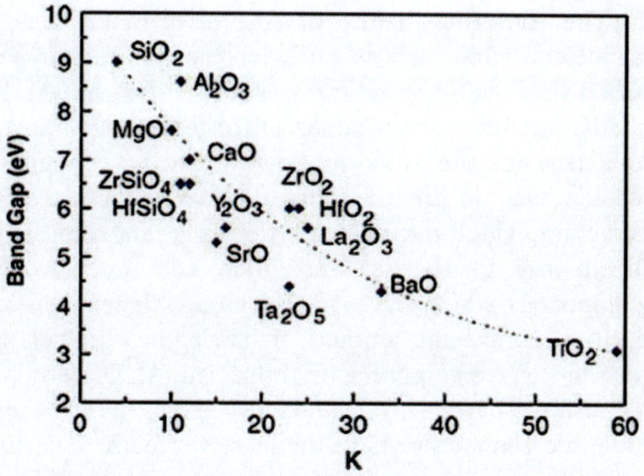

Fig. 6.12 Dielectric constants versus band gap for potential gate oxide materials (From Ref. [31])

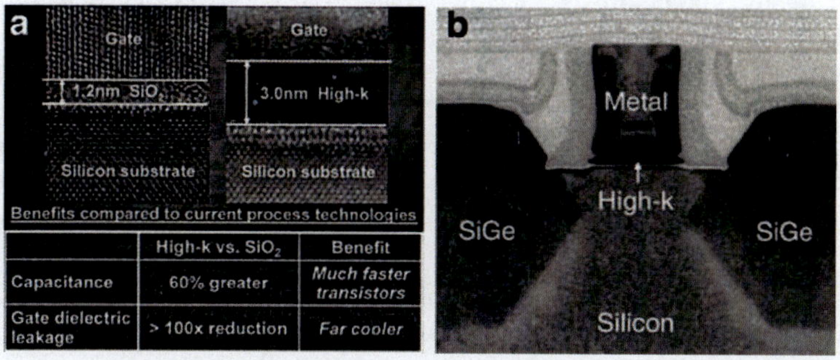

Fig. 6.13 (**a**) Comparison of high-k HfO_2 with SiO_2 [32], (**b**) 45 nm transistor with HfO_2 gate oxide

Hafnium dioxide was chosen by IC companies to replace SiO_2 as the gate dielectric at the 45 nm node due to several performance benefits as shown in Fig. 6.13a. Hafnium dioxide has a dielectric constant 6 times greater than SiO_2, but the band gap is approximately 6 eV compared to 9 eV for SiO_2. A cross section of a 45 nm high transistor is shown in Fig. 6.13b. It is clear from Figs. 6.11b and 6.12 that moving beyond HfO_2 will be difficult without scaling of the power supply voltage.

The selection of HfO_2 was based on many factors that are beyond the scope of this chapter. However, it is clear why ALD was chosen as the deposition method for HfO_2. From (6.5), it is clear that any variation in the thickness of the deposited oxide, within die, within wafer, or wafer to wafer, would lead to unacceptable variation in transistor performance. ALD is the only technology that can deliver the required uniformity as well as defect free films and precise thickness control. The precursors for HfO_2 have been discussed previously. They permit deposition of HfO_2 at low enough temperatures that the film is amorphous and has an exceptional interface with silicon [31].

The conformal nature of ALD also makes it ideal for other applications in nanoelectronics such as diffusion barriers for copper interconnects. Copper is a fast diffuser in SiO_2 and forms a deep trap in silicon. It is therefore necessary to use a diffusion barrier for copper. Diffusion barriers must be thicker than the minimum thickness required to slow down diffusion enough so that it will not impact the device over its lifetime. Copper interconnects are fabricated using a damascene structure, which means that a trench is fabricated in the interlayer dielectric, then lined with the barrier and filled with copper. Barriers are typically nitride compounds which have 50–100 times higher resistivity than copper, so the goal is to minimize the amount of barrier material in the trench so that the overall resistance is not significantly impacted. ALD accomplishes this well because it can deposit the barrier layer uniformly and at the minimum thickness, thereby permitting interconnects to have the lowest effective resistance.

ALD oxide layers tend to be hermetic and have been used to prevent the degradation of the photoluminescence (PL) of organic nanowires when they are exposed to air. GaQ3 organic nanowires made from tris(8-hydroxylquinoline) gallium(III), are of interest for application to organic light emitting diodes, but the PL emission decreases rapidly when exposed to air. The GaQ3 nanowires were coated with ALD alumina deposited at 278 K, a deposition rate of 0.068 nm per cycle, and thickness between 3.4 and 13.6 nm [33]. For all alumina thicknesses there was no difference in PL emission spectra compared to uncoated nanowires, and there was no degradation at any thickness when exposed to air.

6.2.3 Deposition of Nanoparticles

ALD of some metals onto oxides, such as W on SiO_2, have difficulty nucleating [34] and this is manifested by a delay in film growth once ALD pulses have started, as shown schematically in Fig. 6.14a. Here an ideal ALD growth curve is shown for comparison with a non-ideal curve. During the delay in film growth, metal nuclei

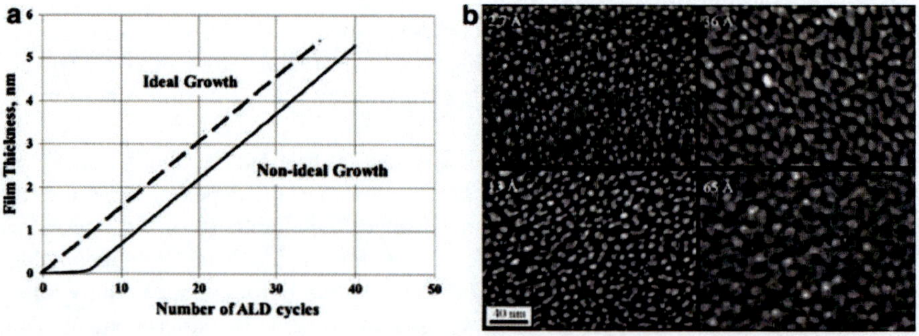

Fig. 6.14 (a) A comparison of ideal and non-ideal ALD growth due to difficult nucleation. (b) Growth of palladium (Pd) nanoparticles during nucleation on Al_2O_3 for different equivalent Pd film thicknesses (From Ref. [35])

form as islands on the oxide surface as shown in Fig. 6.14b for the deposition of Pd on Al_2O_3. The nanoparticles grow larger as the number of ALD pulses increase and finally the particles coalesce into a thin film. The number of pulses required to form a continuous film depends on the width of precursor pulses, the longer the pulse the fewer the number of pulses required. For Pd particle deposition from hexafluoroacetylacetonatepalladium (Pd(II)) and formalin, 1 second wide pulses required a pulse count of approximately 85 to form a monolayer (0.24 nm), while 10 second pulses only required a pulse count of approximately 40 pulses [35].

This difficulty in nucleation has developed into a research area for the deposition of metal nanoparticles on surfaces and other particles with application to catalysis. Platinum (Pt) and Pd are of particular interest in catalysis. Earlier work on the ALD of Pt has relied on unstable precursors and the first successful ALD deposition of Pt was reported in 2003 [36, 37] for the growth of Pt thin films using (methylcyclopentadienyl) trimethylplatinum [$CH_3C_5H_4Pt(CH_3)_3$] and oxygen precursors. The half reactions proposed for these precursors in the temperature range 573–623 K is based on a CVD process [37],

$$Pt + O_2 \rightarrow Pt - O_x \tag{6.6}$$

$$CH_3C_5H_4Pt(CH_3)_3 + Pt - O_x \rightarrow Pt + CO_2 \uparrow + H_2O \uparrow + \text{fragments} \tag{6.7}$$

The absorption of oxygen molecules (or atomic oxygen radicals or ions) on the Pt surface is given by (6.6) and the oxidation of the platinum precursor is given in (6.7). This latter process is the rate limiting step in Pt deposition [38].

These precursors have also been used by other groups to do thin film Pt research [39, 40]. While the ALD curves (film thickness versus number of pulses) shown in this work appeared to be ideal, the density of the films did not match bulk values until the film reached 18.3 nm thickness, indicating that films below this thickness are discontinuous [40]. However, more recent work has shown that Pt nanoparticles are deposited in the early stages of deposition and can be deposited on carbon materials if they were given an oxygen plasma [41] or acid pretreatment [42]. The dispersion of Pt on CNT's from this work is shown in Fig. 6.15.

Platinum has been deposited on strontium titanate ($SrTiO_3$) substrates and the Pt growth rate appears to be catalyzed by $SrTiO_3$ [43]. Nanoscale Pt particles (approximately 0.7–3 nm diameter) have also been grown on strontium titanate ($SrTiO_3$)

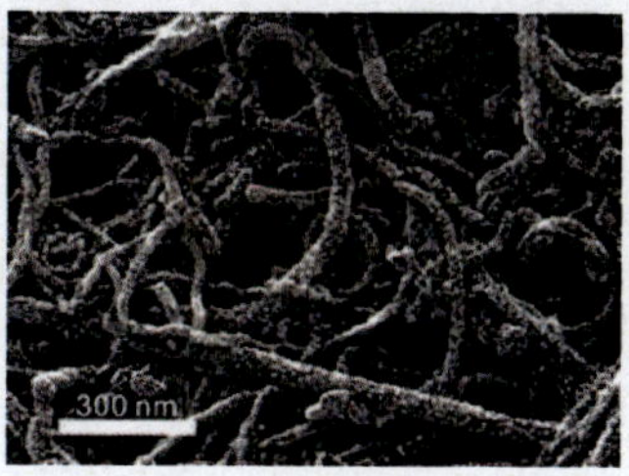

Fig. 6.15 ALD of Pt nanoparticles on carbon nanotubes (From Ref. [42])

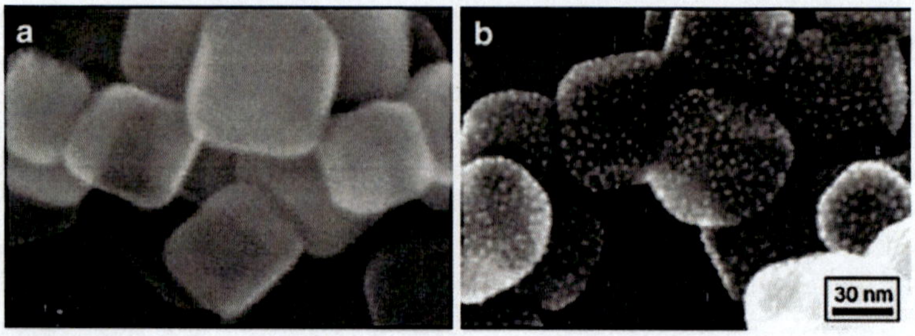

Fig. 6.16 High resolution SEM images of strontium titanate nanoparticles (cubes) (**a**) before and (**b**) after Pt nanoparticle deposition (three ALD cycles) (From Ref. [44])

nanoparticles, a high surface area support, as shown in Fig. 6.16 [44]. In this investigation the same Pt precursors were used as discussed earlier (6.6)–(6.7), but this research focused on the early stages of deposition, not the steady state deposition conditions that prior research investigated. Platinum loading was approximately 1 μg cm^{-2} and uniformly dispersed. At the initiation of deposition, between 1 and 5 ALD cycles, the Pt went from 90% oxidized (Pt-O_x) to 43% oxidized.

One of the biggest issues with metallic nanoparticles is the fact that at high temperature diffusion occurs and particle size increases and the number of particles decreases, thereby reducing catalytic activity. Novel research on 1–2 nm diameter Pd nanoparticles, deposited by ALD, has shown that a thin overcoat of ALD Al_2O_3 stabilizes particle size and distribution at 773 K, and in some cases, even enhances catalytic activity [45]. The support was silica gel (99.6 m^2/g), which was coated by ALD Al_2O_3, then ALD nanoparticles, and then an Al_2O_3 overcoat with various thicknesses. The effect of annealing at 773 K for 6 h on the uncoated and coated Pd nanoparticles is shown in Fig. 6.17a, b, respectively. It is clear that the impact of the Al_2O_3 overcoat is significant. Further studies showed that overcoated ALD Al_2O_3 preferentially nucleates at the edges and corners of the Pd nanoparticles leaving the Pd(111) face available for catalytic activity. Over coated Pd catalysts were tested for performance using the decomposition on methanol. At 543 K the methanol conversion jumped from 70% to 100% conversion within the addition of between 1 and 16 ALD Al_2O_3 cycles, there was a large drop in efficiency with more than 16 cycles.

While metals have difficulty nucleating on metal oxide surfaces, ALD metal growth on metal substrates is ideal. This means that ALD can be used to fabricate bimetallic nanoparticles, by first depositing metal nanoparticles, and then selectively coating those particles with another metal. Ruthenium(Ru)-Pt nanoparticles have been grown using sequential ALD deposition of Ru and Pt on spherical alumina powder [46]. The Ru cycle consisted of 2,4-(dimethylpentadienyl) (ethylcyclopentadienyl) ruthenium precursor and oxygen, while the Pt cycle is the same as discussed earlier in this section, (6.6)–(6.7). With a 1:1 ratio Ru and Pt cycle, Ru deposition rate was 0.031 nm per cycle, while Pt was 0.074 nm per cycle.

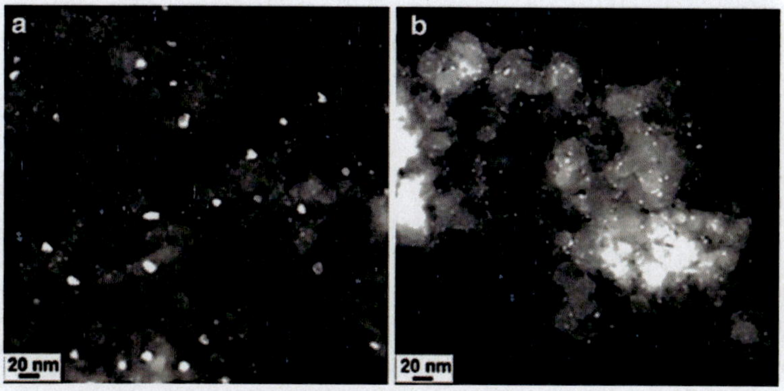

Fig. 6.17 The effect of annealing of Pd nanoparticles (**a**) uncoated, and (**b**) 16 ALD cycles of Al_2O_3 over coating (From Ref. [45])

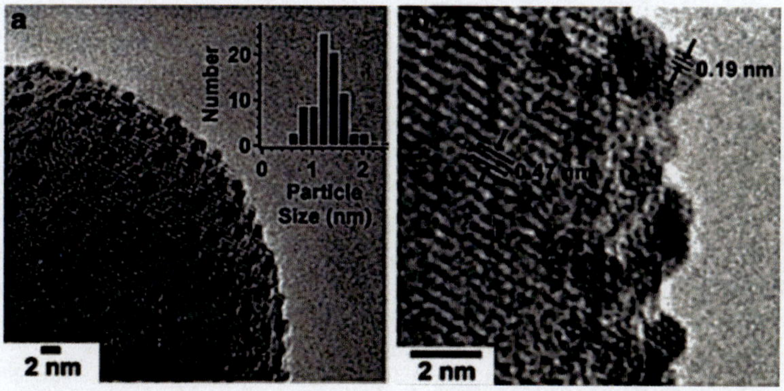

Fig. 6.18 Transmission electron microscopy images of (**a**) Ru-Pt nanoparticles (average diameter = 1.2 nm) on the surface of a spherical Al_2O_3 particle; (**b**) high resolution images showing lattice images for the constituent materials (From Ref. [46])

The Ru-Pt particle size was between 0.9 and 1.3 nm, and the Ru–Ru bond distance was 3% less than bulk values, and the Ru-Pt bond distance was contracted 1.1% with respect to bulk values. These contractions in bond distance are consistent with the contraction normally found in metallic nanoparticles [47, 48]. The bimetallic nanoparticles improved methanol conversion from 40% to 80% at 523 K compared to a physical mixture of Ru and Pt nanocatalysts (Fig. 6.18).

Conclusions

Atomic layer deposition is a very important technology to many current and future nanotechnologies. In this chapter we have shown that ALD is important for any application that requires exceptionally good conformity and uniformity. In addition, the recent discovery of the ability on ALD to deposit metallic and bimetallic nanoparticles has opened up a new avenue for research and development.

References

1. Puurunen RL. J Appl Phys. 2005;97:121301.
2. George SM. Chem Rev. 2010;110:111.
3. Delabie A, Puurunen RL, Brijs B, Caymax M, Conard T, Onsia B, Richard O, Vandervorst W, Zhao C, Heyns MM, Meuris M, Viitanen MM, Brongersma HH, De Ridder M, Goncharova LV, Garfunkel E, Gustafsson T, Tsai W. J Appl Phys. 2005;97:064104.
4. Deshpande A, Inman R, Jursich G, Takoudis C. J Vac Sci Technol A. 2004;22:2035.
5. Lee F, Marcus S, Shero E, Wilk G, Swerts J, Maes JW, Blomberg T, Delabie A, Gros-Jean M, Deloffre E, 2007 IEEE/SEMI advanced semiconductor manufacturing conference (2007) 359.
6. Leskelä M, Ritala M. Angew Chem Int Edit. 2003;42:5548.
7. No SY, Eom D, Hwang CS, Kim HJ. J Electrochem Soc. 2006;153:87.
8. Widjaja Y, Musgrave CB. Appl Phys Lett. 2002;80:3304.
9. Ka¨a¨ria¨inen TO, Cameron DC. Plasma Process Polym. 2009;6:S237.
10. Wilson CA, Grubbs RK, George SM. Chem Mater. 2005;17:5625.
11. Niskanen A, Arstila K, Ritala M, Leskelä M. J Electrochem Soc. 2005;152:F90.
12. Hausmann D, Ph.D. Thesis, Harvard University (2002).
13. Kukli K, Pilvi T, Ritala M, Sajavaara T, Lu J, Leskelä M. Thin Solid Films. 2005;491:328.
14. Kukli K, Ritala M, Sajavaara T, Keinonen J, Leskelä M. Chem Vap Depos. 2002;8:199–204.
15. SAFC Hitech Technical Bulletin "New hafnium oxide ALD precursors", http://www.safcglobal.com/safc-hitech/en-us/home/overview/technical-library. html
16. Kim H, McIntyre PC. J Korean Phys Soc. 2006;48:5.
17. Seidel T, Dalton J, Karim Z, Lindner J, Daulesberg M, Zhang W, 8th International conference on solid-state and integrated circuit technology (2007) 436.
18. Leskelä M, Ritala M. Thin Solid Films. 2002;409:138.
19. Niinisto L, Paivasaari J, Niinisto J, Putkonen M, Nieminen M. Phys Status Solid A. 2004;201:1443.
20. Martin CR. Science. 1994;266:1961.
21. Ott AW, Klaus JW, Johnson JM, George SM, McCarley KC, Way JD. Chem Mater. 1997;9:707.
22. Elam JW, Routkevitch D, Mardilovich PP, George SM. Chem Mater. 2003;15:3507.
23. Pellin MJ, Stair PC, Xiong G, Elam JW, Birrell J, Curtiss L, George SM, Han CY, Iton L, Kung H, Kung M, Wang H-H. Catal Lett. 2005;102:127.
24. Mertinson ABF, Elam JW, Hupp JT, Pellin MJ. Nano Lett. 2007;8:2183.
25. Comstock DJ, Christensen ST, Elam JW, Pellin MJ, Hemson MC. Adv Funct Mat. 2010;20:3099.
26. Xiong G, Elam JW, Feng H, Han CY, Wang H-H, Iton LE, Curtiss LA, Pellin MJ, Kung M, Kung H, Stair PC. J Phys Chem B. 2005;109:14059.
27. Banerjee P, Perez I, Henn-Lecordier L, Lee SB, Rubloff GW. Nat Nanotechnol. 2009;4:292.
28. Yeo Y-C, King T-J, Hu C. IEEE Trans Electr Dev. 2003;50:1027.
29. Moore GE, Electronics (1965) 38/8, April 19.
30. Taur Y, Ning T. Fundamentals of modern VLSI devices. New York: Cambridge University Press; 1998. ISBN 9780521559591.
31. Robertson J. J Vac Sci Technol B. 2000;18:1785.
32. Intel Corporation, High K metal gate press foils, Nov. 2003.
33. Wang C-C, Kei C-C, Perng T-P. Electrochem Sol State Lett. 2009;12:K49.
34. Elam JW, Nelson CE, Grubbs RK, George SM. Thin Solid Films. 2001;386:41.

35. Elam JW, Zinovev A, Han CY, Wang HH, Welp U, Hryn JN, Pellin MJ. Thin Solid Films. 2006;515:1664.
36. Aaltonen T, Ritala M, Sajavaara T, Keinonen J, Leskelä M. Chem Mater. 2003;15:1924.
37. Aaltonen T, Ritala M, Tung YL, et al. J Mater Res. 2004;19:3353.
38. Hiratani M, Nabatame T, Matsui Y, Imagawa K, Kimura S. J Electrochem Soc. 2001;148: C524.
39. Zhu Y, Dunn KA, Kaloyeros AE. J Mater Res. 2007;22:1292.
40. Jiang XR, Huang H, Prinz FB, Bent SF. Chem Mater. 2008;20:3897.
41. Hsueh Y-C, Hu C-T, Wang C-C, Liu C, Perng T-P. ECS T. 2008;16:855.
42. Liu C, Wang C-C, Kei C-C, Hsueh Y-C, Perng T-P. Small. 2009;5:1535.
43. Christensen ST, Elam JW, Lee B, Feng Z, Bedzyk MJ, Hersam MC. Chem Mater. 2009;21:516.
44. Christensen ST, Elam JW, Rabuffetti FA, Ma Q, Weigand SJ, Lee B, Seifert S, Stair PC, Poeppelmeier KR, Hersam MC, Bedzyk MJ. Small. 2009;5:750.
45. Feng H, Lu J, Stair PC, Elam JW, Catal Lett, Published online 25 January 2011.
46. Christensen ST, Feng H, Libera JL, Guo N, Miller JT, Stair PC, Elam JW. Nano Lett. 2010;10:3047.
47. Miller JT, Kropf AJ, Zha Y, Regalbuto JR, Delannoy L, Louis C, Bus E, van Bokhoven JA. J Catal. 2006;240:222.
48. Setthapun W, Williams WD, Kim SM, Feng H, Elam JW, Rabuffetti FA, Poeppelmeier KR, Stair PC, Stach EA, Ribeiro FH, Miller JT, Marshall CL. J Phys Chem C. 2010;114:9758.

Surface Functionalization in the Nanoscale Domain

7

Richard L. McCreery and Adam Johan Bergren

Abstract

This chapter discusses the modification of surfaces using thin layers of organic molecules to control or modify the properties of the surface. Methods for placing monomolecular adlayer films on various surfaces are outlined, as are methods that result in nanoscopic multilayered films. First, the general properties of surfaces are discussed in the context of how a thin organic film can be used to change their properties and behaviour. Then, an overview of methods for coating various surfaces with nanoscopic organic films (i.e., less than 10 nm thick) is given. Finally, methods for generating thicker films are discussed. Throughout this chapter, illustrative examples are given to show the structure of the film in question or to show how the properties of the modified surface can be dictated by the choice of the specific chemical composition of the adlayer.

7.1 Introduction and Scope

It is a well known consequence of geometry that the surface/volume ratio of an object increases as the object becomes smaller. The progression of many aspects of science and technology into the nanoscale realm therefore inherently results in the greater importance of surfaces in determining the behaviour of nanoscale devices and phenomena. The relentless increase in device density in the microelectronics industry provides a very prominent example of the relative importance of surfaces to device fabrication and performance, as the feature size is currently a few tens of

R.L. McCreery (✉)
Department of Chemistry, University of Alberta, Edmonton, AB, Canada
e-mail: richard.mccreery@ualberta.ca

A.J. Bergren
National Institute for Nanotechnology (National Research Council Canada), Edmonton, AB, Canada

M. Stepanova and S. Dew (eds.), *Nanofabrication*,
DOI 10.1007/978-3-7091-0424-8_7, © Springer-Verlag/Wien 2012

nanometers. If the "surface" is arbitrarily considered to encompass the typical escape depth of an excited electron (1–10 nm), then the progression of microelectronic feature sizes from 65 to 32 nm and below is rapidly approaching the point where the finished device is "all surface". In addition to occupying a greater fraction of device volume for smaller devices, surfaces also provide the opportunity to create unusual arrangements of atoms and molecules. The bulk structure of a given material is interrupted at its surface, and it is often possible to orient and/or bond materials with quite different properties at an interface between dissimilar materials.

Nanofabrication involves a wide range of surface modification procedures that are important in many different areas of nanotechnology. This chapter will emphasize mainly chemical methods for surface modification at the monolayer and near-monolayer level (generally, we consider here films that are less than 10 nm thick), with particular attention to the interactions between the bulk substrate material and the surface modifier occurring at their interface. Spin-coating and vapour deposition are commonly used for surface films substantially thicker than a monolayer, and will not be discussed in this chapter. Atomic Layer Deposition is a monolayer deposition method that is discussed in Chap. 6. Certain self-assembly techniques such as block co-polymer deposition and ordering have structure at the monolayer level, and are discussed Chap. 6. Here, we will focus on methods for orienting and bonding surface layers on solid substrates such as silicon, metals, carbon, and oxides. Such interactions are often classified into two categories: "physisorption" involving relatively weak attractions between the substrate and surface modifier (e.g., electrostatic attraction), and "chemisorption" involving strong surface bonds, often covalent. Of particular interest will be the orientation and structure of the interfacial region resulting from various surface modification procedures. The introduction of order at the substrate/modifier interface often distinguishes the methods described in this chapter from the comparatively disordered interfaces resulting from spin coating and vapour deposition.

It is common to classify surface modification techniques according to their surface chemistry. For example, Au/thiolate self assembled monolayers (SAMs) are considered as one class, while silanes on SiOx represent another. In this chapter, we will instead focus on *function* rather than *chemistry*, since usually the function of the nanostructure is more important to potential applications than its chemical structure and bonding. Overall, we intend the chapter to provide a "toolbox" of surface modification methods to aid nanofabrication and produce particular behaviours, be they electronic, improved environmental tolerance, chemical reactivity, etc.

7.2 A Functional View of Surface Modification

Before considering specific surface modification techniques, it is useful to summarize common objectives underlying the approach. The list below is illustrative rather than comprehensive, but it does cover many of the end results important in nanofabrication. It will become apparent that many of these objectives are controlled by the strength of the interaction between the modifier and the substrate surface, which is in turn controlled by the surface bonding chemistry.

7.2.1 Wettability

A very familiar and readily observed surface property is controlled by the interaction of a liquid, often water, with the surface. Water spreads out evenly on a "hydrophilic" surface, since the interaction between the surface and water is stronger than the intermolecular interactions in bulk water. "Hydrophobic" surfaces, on the other hand, weakly attract or even repel water to produce droplets or "beads" of water on the surface due to much weaker surface interactions. Thus, the wettability of a material can be quantified by measuring the contact angle of a water droplet on the surface of the material. The range of values for this angle from 0° (completely wetted) to above 150° (a "superhydrophobic" surface) can also be related to the surface-free energy [1], as discussed in more detail below. An example of how the surface chemistry of a single molecular layer can dramatically alter wettability is shown in Fig. 7.1, which shows a gold surface coated with two different monolayers: one presenting a hydrophilic –OH end group (left side) at the surface and one having a hydrophobic $-CH_3$ end group (right side) [2]. The relative concentration of the two layers varied from left (–OH groups only) to right (enriched in $-CH_3$ groups) across the figure. This figure illustrates that the surface chemistry will have a significant effect on any nanofabrication process involving water or that depend on the interfacial surface free energy.

The concept of wettability may also be applied to many materials other than water and can be used to describe the tendency for a material to coat a surface. For example, vapour deposited metals may form quite uniform films on a substrate in which the metal-surface interaction is attractive, but will form beads if the attraction is weak relative to the metal-metal interatomic forces. Since solvents themselves can be either hydrophobic or hydrophilic, the wettability of a particular solvent on a surface can vary significantly with the nature of the surface. The basic principle underpinning wetting applies to water as well as all other cases: the

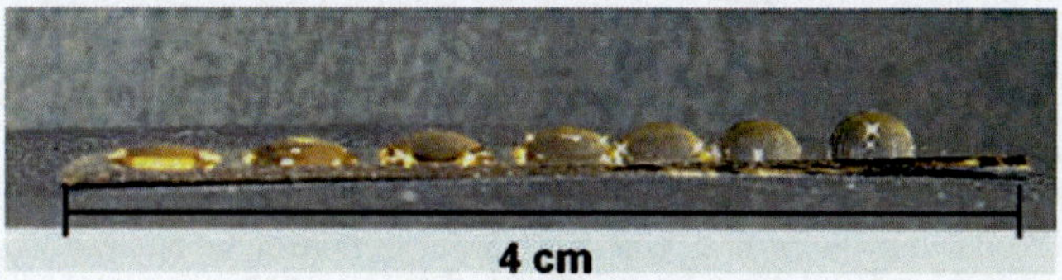

Fig. 7.1 Drops of water on the surface of a gold sample coated with pure ω-hydroxyundecanethiolate (*left*-most drop showing a small contact angle and therefore a high surface energy) and an increasing concentration of methyl-terminated dodecanethiolate (moving to the *right*, where the contact angle increases, indicating lower surface energy) [2]. This figure illustrates that surface energy, as measured using contact angle, is highly sensitive to the chemical groups present at the surface (Reproduced with permission from Ref. [2])

balance of *adhesion* (the attraction of modifier to the surface) and *cohesion* (the attraction between modifier atoms or molecules) acts to determine the wetting characteristics for a particular solid–liquid pair. This balance is the basis of "surface energy", which indicates the propensity of a material to form a surface rather than a particle. Materials with a high surface energy favor particle formation, as the energetic cost of breaking bonds in the bulk materials is large. The contact angle is measured between a line that lies in the plane of the solid surface and the line on the droplet that meets the solid near its edge (usually measured through the liquid). Thus, a contact angle approaching 0° occurs on a wettable surface, while a contact angle above 90° indicates a hydrophobic surface with a weakly interacting surface.

7.2.2 Uniformity and Pinholes

Surface coverage and thickness uniformity are obviously major concerns when coating any surface with an overlayer, and become significantly more demanding when the film thicknesses decreases below 50 nm. This issue is well illustrated by considering the "aspect ratio", i.e., the dimension of the surface film parallel to the surface divided by its thickness. Spin coating a 1 μm thick film on a 100 × 100 μm surface yields an aspect ratio of 100, while a monolayer film of a 1 nm length molecule on the same surface has an aspect ratio of 10^5. Thus, ensuring high coverage and uniformity essentially requires the monolayer film to be 1,000 times more exacting in its spreading over the surface, clearly indicating that successfully fabricating a high coverage, uniform monolayer over even relatively small areas is demanding. For example, significant problems have been encountered when attempting to make molecular electronic devices from single molecular layers due to the presence of a minute population of pinholes in the monolayer that can allow direct electrical contact between the substrate and any top electrical conductors used to complete the circuit [3–7]. Furthermore, techniques such as profilometry, ellipsometry, and interference microscopy are commonly used to determine the uniformity of films with thicknesses greater than ~10 nm, but such methods are generally unsuitable for molecular monolayers. Uniformity for nanoscale surface modifications is largely a function of the relative kinetics of surface bonding, intermolecular interactions, and the reversibility of the surface bond [7, 8]. If adhesion and cohesion are used to describe the energetics of the modifier-surface interaction, the uniformity of the layer and pinhole formation can be directly related to the kinetics of the interactions. For example, consider the commonly used "parylene-N" coating which is applied by polymerization of xylene radicals thermally desorbed from a precursor solid [9, 10]. The reactive xylene radicals can bond to many surfaces, but also to themselves to generate a highly cross-linked and hydrophobic protective film. If the surface bond forms rapidly compared to dimerization reactions, the nucleation rate will be high and the initial adlayer will resemble a "wheat field". Unmodified surface areas will be rapidly coated with additional modifier such that the density of pinholes will be small. On the

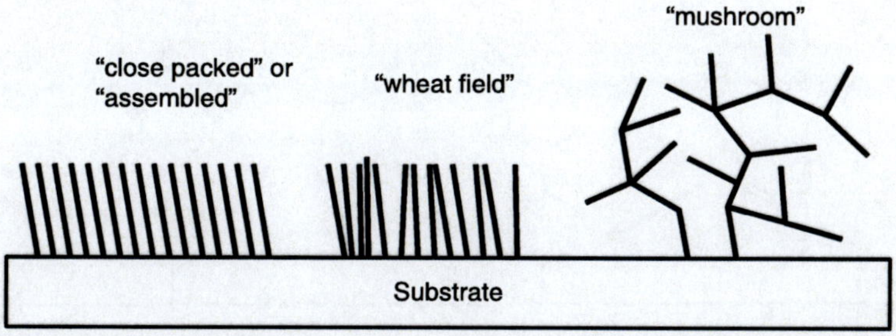

Fig. 7.2 Examples of surface modifying layers that self-assemble to form a close packed structure with highly ordered domains (*left*), less ordered structures that remain oriented and resemble wheat fields (*center*), and disordered layers that result from uncontrolled growth to produce mushroom-like clusters of molecules (*right*)

other hand, if the dimerization (and eventually polymerization) rate is fast compared to the surface reaction, "mushrooms" will result, shown schematically in Fig. 7.2 [8, 11]. Many readers will recognize this issue as an example of "nucleation and growth" kinetics, and it affects a wide range of surface modifications. In the case of parylene, the coatings are usually thick enough (>100 nm) that pinholes are encapsulated by the subsequent layers of the film. But for films of <10 nm thickness, particularly monolayers, the eradication of pinholes is often difficult [3, 12].

A related issue is the process of "self-assembly," which refers to the formation of an ordered layer of molecules by equilibration to a structure of minimum free energy. The Au/thiol SAM and Langmuir-Blodgett films are the most prominent examples, and are discussed in more detail below. In order to "assemble", the surface-modifier bond must be at least partially labile, so it can break and form again to reach the highly ordered structure. In contrast, modification methods which form irreversible surface bonds are unable to "assemble", but have the attraction of relatively high thermal stability. As always, the choice of surface modification procedures is governed by the eventual application, and the relative importance of stability vs. order.

7.2.3 Chemical Reactivity

Wettability, as discussed in Sect. 7.2.1 is an example of the broader description of a surface in terms of chemical reactivity. There are many examples in the literature of a surface modification designed to enable a subsequent interaction between the surface and additional modification, analogous to a "primer" between a coating and a solid surface. For example, a monolayer of an organic compound bonded to a metal may completely change the chemical characteristics of the surface, depending on the identity of the terminal group of the organic modifier. The molecule can be terminated by a functional group that is hydrophilic, acidic, chelating, or that has

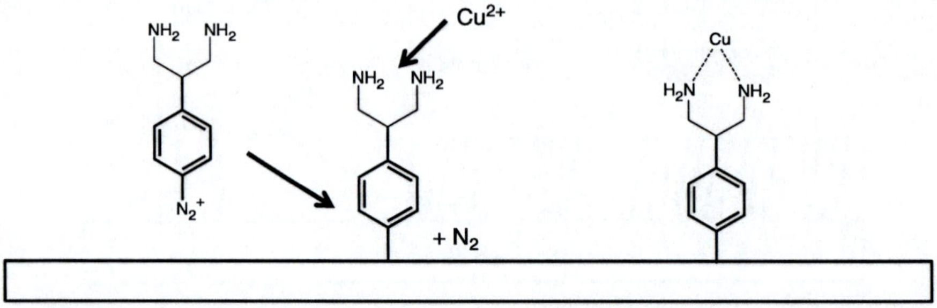

Fig. 7.3 Hypothetical example of a "seed layer" deposited onto a surface using a chelating molecule that is grafted to a surface. The chelating agent acts to bind Cu^{2+} ions from solution to result in a seed layer that promotes uniform plating of copper. Note that this is an illustrative example only; actual molecular structures are generally proprietary

other chemical properties which are quite different from those of the original metal. A solution or gas in contact with the modified metal surface will "see" the layer of organic molecules terminated by a range of possible functional groups first, and the properties of the metal (such as possible high reactivity) may become irrelevant (or at least secondary to the properties of the adlayer).

A prominent example from the semiconductor industry is the "seed layer" for the Damascene copper plating process. Effective electroplating of Cu requires nucleation sites on the surface to be plated, which typically consist of a sputtered layer of Cu applied in a vacuum. More recently, "primers" made from organic molecules with binding sites for Cu^{+2} ions have been used to provide seed layers in regions not accessible by sputtering, as shown schematically in Fig. 7.3. The reactive diazonium reagent bonds to the surface to be plated irreversibly, and contains a functional group that attracts Cu^{+2} ions from the electroplating solution. The diazonium surface modification is aggressive, and forms a high density of nucleation sites on the surface for the Damascene process. Primer layers based on organic modification of conducting and non-conducting surfaces are available commercially for use in semiconductor manufacturing (e.g., www.alchimer.com/technology/index2; www.zettacore.com/molecularinterface.html)

As already noted, "primers" are possible for a variety of additional surface modifications with many more end uses than copper plating. Examples include silane bonding of thiophene precursors for electropolymerization on tin oxide [13], surface epoxy groups on carbon fibers to initiate epoxy curing [14], surface-bound ligands for metal complexes [15] and binding of biomolecules to a silicon surface modified with aromatic amines [16]. There is an extensive literature on surface modification by a variety of methods to impart particular reactivity on electrodes for electrochemistry [17–22]. Adhesion between two materials is a special case of surface reactivity, and is governed by some of the same principles as wettability. Strong adhesion generally involves strong bonds between the two materials, with covalent bonds generally much stronger than typical physisorption governed by electrostatic or dipole–dipole interactions.

7.2.4 Protection

Prominent examples of surface modifications designed to stabilize surfaces toward environmental exposure and prevent corrosion include chemical modification of silicon and chromate conversion coatings on aluminum alloys. Such treatments can passivate a reactive native solid surface with more stable materials, often covalently bonded to the surface. These two example reactions are shown schematically in Fig. 7.4. The protection of aircraft alloys by reduction of chromate ions to an inert, insulating Cr^{III} oxide is essential to the aerospace industry to stabilize quite reactive aluminum/copper alloys [23–26]. The propensity of silicon to form surface oxides is well known, and bonding of a methyl group to the Si surface greatly enhances its resistance to photochemically induced corrosion [27–29].

7.2.5 Electronic Interactions

Of the many known electronic phenomena in solid state structures and in solutions, several of direct relevance to surface modification deserve special note. Rather than considering "bulk" properties such as dielectric constant, conductivity, and mobility, we will focus on electronic effects present at surfaces, many of which are strongly affected by the structure and orientation of surface layers. A relatively simple example is the effect of surface dipoles on the apparent work function of a solid substrate. Bonding of a molecule containing a finite dipole to a surface by a

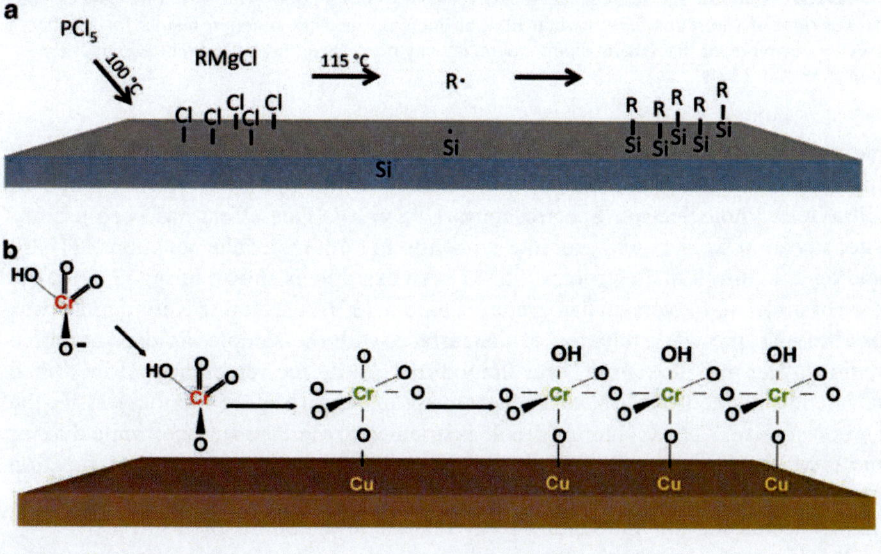

Fig. 7.4 (a) Passivation of a silicon surface by chlorination followed by reaction with a Grignard reagent to bind R groups at the surface [29]. (b) Passivation of a Cu surface by a chromate conversion coating

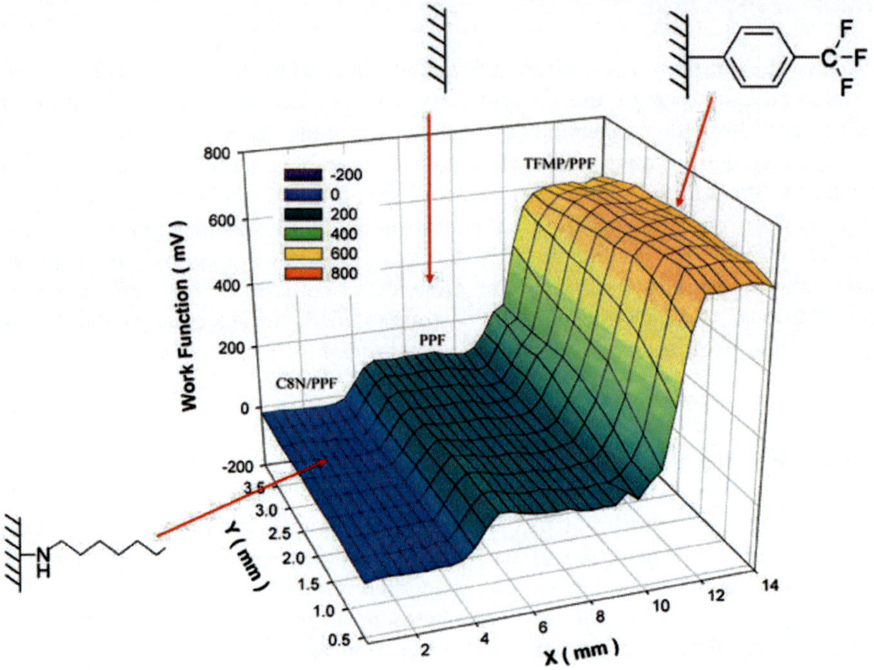

Fig. 7.5 Modification of the work function of a carbon surface (PPF, or pyrolyzed photoresist film, *center* region) by molecular layers with different dipole orientations. The alkylamine dipole is oriented toward the surface, causing the decrease in work function to the left of the bare carbon. To the right of the unmodified carbon film, an increase in work function results for a trifluoromethylphenyl layer that has its dipole oriented away from the surface (Reproduced with permission from Ref. [34])

method which orients the dipoles relative to the surface produces a shift in the work function of substrate/monolayer combination, as judged from Kelvin probe or Ultraviolet Photoelectron Spectroscopy (UPS) [30]. This effect has been used to alter injection barriers and promote tunneling in both molecular junctions [31–35] and organic thin film electronics [36, 37]. An example is shown in Fig. 7.5, for the case of molecular layers on flat, graphitic carbon [34]. A scanning Kelvin probe was used to map the work function of the surface, with the sample divided into three regions. The middle "stripe" was unmodified, while the remaining "stripes" had covalently bonded $C_8H_{17}N$ or trifluoromethylphenyl (TFMP) monolayers on the carbon surface. The alkylamine dipole is oriented toward the surface, while that for the TFMP is oriented in the opposite direction. Note that the apparent work function varies by >600 mV due to the surface modifier, with the electron withdrawing TFMP decreasing the interfacial energy relative to the unmodified surface.

It is generally recognized that bonding a molecule to a surface can change the electronic properties of both components in more respects than the surface work function. Strictly speaking, a covalently bonded molecule on a conducting surface is a new "molecule" with some degree of electronic coupling between the orbitals of the molecule and those of the surface. There is strong theoretical and experimental support for the conclusion that the modified surface should be considered a "system" with possibly quite distinct properties from the component parts [38, 39]. The energies and appearances of the molecular orbitals of the surface bound molecule change upon surface bonding, but the molecular orbitals may also mix with those of the graphite support [40].

An interesting consequence of orienting molecules by bonding to surfaces is the effect on the dielectric constant and polarizability of the molecular layer [41–43]. Although dielectric constants are often considered for bulk materials to be isotropic, they are in fact tensors which depend on molecular orientation. For example, the delocalized electrons in anthracene cause the polarizability along the longer Z axis of the molecule to be higher than that along either the X or Y axes. Furthermore, the polarizability and resulting dielectric constant can increase as the length of a conjugated molecule increases [42]. A related phenomenon relevant to electron transport is the electronic coupling between the surface layer and the substrate. In the limit of strong electronic coupling, the electrons can be considered to be delocalized over a region including both substrate and molecule orbitals, with the combination behaving as one electronic system [41].

Electron transport in thin films has been studied extensively, of course, due to its importance in the behavior of microelectronic devices. Tunneling and field emission of electrons through thin films of SiO_2 are major considerations in widely used devices such as field effect transistors and floating-gate "flash" memory. The investigation of electron transport through single molecules and molecular monolayers has been stimulated in the past decade by "molecular electronics", in which molecules become circuit components. Since the great majority of non-polymeric molecules have dimensions of less than a few nm, transport by quantum mechanical tunneling can contribute significantly, or completely control charge transport through surface modification layers.

It is useful to consider electron transport in surface layers as a function of thickness and transport mechanism. Classical transport in bulk conductors occurs by a series of scattering events, in which the electron travels a short distance, usually a few tens of nm, then scatters and changes direction. Progress through the conductor is diffusive, with the net motion of charge carriers along the conductor driven by the applied electric field. For thick films on surfaces, i.e., >100 nm thick, transport across the layer is also by a series of small steps, either by scattering or "hopping". "Hopping" is a rather general term, which usually involves "sites" where an electron (or hole) can reside temporarily during its progress through the material [44–47]. For organic semiconductors, such sites are cation or anion radicals, for hole or electron transport, respectively. The low mobility and temperature dependence of most organic semiconductors are due to the energy required to form and eliminate these radical sites during transport, as well as the usually limited delocalization of electrons or holes in the materials. Another term for transport between radical sites is "redox exchange" with the associated activation barrier equal to the "reorganization energy" [48–50]. In conducting polymers, the "doped" forms contain delocalized radical sites, often referred to as "polarons", which exhibit band-like transport over distances of up to ~10 nm. Unless the polymer is heavily doped, electrons still must hop between polarons, leading to a positive dependence of conductivity on temperature.

When the thickness of a surface layer is in the range of <1–10 nm, electron tunneling becomes possible, and transport behavior may change dramatically. Tunneling through such layers has been studied in electrochemistry, in which the electron tunnels through the layer to react with a molecule in solution, and in molecular electronics, where a 1–10 nm thick molecular film separates two solid conductors. The quantitative dependence of tunneling rate on film structure and thickness is often complex, but some useful generalizations are possible in the context of the early "Simmons" model [41, 51, 52] (7.1).

$$J = \frac{q}{2\pi h d^2}\left(\bar{\phi} e^{-A\sqrt{\bar{\phi}}} - (\bar{\phi} + qV)e^{-A\sqrt{\bar{\phi}+qV}}\right) \tag{7.1}$$

$$A = \left(\frac{4\pi d}{h}\right)\sqrt{2m_e}$$

where m_e is the electron mass, d is the film thickness, ϕ is the tunneling barrier height, q the electron charge, and V the voltage applied across the film.

First, the tunneling rate and therefore conduction through thin films is exponentially dependent on film thickness, d. For ohmic (diffusive) conduction and hopping mechanisms, the current usually scales with d^{-1}. The exponential dependence of the tunnel current leads to vanishingly small conduction by tunneling when $d >$ ~5–6 nm (at which point the hopping current becomes dominant [53, 54]). Second, the tunneling rate is also exponentially dependent on the barrier height, ϕ, usually taken as the offset between the surface Fermi level and a single orbital energy in the film. Usually, a good approximation of the barrier height can be obtained by using the Lowest Unoccupied Molecular Orbital (LUMO) for electron tunneling or the Highest Occupied Molecular Orbital (HOMO) for hole transport, since these lie closest to the Fermi level in energy. The actual barrier is usually more realistically visualized by using a spectrum of orbital energies [55, 56] rather than a single orbital, but this requires significantly more complex methods to determine the barrier. To illustrate how the barrier height correlates with molecular properties, it is known that the molecular frontier orbital energies for conjugated, aromatic molecules are generally closer to the Fermi energy of most contact materials compared to alkanes, indicating that tunneling through aromatic layers is more efficient and will lead to higher tunneling currents. The Simmons model has a variety of enhancements to account for the electric field strength, the effective carrier mass, the tunneling barrier shape, etc., so correlation of theory and experiment is often quite involved [41, 57]. A useful empirical parameter is the attenuation factor, β, with units of nm^{-1}, which is the absolute value of the slope of a plot of ln(J) vs. d, where J is the tunneling current density and d is the layer thickness. For example, $\beta = 1.0\ nm^{-1}$ indicates that each nm of layer thickness decreases the

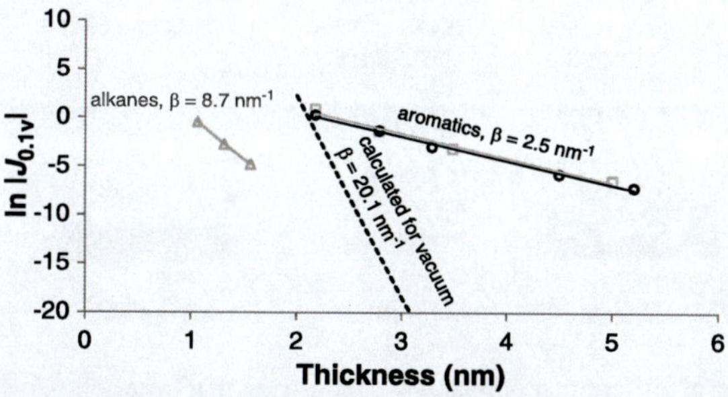

Fig. 7.6 Attenuation plots given as the tunnel current density at 0.1 V as a function of thickness for different molecular structures and a vacuum gap. The alkane series (*triangles* [58]) yielded a value for β of 8.7 nm^{-1}, while two different aromatic layers (*squares* = azobenzene, circles = nitroazobenzene) give a much lower β of 2.5 nm^{-1}. Finally, the dashed line shows an attenuation plot calculated using the Simmons model with image charge effects included [41, 51] for a vacuum gap with a 4 eV barrier height (with dielectric constant and effective mass equal to 1) that gives β = 20.1 nm^{-1}

tunneling rate by a factor of 1/e. Although the Simmons relation predicts that β is proportional to $\phi^{1/2}$, there are often other factors included in the experimental value, notably the effective carrier mass. Figure 7.6 compares experimental and predicted plots of ln(*J*) vs. *d* for several cases, to illustrate the large range of β possible for thin film materials [41, 58].

It is obvious from (7.1) and Fig. 7.6 that tunneling is a strong function of thickness and layer structure, but is usually exponential with layer thickness and weakly dependent on temperature. It should also be noted that this discussion applies to "off resonant" tunneling, in which there is a barrier between a molecular orbital and the substrate Fermi level. "Resonant tunneling" can occur when orbital energies are with *kT* of the Fermi level, and such transport is expected to a lead to a β close to zero [59–61].

7.2.6 Thermal Stability

The thermal stability of chemically modified surfaces is a function both of the bond between the surface and the modification layer, and of the inherent stability of the molecules comprising the layer. While the molecular stability may vary over a wide range for the many possible surface modifications, some generalizations about the surface bond are useful for predicting thermal stability. As noted in Sect. 7.2.2, self-assembly techniques based on Langmuir-Blodgett and Au/thiol chemistry require a relatively weak surface bond to permit the modification layer to "assemble" into an ordered, low energy configuration. The commonly studied Au-S bond for thiols adsorbed to Au surfaces has a bond energy of ~1.6 eV (~40 kcal/mol), while the typical forces that hold L-B structures on a surface are significantly weaker (< 0.5 eV).

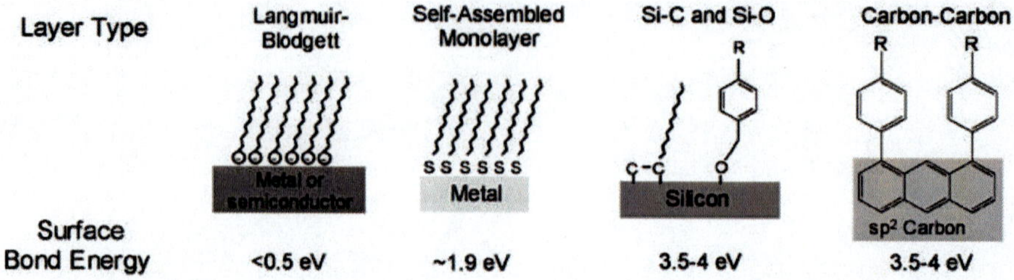

Fig. 7.7 Examples of different modification chemistries that show varying surface-molecule bond strengths, illustrating that the surface chemistry impacts layer stability (Reproduced with permission from Ref. [8])

Irreversible adsorption through C-C, Si-C, and Si-O bonds is mediated by much stronger bonds, as indicated schematically in Fig. 7.7.

Thermal stability of modified surfaces deserves attention in at least two applications. First, subsequent processing of the modified surface in order to incorporate it into a finished product may involve temperature excursions of possibly hundreds of °C, with the possibility of damage or disordering of the modification layer. Second, the end-use of the device may require operation over a much wider temperature range than that of the laboratory or fabrication line. Examples of thermal stability issues in molecular electronics are the failure of Au/thiol surface modifications at 50–100°C which resulted in device failure [62, 63]. Similar molecular junctions based on C-C surface bonds survived 40 h at 150°C [64] or brief exposure to 250°C in vacuum.

7.3 Nanomolecular Layers

As noted above, derivatizing a surface with a nanoscopic layer of organic molecules can alter the properties of the interface in a controllable fashion. The modifying layer can impart some chemical or physical property that is required in order to meet the demands of the end-application, or it can be used to study fundamental phenomenon. For example, the surface of a medical device may need to be rendered bio-compatible, or a systematic study of electronic and/or optical interactions of a solution phase molecule with a particular substrate can be carried out as a function of modifier hydrophobicity, length, etc. As described in this Section, there are many different ways to modify the surface of a substrate material, and the choice of the method that is used usually depends on more than the desired properties of the interface imparted by the modifier. For example, if a hydrophobic surface is desired for the application of a hydrophilic metal, there are methods available to achieve a hydrophobic adlayer. However, one method may require a complicated, time-consuming procedure but result in a highly stable layer, while another may provide a suitable hydrophobic layer with ease that is less robust. Thus, the application should be considered and balanced against these other factors when choosing a modification method.

The thicknesses of the films discussed in this Section generally reside in the nanoscale domain, having a total thickness of less than 10 nm, with thicker films discussed in Sect. 7.4. The reason for this distinction lies partially in how the properties of a material vary with length, and is somewhat arbitrary. As noted in the introduction, the surface of a material becomes increasingly important as the size of an object is reduced. Partially a consequence of this phenomenon, the properties of a material often undergo significant changes as the size approaches the nanoscale. Thus, a sample of bulk material has the same properties when divided into two separate pieces. However, if this division is continued until at least one dimension approaches the nanoscale (roughly defined as less than 100 nm), the properties begin to depend quite sensitively on thickness. That is, a 20 nm slice of the material can have vastly different properties than a 5 nm slice, and both of these slices can have different properties than the bulk material even for identical molecular structures. Since these effects underlie the emerging field of nanoscience and nanotechnology, we refer to modifying layers with thicknesses less than 10 nm as *nanomolecular* layers.

There are numerous methods for placing a thin layer of organic molecules on the surface of a material. In general, the substrate that is modified determines in large part the types of modification methods that can be used. There are specific methods that rely on interactions or bond formation between a substrate material and the modifying film and non-specific methods that can be generally applied. The former case is less versatile, but usually more stable. For each type of nanomolecular layer, a general description of the layer structure is given, along with commonly used preparation procedures and some brief example applications.

7.3.1 Langmuir-Blodgett Layers

Katherine Blodgett described the deposition of monolayer films on glass in 1934 [65] and provided a description of multiple layers of monomolecular films on solid surfaces in 1935 [66]. This remarkable feat of nanotechnology over 75 years ago laid the foundation for the application of Langmuir-Blodgett layers (L-B layers) [67] in numerous applications, although this did not happen for several decades following Blodgett's original papers. The technique relies on the interactions between amphiphilic molecules and water to create an ordered layer of molecules at the liquid–gas interface [68]. The oriented molecular layer is compacted by applying lateral pressure using a barrier, causing an increase in order and packing density to form a compact molecular layer at the air-water interface (a Langmuir film, Fig. 7.8a). To form an L-B layer, the Langmuir film is transferred to a solid support simply by passing the substrate slowly through the film, as shown in Fig. 7.8b, c. Repeatedly dipping the substrate can yield multilayer films, an example of which is shown in Fig. 7.8d. The details of layer orientation initially depend on the substrate surface chemistry, the type of molecule that is spread on the surface of the water, the orientation of the solid, and whether the support is lowered or raised through the interface (extending or receding deposition).

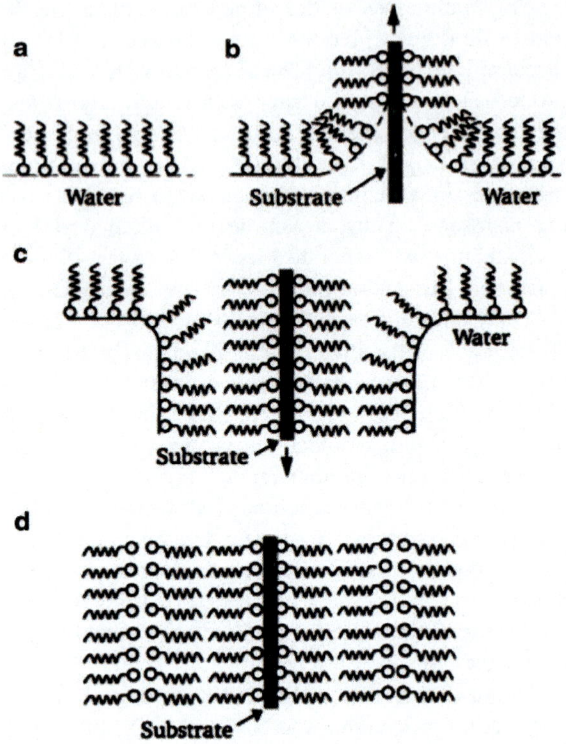

Fig. 7.8 Schematic of the process used to create an L-B layer. (**a**) First, a Langmuir film is formed at the air–water interface. (**b**) Next, a substrate is passed through the Langmuir film to create an L-B film at the surface of the substrate. (**c**) Additional steps can results in a multilayer film, shown in (**d**) (Reproduced from Ref. [67])

There are numerous molecules that can be used to create L-B films, with the only requirement being a dipole that imparts amphiphilicity. However, the ordering of the layer depends on intermolecular interactions between the molecules so that the most well-defined layers result from the use of regular repeating units (such as long-chain alkanes) with an appropriate functional group at one end to create a dipole. Thus, films formed using fatty acids have set the standard for quality and coverage [67].

L-B films have been extensively applied as nanomolecular layers in a large number of applications. The versatility of the structures that can be used and the flexibility in the number of layers makes them attractive for applications in which molecular scale precision is desired. Reported examples include molecular electronics [69, 70], biological sensing [71], scanning probe microscopy [72], organic electronics [73], and many other areas including electrochemistry and non-linear optics [74].

7.3.2 Self-Assembled Monolayers

Using specific chemical interactions, organic molecules can be spontaneously assembled onto many types of surfaces. Usually the organic molecule contains a functional group at one end that has an affinity for the surface to be modified. This affinity leads to the formation of a chemical bond between the functionality and the surface such that the molecule is retained at the substrate surface and the terminus of the molecule is oriented outward from the surface. The fact that the adsorption is driven by a specific surface-molecule interaction limits the coverage to a single molecular layer, and the term *self-assembled monolayer* (SAM) has been widely adopted to describe these layers. In addition, the type of bond formed between the surface and modifier is usually labile: molecules have some mobility on the surface after adsorption. Although this limits the stability of the layer somewhat, it enables the formation of highly ordered structures with full monolayer coverage since adsorbed molecules can move to accommodate additional adsorbates. Thus, a SAM is stabilized by intermolecular interactions between molecules in the film, which serve to orient the layer. This driving force, combined with the surface mobility, enables molecules in low-coverage areas to migrate to more ordered domains, resulting in a well-ordered monomolecular film. The reversible nature of SAM-surface bonding is one element that distinguishes these types of layers from other methods that rely on permanent, irreversible bond formation, as will be discussed in more detail below.

Noble metals are a very common surface for SAMs [75] due to their ease of handling in laboratory ambient. In addition, many other surfaces serve as supports for SAMs [76], including a variety of semiconductors and metal oxides [77]. To illustrate the general features of a SAM, we will use the example of an alkanethiolate monolayer on gold, which has been studied in depth so that many of its structural and formation characteristics are well established [68, 75, 78–80].

A schematic of a self-assembled monolayer composed of alkanethiolate molecules at gold is shown in Fig. 7.9 [81]. As shown, a head group (in this case, the sulfur atom) acts to anchor the molecules onto the gold substrate by a specific interaction that leads to some form of bonding. Although this bond can be quite

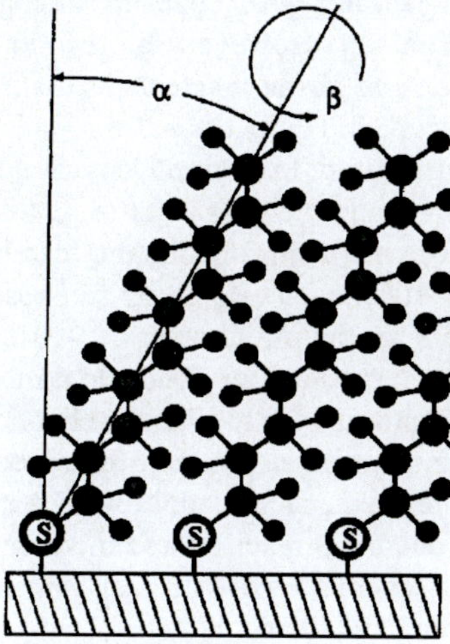

Fig. 7.9 Schematic of a SAM composed of alkanethiolates chemisorbed at Au [81]. The tilt (α) and twist (β) angles are shown here to define the precise orientation of the molecules on the surface (Reproduced with permission from Ref. [81])

strong, it is also reversible. The alkyl chains of the molecules pack into a low free-energy state characterized by van der Waals interactions between the chains. Finally, a tail group is present at the chain terminus. Often, the chemistry of the tail group can be used to control the properties of the interface. Note the that fine structure of the adlayer, including the orientation and packing density, evolve during the formation of the layer [82, 83]. However, very dense layers can be readily produced with a surface coverage that closely corresponds to the closest-packed value expected for a $(\sqrt{3} \times \sqrt{3})R30°$ overlayer at Au (111) (7.7×10^{-10} mol cm^{-2}) [79]. It is this relative ease of preparation of reasonably well-defined molecular monolayers that has made this method an extremely popular surface modification technique in the last few decades.

There are numerous examples of uses for self-assembled monolayer structures [75, 79, 80] including (but certainly not limited to) growth of nanoparticles, electron transfer studies, biological sensing, and uses as etch resists. An important point for many applications is the ease with which the chemistry and function of the surface can be controlled simply by variation of the tail group. Although SAMs can display good stability to some degree of ambient exposure, their stability is limited when subjected to more extreme conditions. Thus, surface modification procedures that utilize stronger covalent bonds have been developed, as described below.

7.3.3 Layers Anchored with Covalent Bonds

The use of irreversible chemical bonds to modify the surface of metals, semiconductors [28], carbon [84], and other surfaces produces stable modifying layers and has been growing in popularity in recent years. This demand is driven by the need for highly stable modified surfaces that can be exploited in an extensive range of chemical environments and physical conditions. Many of the methods used to modify a surface with covalent linkages stems from classical organic and inorganic chemistry. However, the reaction conditions often need to be adjusted to account for one of the reactants as a surface rather than a distinct chemical or functional group, and quite often the surface needs to be activated in order to obtain efficient modification. In addition, covalent chemical bonds can be formed between a pre-existing nanomolecular layer (e.g., SAM or L-B film) and a secondary layer. In this way, the intended functionality can be achieved using the covalent linkage, but with the advantages offered by self-assembly (high ordering, ease of preparation) available for the initial layer.

There are a large number of covalent modification schemes. Often, they rely on some external stimulus to initiate the chemical reaction that results in the formation of a new chemical bond between the modifier and the surface. Electrochemistry, light exposure, heat, or the addition of a chemical catalyst is often employed to activate the chemical reaction and increase efficiency. To illustrate the variety of modification schemes, we will give some specific examples for modification of metals, carbon, and silicon.

7.3.3.1 Covalent Modification of Metals

The covalent modification of metal surfaces generally follows two approaches: (1) treatment of the metal surface to contain a high density of surface groups such as oxides or hydroxyl groups that will serve to form a bond with the modifying layer [85, 86] or (2) direct reaction of the metal surface with a chemical agent. In the first case, conventional organic chemistry is used treating the surface as a hydroxyl moiety, while in the second case, the direct formation of a metal-modifier bond results. However, in either case, one reactant is confined to the surface of a material and steric hindrance and configurational limitations can pose significant restrictions on the yield and efficiency of the reaction. Thus, in order to obtain dense nanomolecular layers with high coverage, the more aggressive chemical reactions are often favoured.

For the purposes of this Section, we consider self-assembled monolayers to be a distinct class of nanomolecular layers due to the spontaneous nature of formation and the labile nature of the substrate-molecule bond. Instead, we focus here on targeted chemical reactions that result in the irreversible formation of a new covalent bond that anchors a nanomolecular layer on the metal. While some degree of modification can sometimes be achieved by the use of a spontaneous chemical reaction using these methods, most often they greatly benefit from some external assistance. For example, the desired reaction can be initiated by heat, light, an added reagent (such as a radical initiator), or an electron transfer step induced by controlling the substrate potential electrochemically. In the literature, the distinction between covalent modification and self-assembly is often not clear; spontaneity, the energy of the resulting anchoring bond, the number of steps required to prepare the surface, and other, more arbitrary criteria are often used to classify these films.

The modification of surfaces using diazonium chemistry involves the formation of the highly reactive $-N_2^+$ group on the molecular species to be anchored, and can generally begin from primary aromatic amine compounds. Diazotization can either be carried out during the modification (in-situ) [87–95] or separately by organic synthesis to yield an isolated diazonium compound [96]. Modification of a metal surface is then carried out by the reduction of the diazonium moiety, which leaves as N_2 gas and generates an aromatic radical species that can bond to the metal surface. The reduction of diazonium reagents is discussed below for carbon, noting that a similar reaction pathway has been shown for many different surfaces, including a wide variety of metals [97–105]. There may be significant differences, however, between the specific characteristics of diazonium-derived molecular layers on metals compared to those deposited on carbon.

7.3.3.2 Covalent Modification of Silicon

Modification of semiconducting surfaces has been gaining in popularity due to the well-defined nature of these surfaces and the excellent control of the surface structure and crystallinity that is available. While many semiconductors have been modified, the most prominent example is silicon [28] due to its ubiquitous use in the semiconductor industry. A variety of methods is available to form Si-C,

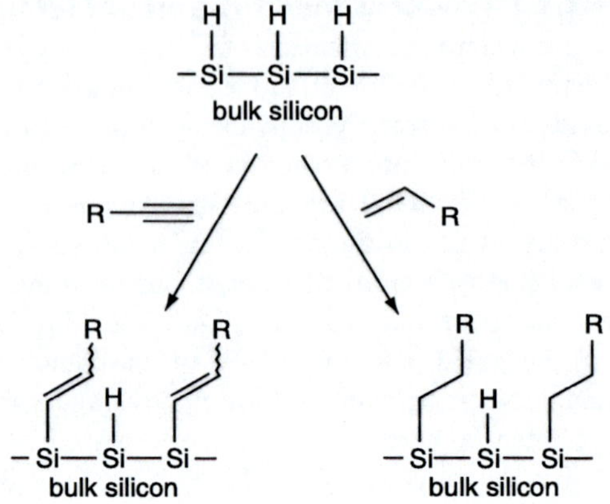

Fig. 7.10 Covalent modification of Si: using either an alkyne (*left*) or alkene (*right*), a covalent bond results between the Si surface atoms and the modifying layer. These reactions can be activated with heat, light, or an added radical initiator (see text) (Reproduced with permission from Ref. [28])

Si-O, and other covalent bonds between Si and a modifying molecule. While a full description of all available methods is beyond the scope of this chapter, we will give an example that illustrates covalent modification of silicon.

Figure 7.10 shows a popular covalent modification scheme that begins with a hydrogen-terminated silicon surface and uses an alkene, alkyne or alcohol functional group to bond the molecular layer [106–108]. Depending upon the specific surface and modifier involved, the reaction can be initiated in at least three ways, all of which involve a radical-mediated modification pathway: through the use of a chemical radical initiator, heat, or UV light. Diazonium reagents have also been demonstrated for forming surface bonds to silicon surfaces [109–112], as well as GaAs and semiconducting carbon nanotubes [113, 114]. The details of the reaction mechanisms and the resulting quality, packing density, surface termination, etc., of the nanomolecular layer are determined by the way the Si surface is prepared, the exact nature of the modifying molecule used, and the initiator. Thus, the choice of method depends on the demands of the application of the resulting modified surface and any limitations on the initial choice of the Si surface.

Many of the applications for modified Si surfaces are similar to that for other types of surfaces, including electronics [31] and biosensors [115].

7.3.3.3 Covalent Modification of Carbon

Carbon surfaces can be modified using a wide variety of chemical reactions. Classical organic chemistry can be carried out using a graphitic carbon surface as the reactant and reactions designed to target the edge plane or basal plane of the substrate. Furthermore, functional groups (such as oxygen containing –OH, –COOH, etc.) are often found at the surface of carbon materials, and these can be used to form chemical bonds. However, for methods that rely on a specific functional group, its concentration is usually increased through pre-treatment prior to modification procedures [11]. On the other hand, some methods do not require any additional specific functional groups to be present, but instead rely on a reactive solution phase molecule to modify the carbon surface in its native form.

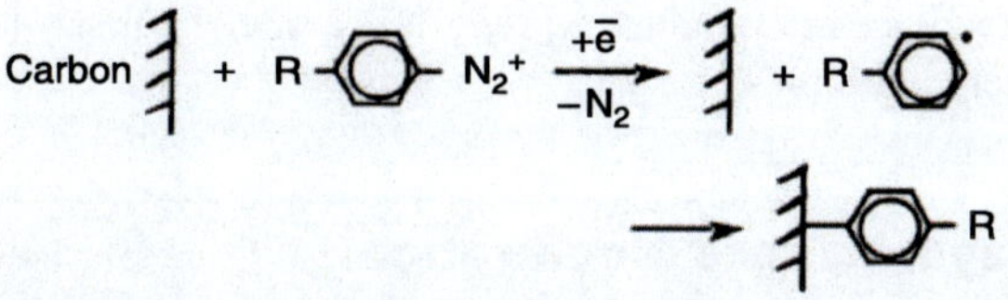

Fig. 7.11 Electrochemical reduction of aromatic diazonium reagents leading a covalently bonded layer, anchored by a carbon-carbon bond. This scheme can also be applied to modify many other surfaces, including metals and semiconductors (Reproduced with permission from Ref. [84])

A very common method for covalent attachment of nanomolecular layers to carbon makes use of diazonium chemistry [8, 11, 105]. The carbon surface to be modified is used as an electrode in a conventional three-electrode electrochemical cell containing a dilute solution of an aromatic diazonium reagent. The carbon electrode potential is controlled to induce reduction of the diazonium reagents, producing a carbon radical that attacks the surface and then bonds to the electrode (see Fig. 7.11). The use of electrochemistry is an advantage for several reasons. First, the electron transfer event that generates the radical species occurs selectively at the surface of the carbon electrode, which confines the reactive species to a thin diffusion layer adjacent to the surface being modified. Second, the number of radical species generated in a given time may be controlled by monitoring the electrolysis current. By carefully controlling the conditions during electrolysis, fine control of the layer thickness can be obtained [41, 116]. Finally, a wide variety of structures can be used, and the conditions during electrochemical depositions can be varied in order to obtain comparable layer characteristics (e.g., thickness) by accounting for the differences in reduction potentials of the diazonium reagents.

Diazonium chemistry is capable of producing highly stable nanomolecular layers that have very high coverage and excellent packing density. This is due, in part, to the aggressive nature of the radical-mediated bonding mechanism and the generation of the radicals at the surface of an electrode. This also, however, can lead to the formation of multilayers for certain molecules under certain conditions, as described in Sect. 7.4.1. Essentially, any site on the surface of the electrode that is capable of reducing the diazonium reagent will produce reactive radical molecules that will find a place to form a bond. This includes the terminus of the first monolayer of bonded molecules, as long as the rate of electrochemical reduction through the monolayer is sufficient to support further radical formation. In practice, some molecules are more prone to form multilayers than others. In addition, for those that do readily form multilayers, the electrolysis conditions can be adjusted to control thickness with excellent resolution (less than 1.0 nm). This can be used to advantage in studies of thickness dependent phenomenon [41]. Diazonium modification may also be used to form patterns on surfaces by micro-contact printing and scanning probe techniques [16, 102, 117–119].

In addition to diazonium chemistry, several other methods have been used to modify carbon surfaces with nanomolecular layers. The electrochemical oxidation of aliphatic amines [84, 120, 121], treatments with alkenes and alkynes (analogous to that for Si surfaces described above) [122], and azide chemistry [123] are among the most prominent examples.

7.4 Multilayer Surface Modification

The approaches described in Sect. 7.3 for producing monolayers on surfaces are either self-limiting at one monolayer thick, or can be controlled to produce a nanomolecular layer. In all cases, the methods discussed excel at producing layers less than 10 nm thick. There are a variety of surface modification approaches which produce "multilayers" with thicknesses greater than one molecular unit, having total thicknesses from 1 nm to >100 nm. As noted in Sect. 7.1, we will exclude spin-coating and vapour deposition, but will consider "reactive" procedures involving chemical interactions between the substrate and modifying molecules. Examples include radical-based reactions which can form multilayers, "layer-by-layer" techniques which alternate electrostatically or covalently bonded layers, and "electropolymerization" in which the modification is driven electrochemically. Schematics and descriptions of these processes are provided in Sections 7.4.1–7.4.3.

7.4.1 Diazonium Reduction and Related Techniques

Although early reports on diazonium reduction to form covalent bonds on C, Si, and metals assumed that the resulting film was a monolayer, there are conditions where multilayers can form, and films up to >20 nm thick can be produced. As described in Sect. 7.3.3.3, reduction of a diazonium ion produces a phenyl radical which rapidly binds to many conducting surfaces to form the first monolayer. Since the monolayer is not a perfect insulator, additional electrons may transfer through the film to reduce more diazonium ions and the associated reactive radicals, as shown in Fig. 7.12, reactions 1 and 2. In the case of phenylacetic acid, nitrophenyl, and dimethylaminophenyl diazonium ions, the process can be continued to produce disordered, 15–20 nm thick multilayers, corresponding to >20 molecular layers [124]. It is likely that the film itself becomes conductive by partial reduction, since the thicknesses achieved are much greater than the tunnelling distance for electrons [116, 125]. In some cases, solvent molecules or ions may be incorporated into the multilayer, which may resemble a porous "sponge" permeable to species in solution [102, 126, 127].

The degree of disorder in diazonium-derived multilayers varies significantly with the molecular structure, the substrate material [103, 105, 118, 129, 130], and the film-formation conditions. Atomic force microscopy shows that film growth can be quite uniform, with only minor increases in surface roughness as the film thickness increases [116]. FTIR and Raman spectroscopy of both mono- and multi-layers on flat carbon surfaces showed tilt angles of 31–44° relative to the surface normal, and this angle was similar for both mono- and multi-layer films

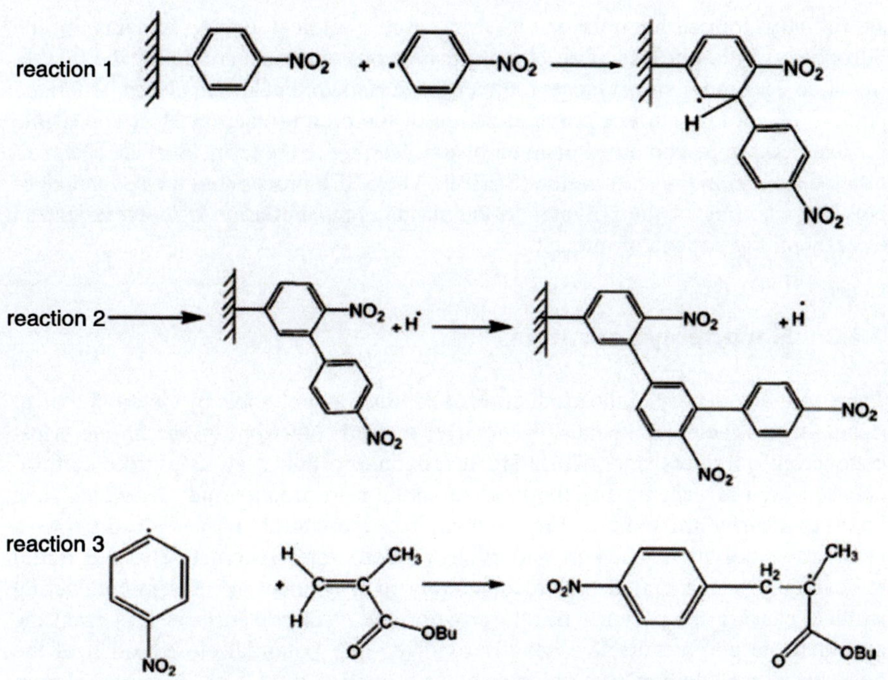

Fig. 7.12 Reactions 1 and 2 show the growth of multilayer films from diazonium precursors. Reaction 3 illustrates the SEEP process, in which olefin polymerization is induced by an electrogenerated diazonium radical. Alternatively, the SEEP process could be initiated from a surface radical (Reproduced with permission from Ref. [128])

[131]. The reader should be cautioned, however, that much of the available characterization data for diazonium-derived multilayers was obtained on carbon surfaces, and the behavior on metal or silicon surfaces may be quite different. Furthermore, the thickness and uniformity of diazonium-derived multilayers should be verified for the particular conditions employed for film formation, usually by AFM [116, 127, 132] and/or ellipsometry [114]. With suitable controls, the ability to grow molecular layers via diazonium reduction provides the benefit of forming films ranging from ~1 to >5 nm thick using similar reagents and procedures. This property of diazonium-derived multilayers was used to construct the attenuation plot of Fig. 7.6 for aromatic films with thicknesses in the range of 2.2–5.2 nm [41].

As noted in Sect. 7.3.3, radical reagents such as those generated by diazonium reduction are quite reactive, and usually form irreversible covalent bonds with the substrate, each other, or both. A beneficial result is high surface coverage, with the more reactive "bare spots" being sites for radical generation and irreversible absorption. A recent development exploits this reactivity in a different manner, by initiating a radical polymerization with a phenyl radical generated from a diazonium ion [128, 133], as shown in Fig. 7.12. Reaction 1 shows formation of a second nitrophenyl layer by reduction and bonding of nitrophenyldiazonium ion to an initially formed monolayer. In this case, a radical center remains in the nitrophenyl film, but loss of an H atom may promote propagation past the bilayer (reaction 2). If an olefin is present, either the surface radical or an electrogenerated phenyl radical may initiate polymerization of the olefin (reaction 3). It is possible for the olefin to be contained in an emulsion, leading to the term "Surface Electroinitiated Emulsion Polymerization (SEEP)". The SEEP process has the advantage of covalent bonding of the polymer to the surface, and initiation of polymerization directly on the surface of interest.

7.4.2 Electropolymerization

There is a long history of the production of conducting polymers by electrochemical rather than chemical means, generally termed *electropolymerization*. Most conducting polymers start with a small monomeric unit such as pyrrole or thiophene, which are chemically oxidized in solution to produce radicals which then bond to nearby monomers. Free radical polymerization ensues, leading to a conducting polymer, often in a partially doped state. Alternatively, the initial radical may be generated electrochemically at a conducting or semiconducting surface, causing the polymer film to grow on the electrode surface. It is often the case that the polymer itself is readily oxidized at a potential close to that of the monomer, thus "doping" the polymer to a conductive state. The growing polymer surface then acts as its own electrode, permitting possibly thick films, readily >100 nm thick. Unlike chemical oxidation, the current and total charge used to oxidize the monomer may be controlled accurately, and initiation of the polymerization at the conducting or semiconducting surface assures a conformal and quite uniform film. Examples of the procedure for electrochemical formation of polypyrrole [134–137], polythiophene [138–140], and polyaniline [141, 142] show that the composition, morphology and thickness of the polymer film are strong functions of the formation conditions, notably the magnitude and form of the potential program used to oxidize the monomer. It is often advantageous to provide an "initiator" on the surface to act as a sight for initiating polymerization. For example, a chlorosilane derivative of thiophene bonded spontaneously to indium-tin oxide can serve as a nucleation layer for polymer growth [13], and "click" chemistry has been used to form a thiophene "termination" layer on conducting diamond [143].

In both cases, the resulting polymer film was more uniform than "random" nucleation on an unmodified surface. A scheme for providing a polyaniline seed layer using diazonium chemistry is shown in Fig. 7.13, in which diazonium surface modification was used to create a monolayer of diphenylamine on a carbon surface [144]. Subsequent oxidation of aniline produced a polyaniline film similar to that formed without the seed layer, and which was more chemically and thermally stable.

Electrochemical methods may also be used to form "redox polymers", in which the monomeric unit contains a redox center such a ferrocene [145, 146] or several Ru or Os complexes [147–149]. Electrochemistry may also be used to form patterns on

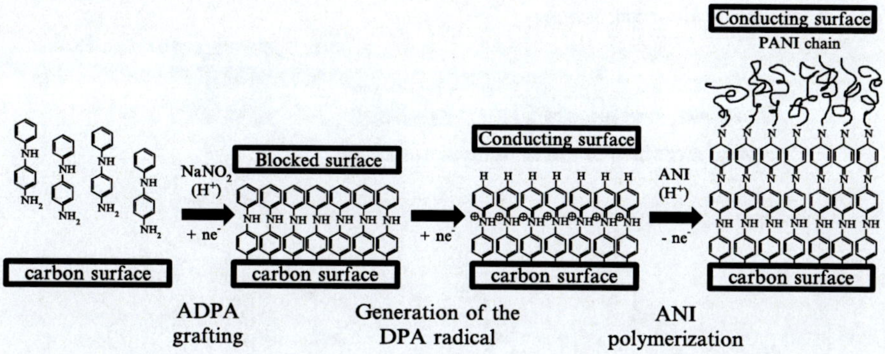

Fig. 7.13 Production of a stable polyaniline layer using a "seed" layer deposited using diazonium chemistry (step 1, resulting in a "blocked" surface) and activated by reduction (step 2) of the surface layer to produce a conductive surface upon which polyaniline can be grown (step 3) (Reproduced with permission from Ref. [144])

surfaces, on the monolayer level as noted in Sect. 7.3.3 above, but also of polymers [150], and to etch patterns on carbon surfaces [151, 152]. A review of electrochemical methods for micro- and nanopatterning has appeared recently [153].

7.4.3 Layer-by-Layer Deposition of Molecular and Atomic Multilayers

Thick films may be generated on surfaces by alternation of two materials, with the general term being "layer-by-layer deposition". A variety of chemical interactions between layers has been exploited to produce possibly thick multilayers, but generally each layer deposition is self-limiting. An example is shown in Fig. 7.14, based on a sequence of reactions between C = C bonds and SH groups, commonly called the "thiol-ene" reaction [154]. A "diene" molecule containing C = C bonds on both ends is initially bonded to silicon, to form a C = C terminated monolayer. Then a "dithiol" with SH groups on both ends is bonded to the terminal C = C bond with UV light and the thiol-ene reaction, resulting in an SH terminated layer. Then the thiol-ene reaction is repeated with alternating layers of diene and dithiol to build up a multilayer. Unlike spin coating or polymerization, the multilayer is formed one layer at a time, so that the total thickness is a linear function of the number of layer-by-layer cycles. In a different approach, formation of a covalent Si-O bond by reaction of a chlorosilane end group with an alcohol or ether can form the basis of a layer-by-layer structure of interest as a dielectric in organic thin-film transistors [155]. Alternating layers containing metal or metal oxide centers and organic molecules have been made, based either on electrostatic bonding or coordination bonds formed between metal ions and electron-rich ligands [156]. A non-vacuum analog to atomic layer deposition has been reported, which alternates electrochemical deposition of single atomic layers [157–160]. With proper conditions, single, epitaxial atomic layers of

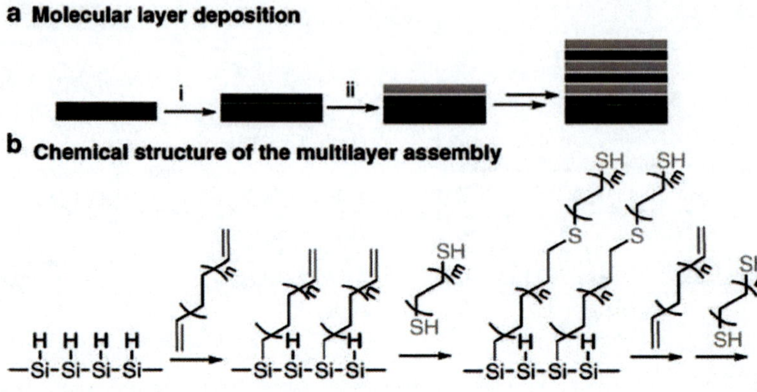

Fig. 7.14 Layer-by-layer growth utilizing the thiol-ene reaction scheme that links together C = C groups with SH groups (Reproduced with permission from Ref. [154])

Cu, Pt, Pb, Se, and other materials may be alternated to produce an ordered multilayer. The approach is a much lower cost alternative to vapor-phase atomic layer deposition (discussed in Chap. 6), and has been shown to produce multilayer structures with interesting electronic properties.

Electrostatic layer-by-layer formation is simpler and more common than "reactive" methods, and exploits the attractions between oppositely charged end groups to stabilize the multilayer [161–165]. Electrostatic interactions have been used to assemble the two components of organic photovoltaic cells, with reported increases in photoconversion efficiency [156, 166, 167].

7.5 Conclusion and Outlook

Although not intended to be comprehensive, this overview of surface modification techniques indicates the great variety of available methods, both in terms of surface chemistry and the properties imparted to the surface. It is noteworthy that many of these modification procedures were developed relatively recently, i.e. in the last two decades. The ability to form covalent bonds between a surface and a molecule with particular chemical properties is a significant advance over the time-tested approaches based on oxide formation, vapour deposition, and spin coating. The advent of molecular electronics was enabled by the ability to make oriented monolayers of molecules on surfaces, and it is now possible to make electronic devices with single-molecular layer active components [3, 8]. It is likely that the initial practical application of surface modification at the monolayer level will involve a process step for making conventional semiconductors, such as a "primer" for Damascene copper plating. However, the growing importance of surfaces to nanofabrication as the feature size continues to decrease provides a strong driving force for more widespread use of surface modification chemistry.

References

1. Israelachvili JN. Intermolecular and surface forces. New York: Academic; 1992.
2. Morgenthaler S, Lee S, Zürcher S, Spencer ND. Langmuir. 2003;19:10459–62.
3. Haick H, Cahen D. Acc Chem Res. 2008;41:359–66.
4. Metzger R, editors. Nano and molecular electronics handbook. Boca Raton, FL: CRC Press; 2007.
5. Metzger RM. Chem Rev. 2003;103:3803.
6. Walker AV, Tighe TB, Haynie BC, Uppili S, Winograd N, Allara D. J Phys Chem B. 2005;109:11263–72.
7. Zhu Z, Daniel TA, Maitani M, Cabarcos OM, Allara DL, Winograd N. J Am Chem Soc. 2006;128:13710–9.
8. McCreery RL, Bergren AJ. Adv Mater. 2009;21:4303–22.
9. Artukovic E, Kaempgen M, Hecht DS, Roth S, Grüner G. Nano Lett. 2005;5:757–60.
10. Fortin JB, Lu TM. Chem Mater. 2002;14:1945–9.
11. McCreery RL. Chem Rev. 2008;108:2646–87.
12. Haick H, Ambrico M, Ligonzo T, Tung RT, Cahen D. J Am Chem Soc. 2006;128:6854–69.
13. Rider DA, Harris KD, Wang D, Bruce J, Fleischauer MD, Tucker RT, Brett MJ, Buriak JM. ACS Appl Mater Interface. 2009;1:279–88.
14. Delamar M, Desarmot G, Fagebaume O, Hitmi R, Pinson J, Saveant J. Carbon. 1997;35:801–7.
15. Kubo W, Nagao M, Otsuka Y, Homma T, Miyata H. Langmuir. 2009;25:13340–3.
16. Flavel BS, Gross AJ, Garrett DJ, Nock V, Downard AJ. ACS Appl Mater Interface. 2010;2:1184–90.
17. Velmurugan J, Zhan D, Mirkin MV. Nat Chem. 2010;2:498–502.
18. Tagliazucchi M, Calvo EJ, Szleifer I. Electrochim Acta. 2008;53:6740–52.
19. Bradbury CR, Zhao J, Fermín DJ. J Phys Chem C. 2008;112:10153–60.
20. Pumera M, Merkoçi A, Alegret S. Electrophoresis. 2007;28:1274–80.
21. Gorodetsky AA, Barton JK. Langmuir. 2006;22:7917–22.
22. Kim Y, Yi J. Langmuir. 2006;22:9805–8.
23. Clark W, McCreery RL. J Electrochem Soc. 2002;149:B379–86.
24. Akiyama E, Markworth AJ, McCoy JK, Frankel GS, Xia L, McCreery RL. J Electrochem Soc. 2003;150:B83–91.
25. Clark WJ, Ramsey JD, McCreery RL, Frankel GS. J Electrochem Soc. 2002;149:B179–85.
26. Zhang W, Frankel GS. Electrochem Solid-State Lett. 2000;3:268.
27. Maldonado S, Plass KE, Knapp D, Lewis NS. J Phys Chem C. 2007;111:17690–9.
28. Buriak JM. Chem Rev. 2002;102:1271–308.
29. Nemanick EJ, Hurley PT, Brunschwig BS, Lewis NS. J Phys Chem B. 2006;110:14800–8.
30. Kim JM, Beebe Y, Jun XY, Zhu XY, Frisbie CD. J Am Chem Soc. 2006;128:4970–1.
31. Vilan A, Yaffe O, Biller A, Salomon A, Kahn A, Cahen D. Adv Mater. 2010;22:140–59.
32. Salomon A, Boecking T, Seitz O, Markus T, Amy F, Chan C, Zhao W, Cahen D, Kahn A. Adv Mater. 2007;19:445–50.
33. Cahen D, Kahn A. Adv Mater. 2003;15:271–7.
34. Yan H, McCreery RL. ACS Appl Mater Interface. 2009;1:443–51.
35. Thieblemont F, Seitz O, Vilan A, Cohen H, Salomon E, Kahn A, Cahen D. Adv Mater. 2008;20:3931–6.
36. Vaynzof Y, Dennes TJ, Schwartz J, Kahn A. Appl Phys Lett. 2008;93:103305–103303.
37. Heimel G, Romaner L, Zojer E, Bredas J-L. Acc Chem Res. 2008;41:721–9.
38. Thygesen KS, Rubio A. Phys Rev Lett. 2009;102:046802.
39. Zahid F, Paulsson M, Datta S. Electrical conduction through molecules. New York: Academic; 2003.
40. Itoh T, McCreery RL. J Am Chem Soc. 2002;124:10894–902.
41. Bergren AJ, McCreery RL, Stoyanov SR, Gusarov S, Kovalenko A. J Phys Chem C. 2010;114:15806–15.

42. Natan A, Kuritz N, Kronik L. Adv Funct Mater. 2010;20:2077–84.
43. Heimel G, Rissner F, Zojer E. Adv Mater. 2010;22:2494–513.
44. Andrews DQ, Van Duyne RP, Ratner MA. Nano Lett. 2008;8:1120–6.
45. Berlin YA, Burin AL, Ratner MA. Chem Phys. 2002;275:61–74.
46. Berlin YA, Ratner MA. Radiat Phys Chem. 2005;74:124–31.
47. Grozema FC, van Duijnen PT, Berlin YA, Ratner MA, Siebbeles LDA. J Phys Chem B. 2002;106:7791–5.
48. Ranganathan S, Murray R. J Phys Chem B. 2004;108:19982–9.
49. Terrill RH, Hatazawa T, Murray RW. J Phys Chem. 1995;99:16676–83.
50. Terrill RH, Sheehan PE, Long VC, Washburn S, Murray RW. J Phys Chem. 1994;98:5127–34.
51. Simmons JG. DC conduction in thin films. London: Mills and Boon Ltd.; 1971.
52. Vilan A. J Phys Chem C. 2007;111:4431–44.
53. Choi SH, Kim B, Frisbie CD. Science. 2008;320:1482–6.
54. Choi SH, Risko C, Delgado MCR, Kim B, Bredas J-L, Frisbie CD. J Am Chem Soc. 2010;132:4358–68.
55. Lahmidi A, Joachim C. Chem Phys Lett. 2003;381:335–9.
56. Soe W-H, Manzano C, Sarkar AD, Chandrasekhar N, Joachim C. Phys Rev Lett. 2009;102:176102.
57. Huisman EH, Guedon CM, van Wees BJ, van der Molen SJ. Nano Lett. 2009;9:3909–13.
58. Bonifas AP, McCreery RL. Nat Nanotechnol. 2010;5:612–7.
59. Mujica V, Kemp M, Ratner MA. J Chem Phys. 1994;101:6849–55.
60. Mujica V, Kemp M, Ratner MA. J Chem Phys. 1994;101:6856–64.
61. Mujica V, Ratner MA. Chem Phys. 2001;264:365–70.
62. Akkerman HB, Kronemeijer AJ, Harkema J, van Hal PA, Smits ECP, de Leeuw DM, Blom PWM. Org Electron. 2010;11:146–9.
63. Coll M, Miller LH, Richter LJ, Hines DR, Jurchescu OD, Gergel-Hackett N, Richter CA, Hacker CA. J Am Chem Soc. 2009;131:12451–7.
64. Ru J, Szeto B, Bonifas A, McCreery RL. ACS Appl Mater Interface. 2010;2:3693–701.
65. Blodgett KB. J Am Chem Soc. 1934;56:495.
66. Blodgett KB. J Am Chem Soc. 1935;57:1007–22.
67. Petty MC. Langmuir-Blodgett films: an introduction. Cambridge: Cambridge University Press; 1996.
68. Edwards GA, Bergren AJ, Porter MD. In: Zoski CG, editor. Chemically Modified Electrodes. Handbook of electrochemistry. New York: Elsevier; 2007.
69. Hussain SA, Bhattacharjee D. Mod Phys Lett B. 2009;23:3437–51.
70. Prokopuk N, Son K-A. J Phys Condens Matter. 2008;20:374116.
71. Siqueira Jr JR, Caseli L, Crespilhoc FN, Zucolottoa V, Oliveira Jr ON. Biosens Bioelectron. 2010;25:1254–63.
72. DeRose JA, Leblanc RM. Surf Sci Reports. 1995;22:73–126.
73. Liu S, Wang WM, Briseno AL, Mannsfeld SCB, Bao Z. Adv Mater. 2009;21:1217–32.
74. Talham DR, Yamamoto T, Meisel MW. J Phys Condens Matter. 2008;20:184006.
75. Love JC, Estroff LA, Kriebel JK, Nuzzo RG, Whitesides GM. Chem Rev. 2005;105:1103–70.
76. DiBenedetto SA, Facchetti A, Ratner MA, Marks TJ. Adv Mater. 2009;21:1407–33.
77. Allara D, Nuzzo RG. Langmuir. 1985;1:52–66.
78. Edwards GA, Bergren AJ, Cox EJ, Porter MD. J Electroanal Chem. 2008;622:193–203.
79. Finklea HO. Electrochemistry of organized monolayers of thiols and related molecules on electrodes. In: Bard AJ, editor. Electroanalytical chemistry. 1st ed. New York: Dekker; 1996.
80. Ulman A. Chem Rev. 1996;96:1533.
81. Walczak MM, Chung C, Stole SM, Widrig CA, Porter MD. J Am Chem Soc. 1991;113:2370–8.
82. Poirier GE. Chem Rev. 1997;97:1117.
83. Poirier GE, Pylant ED. Science. 1996;272:1145–8.
84. Downard AJ. Electroanalysis. 2000;12:1085–96.

85. Murray RW. Acc Chem Res. 1980;13:135–41.
86. Murray RW. Chemically modified electrodes. In: Bard A, editor. Electroanalytical chemistry. New York: Dekker; 1983.
87. Baranton S, Belanger D. J Phys Chem B. 2005;109:24401–10.
88. Baranton S, Belanger D. Electrochim Acta. 2008;53:6961–7.
89. Breton T, Belanger D. Langmuir. 2008;24:8711–8.
90. Han S, Yuan Y, Hu L, Xu G. Electrochem Commun. 2010;12:1746–8.
91. Kullapere M, Seinberg J-M, Maeeorg U, Maia G, Schiffrin DJ, Tammeveski K. Electrochim Acta. 2009;54:1961–9.
92. Liu G, Chockalingham M, Khor SM, Gui AL, Gooding JJ. Electroanalysis. 2010;22:918–26.
93. Noel J-M, Sjoberg B, Marsac R, Zigah D, Bergamini J-F, Wang A, Rigaut S, Hapiot P, Lagrost C. Langmuir. 2009;25:12742–9.
94. Stockhausen V, Ghilane J, Martin P, Trippe-Allard G, Randriamahazaka H, Lacroix J-C. J Am Chem Soc. 2009;131:14920–7.
95. Yesildag A, Ekinci D. Electrochim Acta. 2010;55:7000–9.
96. Pavia DL, Lampman GM, Kriz GS, Engel RG. Organic laboratory techniques: a microscale approach. New York: Saunders College Publishing; 1995.
97. Bernard M-C, Chausse A, Cabet-Deliry E, Chehimi MM, Pinson J, Podvorica F, Vautrin-Ul C. Chem Mater. 2003;15:3450–62.
98. Liu G, Böcking T, Gooding JJ, J Electroanal Chem. 2007;600:335–44.
99. Mahmoud AM, Bergren AJ, McCreery RL. Anal Chem. 2009;81:6972–80.
100. Malmos K, Iruthayaraj J, Pedersen SU, Daasbjerg K. J Am Chem Soc. 2009;131:13926–7.
101. Shewchuk DM, McDermott MT. Langmuir. 2009;25:4556–63.
102. Lehr J, Garrett DJ, Paulik MG, Flavel BS, Brooksby PA, Williamson BE, Downard AJ. Anal Chem. 2010;82:7027–34.
103. Paulik MG, Brooksby PA, Abell AD, Downard AJ. J Phys Chem C. 2007;111:7808–15.
104. Pearson D, Downard AJ, Muscroft-Taylor A, Abell AD. J Am Chem Soc. 2007;129:14862–3.
105. Pinson J, Podvorica F. Chem Soc Rev. 2005;34:429–39.
106. Balakumar A, Lysenko AB, Carcel C, Malinovskii VL, Gryko DT, Schweikart KH, Loewe RS, Yasseri AA, Liu ZM, Bocian DF, Lindsey JS. J Org Chem. 2004;69:1435–43.
107. Li Q, Mathur G, Homsi M, Surthi S, Misra V, Malinovskii V, Schweikart K-H, Yu L, Lindsey JS, Liu Z, Dabke RB, Yasseri AA, Bocian DF, Kuhr WG. Appl Phys Lett. 2002;81:1494–6.
108. Liu Z, Yasseri AA, Lindsey JS, Bocian DF. Science. 2003;302:1543–5.
109. de Villeneuve CH, Pinson J, Bernard MC, Allongue P. J Phys Chem B. 1997;101:2415–20.
110. Allongue P, de Villeneuve CH, Cherouvrier G, Cortes R, Bernard MC. J Electroanal Chem. 2003;550–551:161–74.
111. Scott A, Hacker CA, Janes DB. J Phys Chem C. 2008;112:14021–6.
112. Scott A, Janes DB, Risko C, Ratner MA. Appl Phys Lett. 2007;91:033508–033503.
113. Garrett DJ, Flavel BS, Shapter JG, Baronian KHR, Downard AJ. Langmuir. 2010;26:1848–54.
114. Stewart MP, Maya F, Kosynkin DV, Dirk SM, Stapleton JJ, McGuiness CL, Allara DL, Tour JM. J Am Chem Soc. 2004;126:370–8.
115. Hamers RJ. Annu Rev Anal Chem. 2008;1:707–36.
116. Anariba F, DuVall SH, McCreery RL. Anal Chem. 2003;75:3837–44.
117. Garrett DJ, Lehr J, Miskelly GM, Downard AJ. J Am Chem Soc. 2007;129:15456–7.
118. Downard AJ, Garrett DJ, Tan ESQ. Langmuir. 2006;22:10739–46.
119. Brooksby PA, Downard AJ. Langmuir. 2005;21:1672–5.
120. Deinhammer RS, Ho M, Anderegg JW, Porter MD. Langmuir. 1994;10:1306–13.
121. Barbier B, Pinson J, Desarmot G, Sanchez M. J Electrochem Soc. 1990;137:1757–64.
122. Ssenyange S, Anariba F, Bocian DF, McCreery RL. Langmuir. 2005;21:11105–12.
123. Devadoss A, Chidsey CED. J Am Chem Soc. 2007;129:5370–1.
124. Kariuki JK, McDermott MT. Langmuir. 2001;17:5947–51.
125. Solak AO, Eichorst LR, Clark WJ, McCreery RL. Anal Chem. 2003;75:296–305.

126. Yu SSC, Tan ESQ, Jane RT, Downard AJ. Langmuir. 2007;23:11074–82.
127. Brooksby PA, Downard AJ. J Phys Chem B. 2005;109:8791–8.
128. Deniau G, Azoulay L, Bougerolles L, Palacin S. Chem Mater. 2006;18:5421–8.
129. Brooksby PA, Downard AJ. Langmuir. 2004;20:5038–45.
130. Downard AJ. Langmuir. 2000;16:9680–2.
131. Anariba F, Viswanathan U, Bocian DF, McCreery RL. Anal Chem. 2006;78:3104–12.
132. Brooksby PA, Downard AJ, Yu SSC. Langmuir. 2005;21:11304–11.
133. Tessier L, Deniau G, Charleux B, Palacin S. Chem Mater. 2009;21:4261–74.
134. Sotzing GA, Reynolds JR, Katritzky AR, Soloducho J, Belyakov S, Musgrave R. Macromolecules. 1996;29:1679–84.
135. Liu Y-C, Wang C-C. J Phys Chem B. 2005;109:5779–82.
136. Bof Bufon CC, Vollmer J, Heinzel T, Espindola P, John H, Heinze J. J Phys Chem B. 2005;109:19191–9.
137. Lacroix JC, Maurel F, Lacaze PC. J Am Chem Soc. 2001;123:1989–96.
138. Doherty III WJ, Armstrong NR, Saavedra SS. Chem Mater. 2005;17:3652–60.
139. Marrikar FS, Brumbach M, Evans DH, Lebron-Paler A, Pemberton JE, Wysocki RJ, Armstrong NR. Langmuir. 2007;23:1530–42.
140. Bobacka J, Grzeszczuk M, Ivaska A. J Electroanal Chem. 1997;427:63–9.
141. Wei D, Baral JK, Osterbacka R, Ivaska A. J Mater Chem. 2008;18:1853–7.
142. McCarley RL, Morita M, Wilbourn KO, Murray RW. J Electroanal Chem. 1988;245:321–30.
143. Wang M, Das MR, Li M, Boukherroub R, Szunerits S. J Phys Chem C. 2009;113:17082–6.
144. Santos LM, Ghilane J, Fave C, Lacaze P-C, Randriamahazaka H, Abrantes LM, Lacroix J-C. J Phys Chem C. 2008;112:16103–9.
145. Jureviciute I, Bruckenstein S, Jackson A, Hillman AR. J Solid-State Electrochem. 2004;8:403–10.
146. Kurihara M, Kubo K, Horikoshi T, Kurosawa M, Nankawa T, Matsuda T, Nishihara H. Macromol Symposia. 2000;156:21–9.
147. Manness KM, Terrill RH, Meyer TJ, Murray RW, Wightman RM. J Am Chem Soc. 1996;118:10609–16.
148. Surridge NA, Sosnoff CS, Schmehl R, Facci JS, Murray RW. J Phys Chem. 1994;98:917–23.
149. Surridge NA, Zvanuf ME, Keene FR, Sosnoff SC, Silver M, Murray RW. J Phys Chem. 1992;96:962–70.
150. Zangmeister RA, O'Brien DF, Armstrong NR. Adv Funct Mater. 2002;12:179–86.
151. Kiema GK, Ssenyange S, McDermott MT. J Electrochem Soc. 2004;151:C142.
152. Ssenyange S, Du R, McDermott MT. Micro Nano Lett IET. 2009;4:22–6.
153. Simeone FC, Albonetti C, Cavallini M. J Phys Chem C. 2009;113:18987–94.
154. Li Y-h, Wang D, Buriak JM. Langmuir. 2010;26:1232–8.
155. Yoon M-H, Facchetti A, Marks TJ. Proc Natl Acad Sci USA. 2005;102:4678–82.
156. Zhao W, Tong B, Pan Y, Shen J, Zhi J, Shi J, Dong Y. Langmuir. 2009;25:11796–801.
157. Kim JY, Kim Y-G, Stickney JL. J Electrochem Soc. 2007;154:D260–6.
158. Kim JY, Stickney JL. J Phys Chem C. 2008;112:5966–71.
159. Kim Y-G, Kim JY, Vairavapandian D, Stickney JL. J Phys Chem B. 2006;110:17998–8006.
160. Liang X, Kim Y-G, Gebergziabiher DK, Stickney JL. Langmuir. 2009;26:2877–84.
161. McClure SA, Worfolk BJ, Rider DA, Tucker RT, Fordyce JAM, Fleischauer MD, Harris KD, Brett MJ, Buriak JM. ACS Appl Mater Interface. 2009;2:219–29.
162. Li X, Wan Y, Sun C. J Electroanal Chem. 2004;569:79–87.
163. Tang T, Qu J, Mullen K, Webber SE. Langmuir. 2006;22:26–8.
164. Milsom EV, Perrott HR, Peter LM, Marken F. Langmuir. 2005;21:9482–7.
165. Kim HJ, Lee K, Kumar S, Kim J. Langmuir. 2005;21:8532–8.
166. Mwaura JK, Pinto MR, Witker D, Ananthakrishnan N, Schanze KS, Reynolds JR. Langmuir. 2005;21:10119–26.
167. Huguenin F, Zucolotto V, Carvalho AJF, Gonzalez ER, Oliveira ON. Chem Mater. 2005;17:6739–45.

Nanostructures Based on Self-Assembly of Block Copolymers

8

E. Bhoje Gowd, Mallikarjuna Shroff Rama, and Manfred Stamm

Abstract

Block copolymers have attracted increasing interest for the synthesis of nanomaterials and fabrication of nanostructures due to their ability to self-organize at nanometer scales. A variety of nanoscale morphologies can be obtained by controlling the composition and architecture of the individual block components. In this chapter, we highlight the recent advances in the formation of nanostructures in bulk, generation of block copolymer based nanotemplate ordered thin films and their potential applications in nanofabrication including template nanolithography, deposition of functional nanomaterials as well as nanoporous membranes.

8.1 Introduction

Nanometer scale patterning of surfaces has attracted great attention due to recent applications in interdisciplinary areas including electron beam lithography, x-ray lithography, optical lithography, and imprint lithography [1–8]. However, a feature size less than 30 nm in a large area is hard to achieve with the above-mentioned lithography techniques. Block-copolymer based nanofabrication is an emerging approach for the production of nanostructured materials due to the unrelenting reduction in feature sizes on integrated circuitry [9–15]. Block copolymers are a class of macromolecules that are produced by covalent connection of two or more

E.B. Gowd (✉)
Materials and Minerals Division, National Institute for Interdisciplinary Science and Technology (NIIST), CSIR, Trivandrum, India
e-mail: bhojegowd@yahoo.com

M.S. Rama • M. Stamm
Department of Nanostructured Materials, Leibniz Institute of Polymer Research Dresden, Dresden, Germany

M. Stepanova and S. Dew (eds.), *Nanofabrication*,
DOI 10.1007/978-3-7091-0424-8_8, © Springer-Verlag/Wien 2012

chemically dissimilar homopolymers. The number, composition and connectivity of blocks determine the molecular architecture of block copolymers (e.g. diblock, triblock, multiblock, and star or graft copolymers). Complex architectures have been generated. Since blocks are usually not compatible depending on molecular weight and effective monomer interaction, block copolymers can self-assemble to form a variety of nanoscale periodic patterns including spheres, cylinders, and lamellae with typical dimensions of 5–50 nm [12–23]. In many of the applications block copolymers are used in the form of thin films. The optimal utilization of these phase-separated block copolymer thin films involves several steps, i.e. working with appropriate film thickness, controlling orientation and lateral ordering of the microdomains, and generation of ordered arrays of nanopores [17–19, 21]. Thus obtained block copolymer templates have been used in various nanofabrication processes to generate patterns of interest by further treatment with a variety of different methods including chemical and physical vapor deposition, electrodeposition, incorporation of metal nanoparticles, and chemical reduction techniques [22–31].

In this chapter recent progress in the field of thin film nanostructures based on self assembly of block copolymers is described. The chapter is structured in-to four parts. In the first part, basic principles of microphase separation and morphology in bulk and thin films are discussed. In the second part, we mainly focus on fabrication of nanostructures from hybrid materials where block copolymers serves as structure-directing agent. In the third part, the preparation procedures of ordered nanotemplates are discussed and in the final part, we highlight the applications of block copolymer nanotemplates in the fabrication of functional nanostructures as well as nanoporous membranes for controlled separations.

8.2 Microphase Separation and Morphology

8.2.1 Microphase Separation in Bulk

In bulk, the microphase separation in block copolymers is primarily governed by the mutual repulsion of the chemically dissimilar blocks as well as the packing constraints imposed by the connectivity of each block. The microphase separation in diblock copolymers has been extensively studied from both theoretical and experimental point of view (e.g. [32–35]). The diblock copolymer phase diagram is determined by three independent factors: volume fractions of the two blocks, diblock degree of polymerization, and degree of incompatibility of the two blocks as expressed by the Flory-Huggins interaction parameter (χ). In diblock copolymers, the morphology of the microdomains ranges from spheres and cylinders to lamellae depending on the volume fraction of the blocks. Figure 8.1 presents a theoretical and an experimental phase diagram of the model diblock copolymer poly(styrene-*b*-isoprene). As seen from Fig. 8.1, the phase diagram shows the dependence of morphology on the volume fractions (f_A and f_B) of the two blocks and degree of segregation (χN), where N is the degree of polymerization of the diblock copolymer. In the approximation of mean field theory diblock

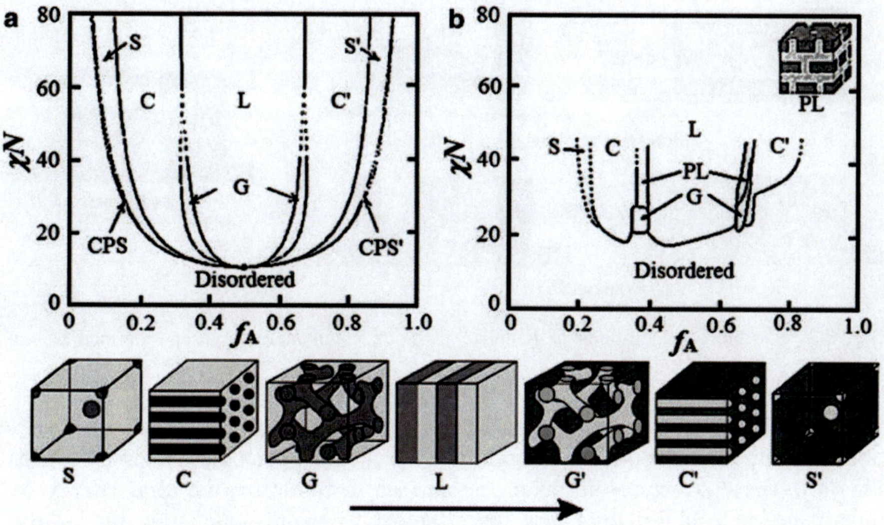

Fig. 8.1 Phase diagram for linear AB diblock copolymers. (**a**) Equilibrium morphologies predicted by self-consistent mean-field theory: spherical (*S*), cylindrical (*C*) gyroid (*G*), and lamellar (*L*) morphologies. (**b**) Experimental phase diagram of poly (styrene-*b*-isoprene) diblock copolymers: same as above plus perforated layer (*PL*). (*Bottom*) A representation of the equilibrium phase structures as f_A is increased for fixed χN (Reproduced from Ref. [34] with permission)

copolymers at $\chi N < 10.5$ are always in disordered state despite the volume fraction [33]. In the case $\chi N \leq 10$, the system is governed by entropic terms. Depending on the temperature dependence of the χ–parameter, this may occur at higher (or in some cases at lower) temperature and lower degree of polymerization. In the strong segregation regime $\chi N >> 10$, enthalpic terms dominate as a function of temperature or pressure, causing an order to disorder transition (T_{ODT}) (also called microphase separation) where the interfaces among the phases tend to be very sharp and dissimilar blocks phase separate into a variety of ordered periodic phase morphologies.

8.2.2 Microphase Separation in Thin Films

In many practical applications, block copolymers are used in the form of thin films, and the structural behavior of BCP thin films compared with bulk materials is often much more complicated due to the interfacial interactions of the blocks with the underlying substrate, the surface segregation of blocks and effects of incommensurability of copolymer periodicity and the film thickness. The block that has the lowest surface free energy will preferentially segregate at the air interface, whereas the block with the lowest interfacial energy, which may be the same block, will segregate at the substrate interface, leading to a preferred parallel orientation of microdomains [36, 37]. However, if the thickness of the film is incommensurate

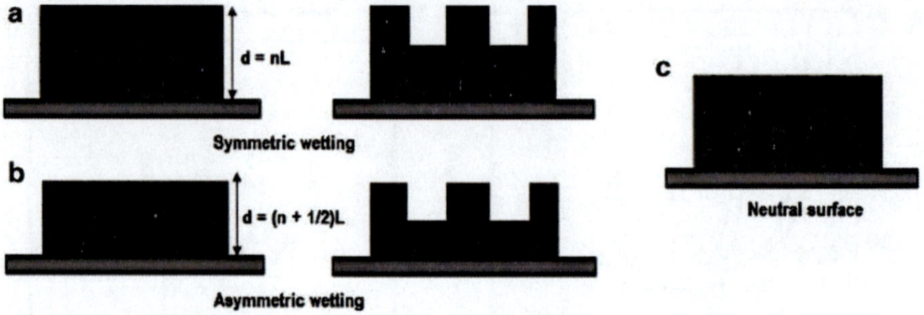

Fig. 8.2 Possible configurations of lamellae in block copolymer thin films confined at one interface

with the microdomain period, the microdomains can be forced to orient perpendicularly or form some other non-equilibrium structures, such as islands and holes [38–40]. These structures form on the surface to minimize the total energy by quantizing the local film thickness. Nevertheless, when an unconfined lamellar film is given the chance to equilibrate, for example by annealing at temperatures above the glass transition temperature of both blocks, it will usually form terraces with thicknesses that are a multiple of the microdomain period. In regions between terraces, the thickness is still incommensurate with the microdomain period, and other non-equilibrium structures are a possibility.

In copolymers that form lamellae parallel to the substrate, a multilayered structure is formed, where the film thickness is quantized in terms of the block copolymer periodicity, L. When one block prefers both the substrate and the air interfaces (i.e., symmetric wetting), the film is smooth if its thickness d is equal to $d = nL$, where n is an integer. Conversely, when one block prefers the substrate and the other block prefers the air interface (i.e., asymmetric wetting), smooth films are attained if the film thickness is $d = (n + 1/2)L$. If the thickness of the prepared film is not commensurable with L, islands or holes with a step height of L form on the top surface. This topography allows the preferred block(s) to be present at both interfaces and also to maintain the characteristic period throughout the whole film thickness. The occurrence of the hole/island morphology has been observed often for lamellar domains and also for cylinder-forming and sphere-forming systems. Some of the possible configurations for a lamellae-forming diblock copolymer confined in a thin film are shown in Fig. 8.2. Only in the case of a neutral substrate, perpendicular lamellae will form spontaneously.

8.2.3 Control of Microdomain Orientation in Block Copolymer Thin Films

Controlling the microdomain orientation is one of the important issues in block copolymer thin films. The parallel orientation of these microdomains is often observed for block copolymer thin films because of the difference in surface energies of the block components and the interaction with the substrate. Although the microdomains that oriented parallel to the substrates (cylinders) are useful to prepare nanowires, the microdomains oriented perpendicular to the substrate are more desirable as templates for the fabrication of high aspect-ratio nanostructures as well as nanoporous membranes. To address this challenge to obtain the perpendicular microdomain orientation, various strategies have been developed. To control microdomain orientations of block copolymer films, for instance solvent annealing [41–45], chemical patterning [5, 46], electric field alignment [24], graphoepitaxy [47, 48], soft lithography [49], shear alignment [50, 51], directional crystallization [52] and zone annealing [53] have been used. In a few cases, combinations of these strategies have proved to be much more effective to induce long-range ordering in microdomains than a single strategy.

In this section we will restrict our discussion to two block copolymer systems, PS-b-PMMA and PS-b-P4VP to explain the perpendicular orientation of cylindrical or lamellar microdomains in a block copolymer thin film. In PS-b-PMMA thin films, perpendicular orientation of cylinders was achieved by balancing the interfacial interactions, i.e., using neutral surfaces. Such surfaces can be obtained by deposition of random copolymers of the same components as in a diblock copolymer [54]. Roughening the substrate can also induce perpendicular orientation because a penalty is imposed for elastic deformation to confirm a rough substrate [55]. Addition of homopolymer to a block copolymer is another simple method to reorient the microdomains. The addition of PMMA homopolymer to a cylinder-forming PS-b-PMMA block copolymer will stabilize the perpendicular orientation of the microdomains in thicker films [56]. Here the domain spacing and/or size is also adjusted according to the amount of the homopolymer added. The other common method used to align cylinders perpendicular to the substrate is application of an electric field. It was demonstrated that annealing of PS-b-PMMA films in an external electric field of high strength (~30 kV/cm) leads to the orientation of PMMA microdomains (lamellae or cylinders) along the field lines, in either normal or in-plane direction depending on the applied electric field direction. Remarkably, the desired perpendicular orientation was achieved for several micrometer thick films [24].

Solvent vapour annealing is another simple and effective method to manipulate the orientation of BC microdomains in thin films. Our group reported solvent induced switching between different alignments of cylindrical microdomains in supramolecular assembly (SMA) based on polystyrene-block-poly (4-vinylpyridine) (PS-b-P4VP) and 2-(4′-hydroxybenzeneazo) benzoic acid [57, 58]. Recently, we also found conditions for switching of P4VP microdomains in PS-b-P4VP thin films [59]. 1,4-Dioxane (preferential solvent to PS block) vapour annealed thin films assemble into hexagonally packed cylindrical microdomains oriented normal to the substrate because the evaporation of the solvent is highly directional. On the other hand, the samples annealed in chloroform (nonpreferential solvent) showed well-developed cylindrical microdomains parallel to the substrate. Recently, Russell and co-workers showed that cylindrical microdomains oriented normal to the film surface could be obtained directly by spin-coating of PS-b-P4VP from mixed solvents of toluene and tetrahydrofuran (THF), and arrays of highly ordered cylindrical microdomains formed over large areas after exposing the films in the vapor of a toluene/THF mixture for a while. This process was independent of substrates, but strongly dependent on the quality of the solvents for each block and the solvent evaporation rate [60].

8.2.4 Long-Range Lateral Order in Block Copolymer Thin Films

In addition to controlling the orientation of block copolymer microdomains, control over the lateral ordering in these oriented microdomains is equally important for many applications, especially in semiconductor and data storage applications. As-cast thin films often reveal lateral domain formation of morphological order where each domain maps a region where the phase separated microstructure has a well-defined arrangement that is highly periodic. The neighboring domains will have the same periodic microstructure but within each domain the director alignment will be random. The presence of domain boundaries can be expected to be due to multiple nucleation and defect formation during microphase separation. These defects destroy the long range orientational and translational order because the domains nucleate independently and grow until a boundary wall is formed. There has been substantial work to minimize domain boundary defects and/or to enhance the long-range lateral order in block copolymer thin films. For more general reviews of the various methods of enhancing order in block copolymers, we refer the reader elsewhere [12, 14, 61–63]. In this section, we highlight some recent advances in improving long-range lateral order in block copolymer microdomains.

Angelescu et al. [50, 64] applied a shearing force to block copolymer thin films with thickness of ~100 nm using a thick sacrificial layer of poly(dimethyl siloxane) to prevent rupture (or breakage) of the thin film during shearing, and obtained the long range ordering of cylindrical nanodomains aligned to the flow direction in the thin film. Hashimoto and co-workers [53, 65] developed a temperature-gradient method to obtain long range ordering of PS-*b*-PI thick films. Russell and co-workers [43, 44] developed a solvent vapour annealing method and obtained perfect long-range ordering of perpendicularly oriented PEO cylindrical microdomains in PS-*b*-PEO. Graphoepitaxy is one of the promising methods to control the lateral ordering of block copolymer microdomains using a topographic pattern. Segalman et al. [38, 47, 66] reported applying this strategy to generate arrays of spherical domains of PS-b-P2VP over large areas with a high degree of perfection. Recently, Ross and co-workers [48] reported the self-assembly of spherical microdomains of poly (styrene-b-dimethylsiloxane) (PS-b-PDMS). In this work, they used surface chemistry to obtain the topographic pattern to match one of the blocks of a spherical block copolymer. The modified hexagonally packed dots interact with the spherical domains and could replace one sphere in the polymer pattern, resulting in a two-dimensional periodic nanostructure array with precisely determined orientation and long-range order.

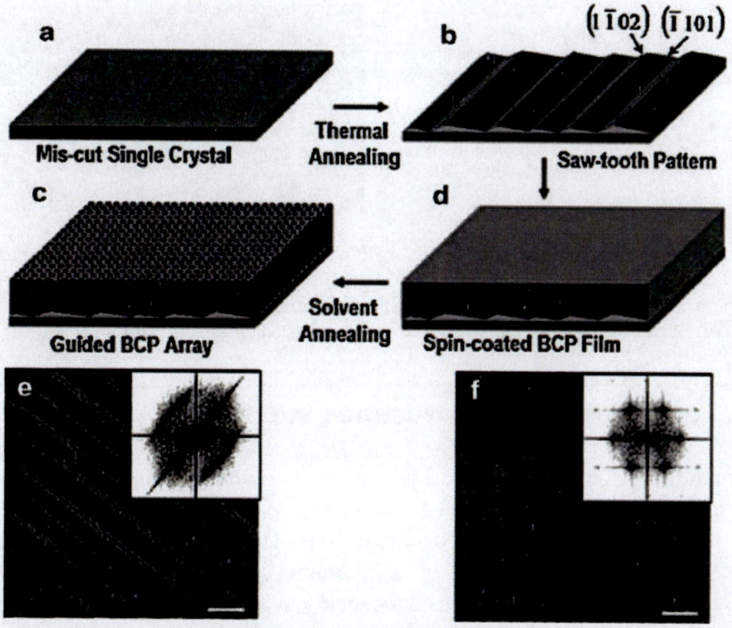

Fig. 8.3 Schematic illustration of the use of faceted surfaces to orient cylinder-forming PS-b-PEO. (**a**) The sapphire surface was mis-cut relative to a specific crystallographic plane and then (**b**) thermally annealed to reconstruct the surface. After (**c**) application of the block copolymer and (**d**) solvent annealing, atomic force micrographs of the block copolymer show excellent registry with the faceted surface for film thicknesses of (**e**) 24 nm and (**f**) 34 nm (scale bar = 200 nm) (Reproduced from Ref. [67] with permission)

Park et al. [67] used surface-reconstructed sapphire wafers to orient perpendicular cylinder-forming PS-b-PEO block copolymers along the saw toothed substrate after solvent annealing. They observed perfect long-range lateral order of PEO microdomains in PS-b-PEO block copolymers, and an ultra high point density of ~10 Tb/in^2 and 3 nm domain spacing was obtained, as shown in Fig. 8.3.

Square arrays of microdomains are particularly attractive, since they are consistent with industry standard design architectures. Park et al. [68] fabricated a square array of vertically oriented cylindrical microdomains of PS-b-PMMA thin film on a chemically nanopatterned surface using a pre-patterning of the square array. Tang et al. [69] fabricated a square array of nanoscale patterns using supramolecular interactions (hydrogen bonding) between A–B and B′–C copolymers, as shown in Fig. 8.4. A-B is poly (ethylene oxide)-*b*-poly (styrene-*ran*-4-hydroxystyrene), and B′–C is poly (styrene-*r*-4-vinyl pyridine)-*b*-polymethylmethacrylate. The stoichiometric ratio of the hydrogen bonding between phenolic and pyridyl units determines the degree of packing of cylindrical domains. By degradation of the PMMA and subsequent CHF_3 RIE with a nanoporous block copolymer mask, a square array of cylindrical pores was successfully transferred on a silicon oxide wafer.

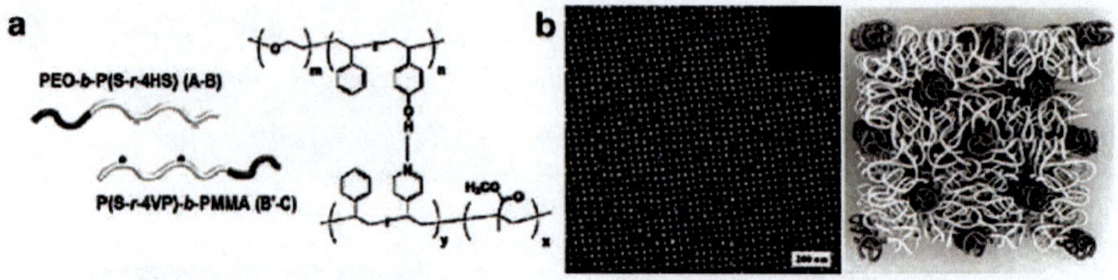

Fig. 8.4 (**a**) Hierarchical self-assembly and target morphology for a blend of supramolecular A–B and B′–C diblock copolymers stabilized by H-bonding. (**b**) TEM image and associated Fourier transform (*inset*) of a solvent-annealed blend film of supramolecular block copolymers. A cartoon (*right panel*) illustrates proposed chain packing (Reproduced from Ref. [69] with permission)

8.3 Formation of Nanostructures in Bulk

Nanostructured polymer-inorganic hybrids are known for their good mechanical and thermal properties. They are often harder and less fragile than the same material without inner structuring. Such hybrid materials can be prepared through different strategies: direct polymerization of a monomer-metallic complex [70], direct assembly of block copolymers with nanoscale metallic species [71] and assembly through chemical coordination with organometallic compounds [72, 73]. Characteristics such as selectivity, particle size and spatial organization of the nanoparticles play an important role when engineering the structure of these polymer-inorganic hybrid materials. Cohen and co-workers [74] used block copolymers as nanoreactors to synthesize metallic nanoparticles in-situ by loading inorganic metal salts selectively into one block of the block copolymers followed by a reduction step to obtain nanoparticles. This concept was expanded to synthesize nanoparticles with controlled size within the preformed micelles of block copolymer [75, 76]. In this section, we highlight the strategies to develop bulk block copolymer hybrid materials.

Wiesner and co-workers [77–79] reported the successful approach to incorporate inorganic materials in bulk. They use an amphiphilic poly (isoprene-b-ethyleneoxide) (PI-b-PEO) block copolymer as a structure-directing agent during the sol-gel synthesis of an organically modified aluminosilicate network based on 3-(glycidyloxypropyl) trimethoxysilane (GLYMO) and aluminum sec-butoxide, $Al(O\text{-}s\text{-}Bu)_3$. The aluminosilicate produced by reaction of GLYMO and $Al(O\text{-}s\text{-}Bu)_3$ preferentially swells the hydrophilic PEO block of PI-b-PEO to form a nanostructured organic–inorganic hybrid materials. A schematic showing the synthetic approach used to prepare these nanoobjects is shown in Fig. 8.5. By dissolving the organic component, spherical, cylindrical and plate-like nanoparticles could be prepared. These are coated with PEO chains (so-called 'hairy' nano-objects), which can be removed by thermal treatment.

Recently, Fahmi and co-workers developed multifunctional hybrid materials in bulk using diblock copolymers as structure-directing agents for the inorganic materials [80–83]. Hybrid materials based on PS-*b*-P4VP in combination with

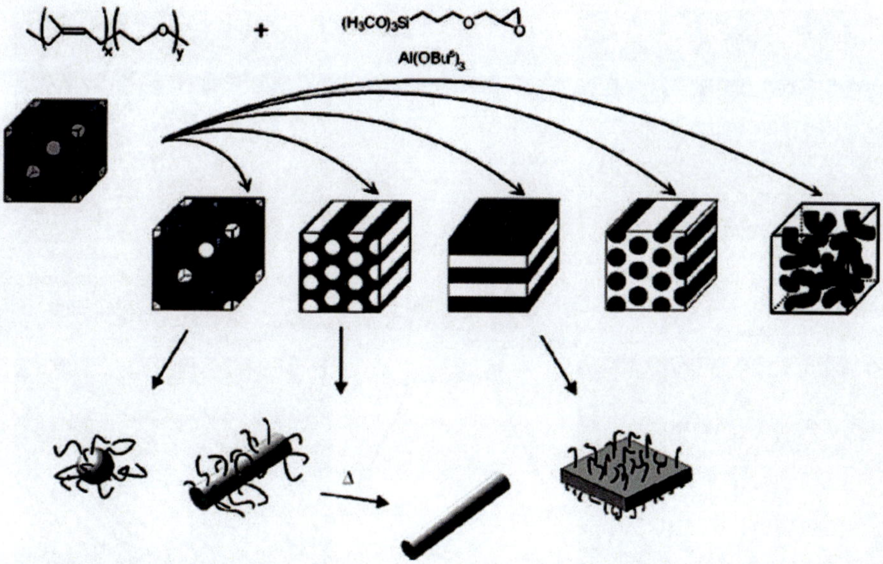

Fig. 8.5 Schematic drawing for the preparation of nano-objects of different shapes using a sol-gel synthesis of metal alkoxide template by use of PI-PEO diblock copolymer (Reproduced from Ref. [77] with permission)

inorganic nanoparticles were synthesized in-situ by incorporating functional elements, such as metallic or semiconductor nanoparticles. The P4VP block has a base character owing to the pyridyl groups selectively coordinating with inorganic precursors in solution. Through evaporation of the solvent, the block copolymer undergoes a process of microphase separation, where the P4VP block carries the precursor to its own nanodomains, separated from the PS domains. Subsequently, the inorganic precursor is reduced to its elemental state in the form of nanoparticles, where the nanoparticles grow while being confined to the P4VP of PS-*b*-P4VP. These blocks act as nanoreactors that control the nanoparticle growth in terms of size and size distribution. Depending on the selected precursor, the inorganic component may exhibit conductive, semiconductive or magnetic properties.

Figure 8.6 shows four hybrid block copolymer morphologies after the incorporation of gold precursor into PS-*b*-P4VP. Varying the volume fraction of P4VP blocks, the hybrid material can adopt different morphologies such as body centered cubic spheres (Fig. 8.6a), hexagonally-packed cylinders (Fig. 8.6b), gyroids (Fig. 8.6c) and lamellar structure (Fig. 8.6d). With characteristic domain spacing between 10 and 100 nm, BCP morphologies can be tuned accurately by changing molecular weight or the volume fraction.

As discussed in the preceding section, it is possible to align block copolymer microdomains by applying external fields like large amplitude oscillating shear (LAOS). Fahmi and co-workers [80] successfully applied this technique to align hybrid materials to improve the macroscopic order. Schematics of the orientation

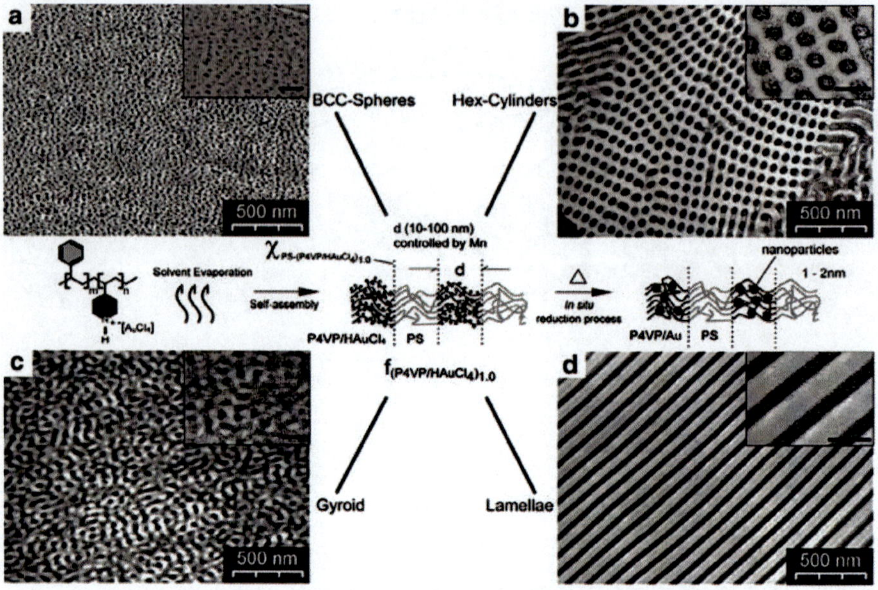

Fig. 8.6 TEM images showing four hybrid diblock-copolymer morphologies. (**a**) Body centered cubic spheres, (**b**) hexagonally packed cylinders, (**c**) gyroids, and (**d**) lamellae. The *scale bars* in all the insets represent 100 nm. *Centre* shows the method to prepare hybrid materials using PS-b-P4VP and a gold precursor (Reproduced from Ref. [83] with permission)

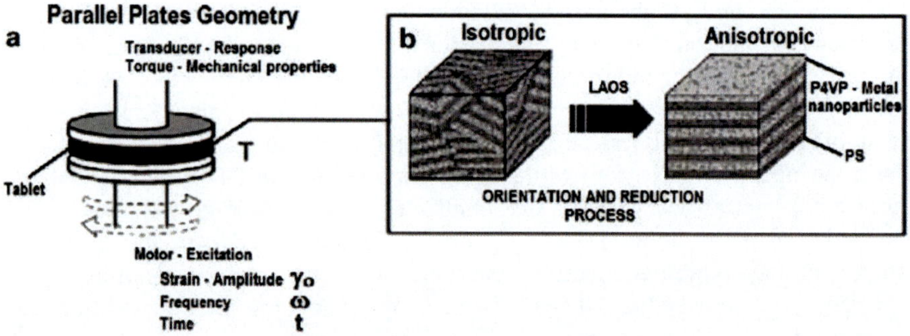

Fig. 8.7 Schematics of the shear orientation process in block copolymers. (**a**) A tablet of the hybrid material is placed in parallel-plate geometry. (**b**) The block copolymer loaded with nanoparticles (which is initially isotropic) is exposed to Large Amplitude Oscillating Shear (LAOS), so that alignment of the polydomain structure takes place (Reproduced from Ref. [80] with permission)

process are shown in Fig. 8.7. The hybrid material (PS-*b*-P4VP + $HAuCl_4$) is placed between parallel plates and mechanical shear is applied by oscillations of the lower plate at temperatures close the glass transition temperature of PS. At these conditions the self assembly of isotropic bulk hybrid materials is enforced and the nanostructures align at a scale up to micrometers. One thus can aim for a single-domain anisotropic material as shown in Fig. 8.7b.

8.4 Generation of Block Copolymer Nanotemplates

Block copolymer films with cylindrical microdomains oriented normal to the surface are attractive for template based applications. The nanoporous templates can be produced by selective etching of the cylindrical domains from the etch resistant majority component. Nanotemplates generated by this method have well defined porosity generally ranging from a few to about 100 nm. These nanotemplates exhibit significant potential for the fabrication of arrays of nanometer-scale structures and for membrane separations. Generation of block copolymer nanotemplates may be roughly categorized into three general methods.

8.4.1 Nanotemplates from Selective Elimination of Sacrificial Microdomain

In this method, the minor phase of the block copolymer microdomains is selectively removed by one of the several methods, such as ultraviolet (UV) irradiation, ozone, oxygen plasma, dry or chemical etching [21, 84–89]. Figure 8.8 shows the schematic representation of this method. Among various nanostructures, perpendicularly oriented cylindrical microdomains of poly (styrene-*b*-methyl methacrylate) (PS-*b*-PMMA) based nanotemplates have been extensively studied. Russell and coworkers [21, 86] proposed the easy removal of PMMA block by exposing the PS-*b*-PMMA thin film under UV radiation, followed by washing with acetic acid. The key advantage of this method is simultaneous crosslinking and degradation of the PS and PMMA, respectively. The nanotemplate obtained by this method is stable even above the glass transition temperature of PS and can be sustained under a very harsh solvent environment.

Chemical etching methods for removing the sacrificial block have also been investigated [88]. For example, hydrolytic degradation of poly (D, L-lactide) from perpendicularly oriented cylindrical microdomains of poly (styrene-*b*-D, L -lactide) (PS-*b*-PLA) leads to uncross-linked nanotemplates. All these methods have the disadvantage to be irreversible.

Fig. 8.8 Schematic representation of the formation of nanotemplates from ordered block copolymers

8.4.2 Nanotemplates from Block Copolymer Thin Film Surface Reconstruction

In this method nanotemplates can be produced without removal of either polymer block by immersing the copolymer film in a selective solvent for the minor component [18, 19, 59, 90]. The solvent selectivity and solubility of the minor component block are crucial to generate nanoporous films with pores that penetrate through the entire film thickness. PS-*b*-PMMA based nanotemplates are generated by immersing the well-ordered PS-*b*-PMMA thin films in acetic acid, a good solvent for PMMA and a non-solvent for PS. Upon drying, the swollen PMMA chains migrate to the air surface and rest on top of the PS matrix, leaving cylindrical nanopores at the positions of the PMMA microdomains [18, 90]. Similar to PMMA-based system, poly (styrene-*b*-4-vinyl pyridine) (PS-*b*-P4VP) thin films are surface reconstructed using ethanol as a preferential solvent (a good solvent for P4VP and a non-solvent for PS) to generate nanoporous films [19, 59]. In the case of PS-*b*-P4VP thin films, the orientation of the cylindrical microdomains can be reversibly switched from the perpendicular to parallel orientation and vice versa upon exposure to appropriate solvent vapor. Dipping these ordered thin films in ethanol, only the P4VP blocks are swollen and move on top of the PS block leaving either nanopores or nanochannels within the PS matrix [59]. Figure 8.9 shows the schematic representation of the formation of nanopores and nanochannels using surface reconstruction method. It should be noted that the surface reconstruction of the film keeps well-developed microdomain structure of the solvent annealed samples without changing the domain size and diameter of the cylindrical microdomains.

This method is fully reversible as the solvent used for surface reconstruction does not alter the chemical structure of the block copolymer. After the formation of nanotemplates, the initial morphology can be recovered by heating the film above the glass transition temperature or annealing the film in an appropriate solvent vapor. The key advantage of this method is the generation of nanoporous materials with functional groups that are readily available for further modification.

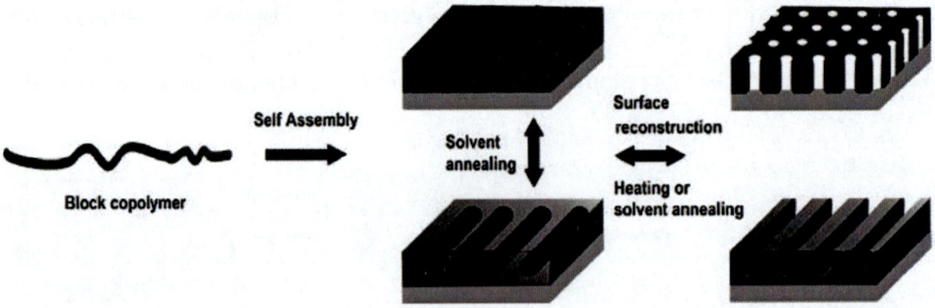

Fig. 8.9 Schematic representation of the generation of nanotemplates from ordered block copolymers using the surface reconstruction method

8.4.3 Nanotemplates from Block Copolymer Supramolecular Assembly

The supramolecular approach (SMA) to block copolymer self-assembly is a simple and powerful technique for fine tuning of the block copolymer morphologies, and has been successfully applied in bulk and in thin films [57, 58, 91–93]. In this approach, a low molar mass additive is associated with one of the blocks by noncovalent interactions. The major advantage of the SMA strategy is that the low molar mass additive can be removed easily from the SMA by selective dissolution to obtain a nanoporous material. Among various types of supramolecular assemblies, those containing hydrogen bonds hold a prominent place in supramolecular chemistry because of their directionality and versatility. The pioneering groups of Ikkala and ten Brinke have demonstrated the preparation of hierarchical polymeric materials through the complexation of 3-pentadecyl phenol (PDP) and PS-*b*-P4VP in bulk by hydrogen bonds [91–93]. The resultant complexes display a structure-within-structure pattern characterized by two length scales, one provided by the copolymer morphology and a smaller one formed by the ordered assembly of PDP.

Our group has demonstrated the formation of smooth thin films from SMA of PS-*b*-P4VP and a non-surfactant molecule, 2-(4′-hydroxybenzeneazo) benzoic acid (HABA) having two different hydrogen-bonding groups [57, 58]. Figure 8.10 shows the schematic representation of the formation of nanotemplates by this methodology. A mixture of a PS-*b*-P4VP block copolymer and HABA containing a one to one molar ratio of acid to pyridine unit was spin cast as a thin film on to a silicon wafer. Depending on the casting solvent, a cylindrical morphology was observed either with the cylinder axis parallel or perpendicular to the substrate. The orientation of the cylindrical microdomains P4VP (HABA), surrounded by PS matrix could be switched by exposure to different solvent vapours. Annealing in chloroform resulted in parallel oriented cylinders, whereas annealing in dioxane resulted in a perpendicular orientation. After fabricating the SMA thin film, HABA can be easily removed by immersing the thin film in ethanol to transform the block

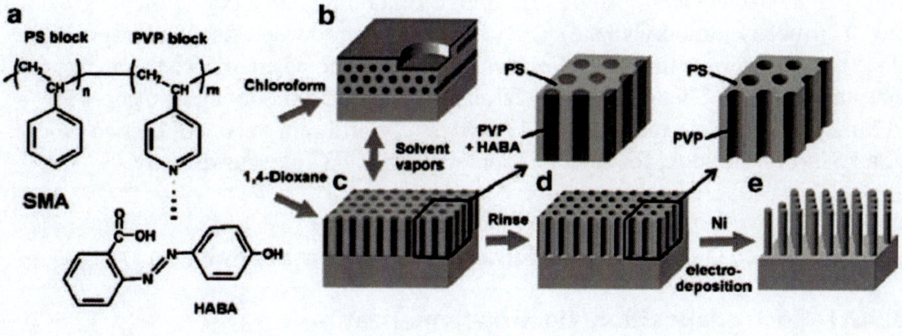

Fig. 8.10 Scheme for the preparation of nanotemplates from PS-b-P4VP: HABA supramolecular assembly (Reproduced from Ref. [57] with permission)

copolymer thin film into a nanotemplate or membrane. These nanotemplates contain functional groups that are readily available for further applications.

8.5 Applications of Block Copolymer Nanotemplates

8.5.1 Nanolithography

Nanoscale structural domain sizes together with differential etch resistance of the polymer blocks are a potentially powerful combination for lithographic technology. Park et al. [94] demonstrated the utilization of poly (styrene-*block*-butadiene) (PS-*b*-PB) or poly (styrene-*block*-imide) (PS-*b*-PIM) block copolymer thin films as lithographic masks to transfer dot and stripe patterns to semiconductor patterns. They demonstrated a periodic array with a density ~10^{11} holes/cm^2 by pattern transfer from spherical microdomains in PS-b-PB or PS-b- PIM to an underlying semiconductor substrate by dry etching, as shown in Fig. 8.11(I). Removal of the PB (or PIM) spheres by ozonation led to the porous structure in a PS matrix, which was further used as a mask to transfer the patterns (pits) into a semiconductor substrate. The inverse structure of semiconductor columns could be prepared by crosslinking the PB (using OsO_4 staining). When plasma is applied, the regions below PB domains are partially masked to generate an array of nanodots.

Later the same group used a trilayer structure to fabricate a metal nanodot array as shown in Fig. 8.11(II) [85]. In this case, the pattern was first transformed into a polyimide-coated silicon nitride film by O_2 RIE to produce an array of nanocolumns. Metal is then deposited on to the pattern where the metal accumulates both on top and bottom of the nanocolumns. By removing the polyimide and silicon nitride, an array of metal nanodots was obtained. This technique has also been used to pattern hexagonally ordered gallium arsenide via a silicon nitride sacrificial layer. The advantage of this trilayer pattern-transfer method is general applicability for nanoscale patterning of different materials on arbitrary substrates.

A simplification of nanolithographic procedures has also been demonstrated by omitting the ozone etching step [95]. Silicon containing block copolymers based on poly (ferrocenyldimethylsilane) (PFS) have been proposed as good candidates, since they can form a thin $Si_xO_yFe_z$ layer on the surface when exposed to an oxygen plasma (O_2-RIE). Cheng et al. [95] fabricated a cobalt nanodot array using a PS-*b*-PFS monolayer. Here the etch rate is sufficiently different between the two blocks of a PS-*b*-PFS diblock, for selective ion etching of PS to occur directly.

8.5.2 Direct Deposition of Functional Nanomaterials

8.5.2.1 Electrodeposition/Electropolymerization

The block copolymer based nanotemplates were used for the fabrication of the ordered array of metallic nanorods by electrodeposition [24, 25, 57]. This process

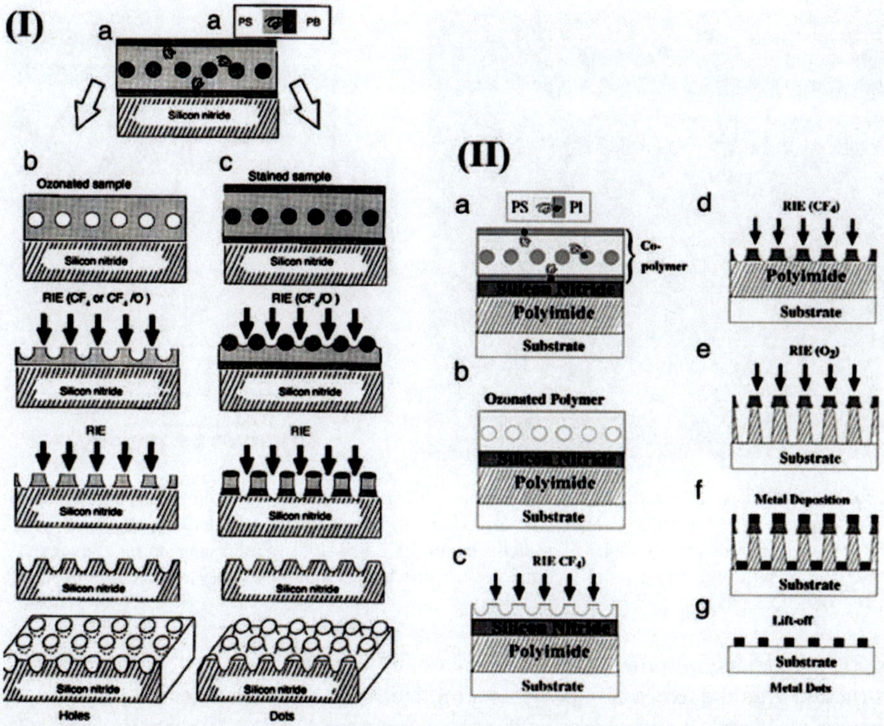

Fig. 8.11 (I) (A) Schematic cross-sectional view of a nanolithography template consisting of a uniform monolayer of PB spherical microdomains on silicon nitride. PB wets the air and substrate interfaces. (B) Schematic of the processing flow when an ozonated copolymer film is used, which produces holes in silicon nitride. (C) Schematic of the processing flow when an osmium-stained copolymer film is used, which produces dots in silicon nitride (Reproduced from Ref. [94] with permission). (II) Schematic of process for production of a metal nanodot array using block copolymer lithography. (a) Cross-sectional view of trilayer structure, (b) selective etching of PIM spherical domains by ozonation, (c) pattern transfer from block copolymer film to silicon nitride by reactive ion etching, (d) reactive ion etching of silicon nitride, (e) further etching, transferring pattern into polyimide, (f) metal evaporation deposition (successively Ti and Au), and (g) polyimide lift-off, along with nitride and metal layers, in a solvent bath (Reproduced from Ref. [85] with permission)

involves the motion of charged nanoparticles in solution under the influence of an electric field and subsequent deposition of the nanoparticles onto an electrode surface. Thurn-Albrecht et al. [24] fabricated ultrahigh-density cobalt (Co) nanowires arrays using PS-*b*-PMMA based block copolymer nanotemplates from a methanol solution. Because of the high aspect ratio and short distance between the adjacent Co nanowires; single magnetic domain behavior was anticipated, with obvious potential for patterned magnetic storage media. In another example, Sidorenko et al. [57] filled SMA based nanotemplates with nickel clusters via electrodeposition (Fig. 8.10). Figure 8.12 shows an AFM height image of the nickel

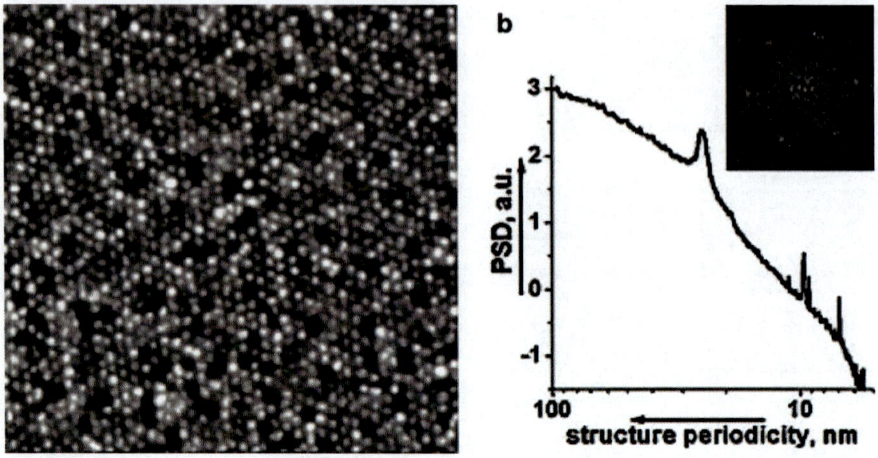

Fig. 8.12 Ni dots electrodeposited throughout 45 nm thick SMA nanotemplates, lateral scale 1 × 1 μm^2: (**a**) topography image, z scale 30 nm, (**b**) power spectrum density, the main peak (24 nm) corresponds to the SMA periodicity (inset, FFT image of (**a**) showing perfect hexagonal ordering of Ni dots). Occasional lacunas appear due to the inhomogeneity of electrodeposition (Reproduced from Ref. [57] with permission)

dots, corresponding power spectral density and fast Fourier transformation of the structure after the removal of polymer template. With 8 nm diameter and an average height of 25 nm, the produced rods provide an aspect ratio of approximately 1:3. Recently Steiner and co-workers demonstrated the generation of freestanding copper oxide nanowire arrays using electrodeposition and subsequent removal of the block copolymer template [25]. They found that the method used for template removal influences the structural stability of the freestanding array: the wires bunch together with solvent dissolution, whereas the UV treatment leads to less clustering of the wires.

High-density arrays of conducting polymer nanorods were also produced inside porous diblock copolymer templates [96, 97]. Russell and co-workers fabricated high-density arrays of polypyrrole nanorods (10^{11} pores/cm^2) by the electropolymerizing of pyrrole inside the pores of the PS-*b*-PMMA nanotemplates [96]. Our group produced dense arrays of polyaniline nanorods (Fig. 8.13) with 10 nm diameter on transparent ITO substrate via electropolymerization using supramolecular assemblies of block copolymers as scaffold material [97]. By washing with chloroform, the template is completely removed, leaving a self-supporting array of conducting polymers oriented normally to the substrate. The *I–V* characteristics of the individual polyaniline nanorods show semiconducting behavior. These ordered arrays of polyaniline nanorods were found to exhibit excellent electrochemical properties with an electrochemical capacitance value of 3,407 F g^{-1}, which is almost 11 times higher than that of the polyaniline thin films (299 F g^{-1}) deposited on bare ITO under identical conditions.

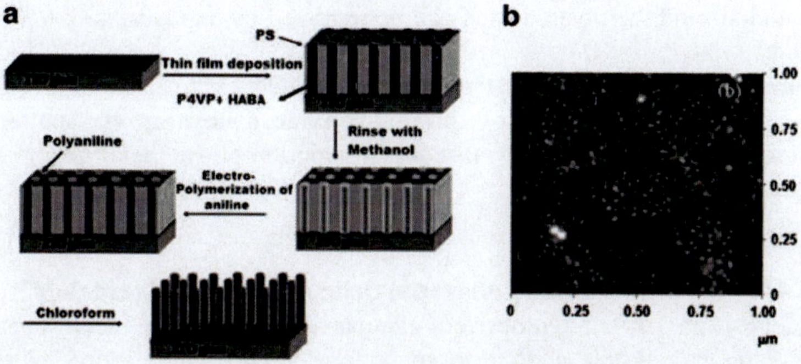

Fig. 8.13 (**a**) Schematic presentation of fabrication of polyaniline nanorods using block copolymer nanoporous template. (**b**) AFM height image of polyaniline nanorods

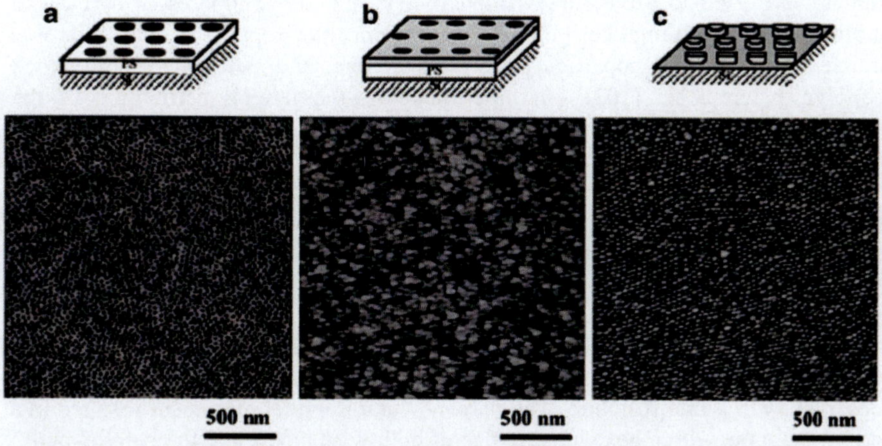

Fig. 8.14 Schematic diagram of the fabrication process for Cr dot arrays (*upper pictures*) and height images of tapping mode AFM of each step (*lower pictures*): (**a**) nanoscopic holes in cross-linked PS matrix, (**b**) evaporated Cr onto the PS template, and (**c**) Cr nanodot arrays. The height range of the AFM images is 10 nm (Reproduced from Ref. [98] with permission)

8.5.2.2 Physical Vapor Deposition

Russell and co-workers [98] fabricated high density arrays of chromium (Cr) and layered gold/chromium (Au/Cr) nanodots by evaporation of the metals onto nanoporous templates produced by the self-assembly of PS-*b*-PMMA diblock copolymers (Fig. 8.14). Evaporation of Cr onto the template followed by sonication and UV degradation of the PS, left Cr nanodots. Subsequent deposition of Au created an array of Cr/Au multi-layered dots. The inverse structure (porous metal arrays) was prepared by using a PS – cylinder forming PS-*b*-PMMA diblock copolymer, etching the PMMA to create PS posts, and then evaporating Cr onto the film and around the posts, which can be removed by exposure to UV followed by rinsing.

Stamm and co-workers used templates fabricated from SMA of PS-*b*-P4VP and HABA for the generation of chromium oxide nanowires via sputtering of chromium onto the template and subsequent oxidation of the metal under ambient conditions [99].

8.5.2.3 Direct Deposition of Nanoparticles/Polymeric Materials

Direct deposition of pre-synthesized nanoparticles onto the block copolymer nanotemplates is a simple method to generate ordered particle arrays. This route is effective for controlling size distribution, shape, and spatial distribution of nanoscopic objects. Russell and co-workers have utilized PS-*b*-PMMA block copolymer thin films, whereby the minority component PMMA was removed by UV photodegradation to form a nanoporous template. In one method, capillary forces were used to drive CdSe nanoparticles into the nanopores of cylindrical diblock copolymer templates [100]. In another one, electrophoretic deposition was utilized to drive the nanoparticles into the nanopores and nanotrenches of diblock copolymer templates [101]. In these methods, the lateral distribution of the nanoparticles into the nanopores was only manipulated by physical forces such as capillary forces or electric fields.

Recently, we demonstrated a simple approach to fabricate highly ordered arrays of nanoscopic palladium dots and wires (Fig. 8.15a) by the direct deposition of pre-synthesized palladium nanoparticles in aqueous solution [59, 101]. In this method, the cylindrical morphology adopted in the thin films of PS-*b*-P4VP could be switched from parallel to perpendicular and vice-versa by annealing in the vapor of appropriate solvents. When these films were immersed into ethanol, a good solvent for P4VP and a non-solvent for PS, a surface reconstruction of the films was observed with a fine structure. The perpendicular cylinder alignment resulted in a nanomembrane with a hexagonal lattice of hollow channels and the parallel cylinder alignment was turned into nanochannels. Figure 8.15b shows the AFM height images of nanopores and nanochannels after surface reconstruction in ethanol. In these templates, the pore or channel walls are formed by the reactive P4VP chains. Hence, two possible driving forces can help filling the nanoparticles inside the pores or channels. The first one is the capillary force which allows the nanoparticle solution to enter inside the pores and the second is the preferential interaction between P4VP chains and Pd nanoparticles, which is the key in holding the nanoparticles tightly inside the pores or channels. A subsequent stabilization by UV-irradiation followed by pyrolysis in air at 450°C removes the polymer to produce highly ordered metallic nanostructures. Figure 8.15c shows the AFM height images of palladium nanodots and nanowires after the removal of the template. This method is highly versatile as the procedure used is simple and provides a facile approach to fabricate a broad range of nanoscaled architectures with tunable lateral spacing. It can be extended to systems with even smaller dimensions.

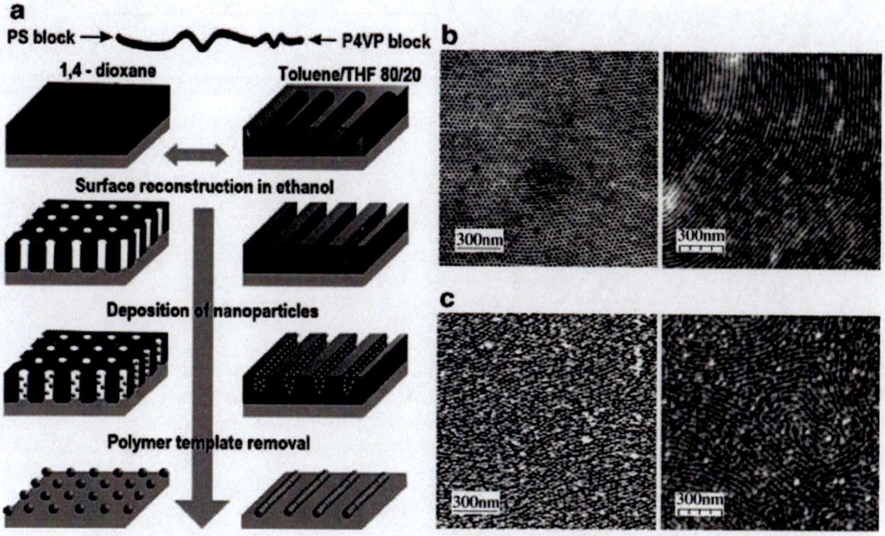

Fig. 8.15 (a) Schematic diagram of the fabrication process of highly ordered arrays of nanoscopic palladium dots and wires using block copolymer templates. (b) AFM height images of PS-*b*-P4VP nanotemplates (nanopores and nanochannels) obtained after surface reconstruction in ethanol. (c) AFM height images of palladium nanodots and nanowires obtained after the polymer removal

We also used a PS-*b*-P4VP copolymer thin film as a template for patterning noble metal nanoparticles such as gold, platinum, and palladium [102]. In this method, hexagonally packed cylindrical morphology with the P4VP microdomain dispersed in a polystyrene matrix was directly immersed in aqueous solution of nanoparticles. The pre-synthesized inorganic nanoparticles selectively bind to the P4VP domains and the nonpolar matrix PS preserves the 3-D structure of the particles. The polymer is removed by pyrolysis or oxygen plasma etching to leave an inorganic nanopattern with identical cylindrical dimensions.

In another example, we have also deposited semiconducting nanoparticles like CdS on the block copolymer templates [103]. In this case, nanotemplates are prepared from supramolecular assemblies (SMA) of PS-*b*-P4VP and 1-pyrenebutyric acid (PBA), consisting of lamellar microdomains formed by P4VP (PBA) surrounded by PS. By annealing the SMA complex in a selective solvent leads to perpendicular orientation of the lamella. The PBA was selectively extracted from P4VP and PBA microdomains with selective solvent to generate SMA templates. These templates were directly dipped into aqueous solution of cadmium acetate for 4 h. The P4VP chains, which are at the pore walls of the templates coordinate with Cd^{2+} ion. These block copolymer templates coordinated with Cd^{2+} ion are directly dipped into aqueous solution of thioacetamide which acts as a source of S^{2-} ions and produce cadmium sulfide nanoparticles stabilized by the pyridine ring of the P4VP chain. Excess of cadmium acetate molecules, which are loosely bound to the surface of SMA template, will go to thioacetamide solution.

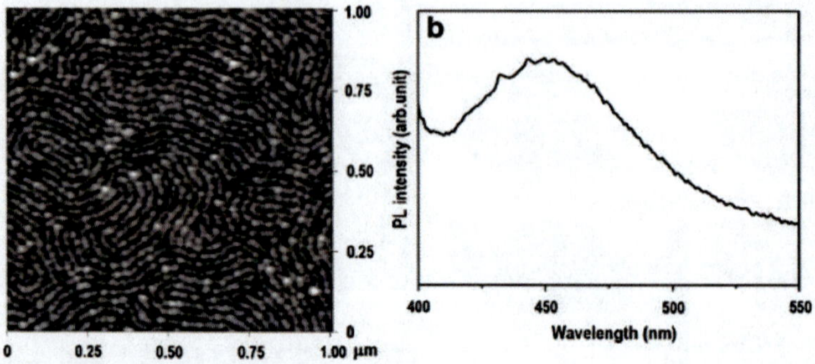

Fig. 8.16 (a) AFM height image of SMA thin film after the deposition of CdS. (b) Photoluminescence spectrum of thin film fabricated from SMA after CdS deposition. The excitation wavelength is 320 nm (Reproduced from Ref. [103] with permission)

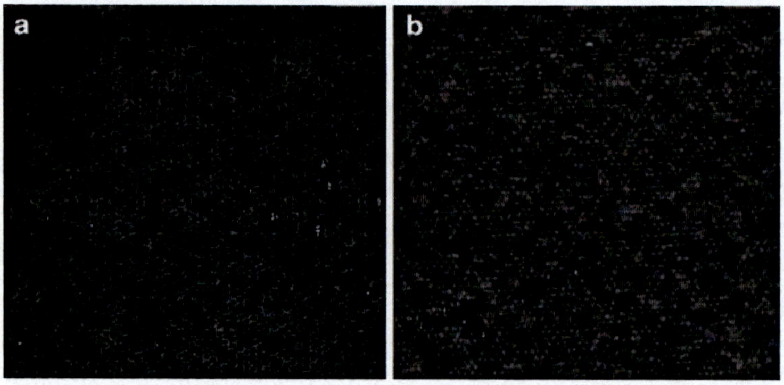

Fig. 8.17 AFM height images of (a) nanoporous SMA thin film (b) highly ordered polymeric nanodots arrays after pyrolysis. Lateral scale 1,500 × 1,500 nm (Reproduced from Ref. [104] with permission)

Figure 8.16a shows the AFM image of the SMA template after CdS deposition. The photoluminescence spectrum of thin film fabricated from SMA after CdS deposition is shown in Fig. 8.16b. The peak at around 450 nm in the spectrum confirms the presence of CdS nanoparticles in the template.

The SMA templates fabricated from SMA of PS-*b*-P4VP and HABA could also be used for patterning polymeric materials [104]. Under influence of interpolymer hydrogen bonding and capillary action of nanopores, a SMA template (Fig. 8.17a) was properly filled with phenolic resin precursor, followed by curing and pyrolysis at medium temperatures to remove the nanotemplate. As a result, polymer nanodot arrays were obtained with spacing below 30 nm (Fig. 8.17b). When additional process steps are added, new structures can be obtained from the templates. For example silica nanodots, which were prepared by pyrolysis of a SMA template loaded with a precursor species, were used to guide the dewetting of a phenolic resin precursor thin film. Curing and calcination of the phenolic precursor, followed

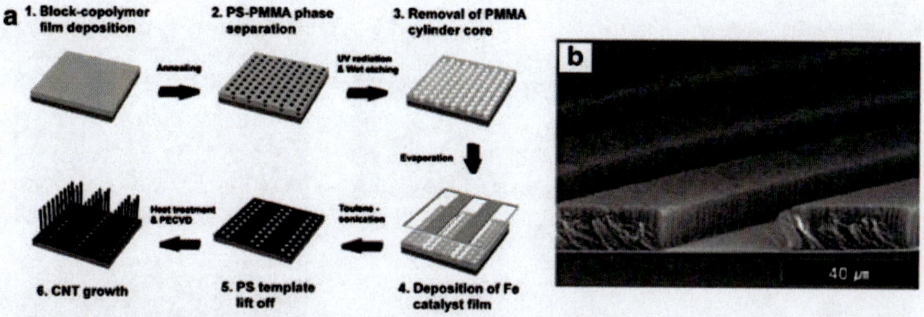

Fig. 8.18 (a) Schematic depiction of the hierarchical organization process used to fabricate vertical CNT arrays. (b) Cross-sectional SEM image for CNT arrays prepared using a parallel grid (Reproduced from Ref. [106] with permission)

by etching of the silica arrays, resulting in large area carbon nanoring arrays with a diameter as small as 25 nm [105].

8.5.2.4 Fabrication of Carbon Nanotube Arrays

Kim and co-workers fabricated hierarchically organized vertical carbon nanotube arrays (CNT) by combining plasma-enhanced chemical vapor deposition (PECVD) with self-assembled block copolymer templates [106]. The cylindrical assemblies of PS-*b*-PMMA have been used to generate a nanoporous block copolymer template. An iron catalyst was deposited within the cylindrical pores of the nanotemplates using copper grid masks and the PS template was subsequently removed to have an array of catalyst particles on the substrate. After catalyst deposition, heat treatment at 750°C has been used to further reduce the size of the catalyst particles. The PECVD growth of CNTs from laterally patterned catalyst arrays yields highly oriented vertical CNTs. The entire fabrication process is schematically depicted in Fig. 8.18a and as an example, the SEM image of hierarchically ordered CNT arrays prepared by employing a parallel grid mask is shown in Fig. 8.18b. The diameter and position of the fabricated CNTs can be determined by the size and lateral distribution of the catalyst particles.

8.5.3 Nanoporous Membranes

8.5.3.1 Filtration of Viruses

Separation and purification of viruses are important processes in the biotechnology industry. Micro- and ultrafiltration have been used for virus separation, but they are not very effective as these membranes still allow smaller-sized virus particles to permeate into a small number of abnormally large-sized pores in the membrane [107]. This broad size distribution of pores in ultrafiltration membranes and the low density of pores in track-etched membranes limit the practical use of virus filtration. Kim and co-workers successfully employed block copolymer based nanoporous films for the filtration of viruses [108]. These membranes exhibit narrow size

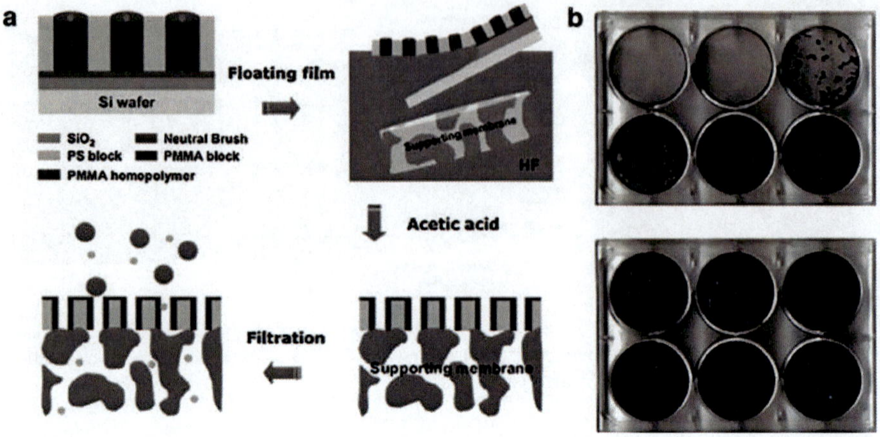

Fig. 8.19 (**a**) Schematic depiction of the procedure for the fabrication of nanoporous membrane consisting of block copolymer thin film and supporting membrane. (**b**) Plaque assays of HRV 14 solution before filtration (*top*) and after filtration through the block copolymer membrane (*bottom*) (Reproduced from Ref. [108] with permission)

distribution of pores but have limitations such as low mechanical and chemical stability in order to be widely used for effective virus filtration. A supporting membrane was used to provide the mechanical strength. Figure 8.19a shows the schematic illustration of the fabrication of nanoporous membranes. The template consists of an 80 nm thick nanoporous layer, prepared from a PS-b-PMMA block copolymer, and a supported microporous polysulfone membrane that provides mechanical strength. This composite membrane did not show any damage or crack even at a pressure of 2 bar, while high selectivity was maintained for the filtration of human rhinovirus type 14 (HRV 14) which has a diameter of ~30 nm and is a major pathogen of the common cold in humans. They also showed via plaque assay test that none of the 2.5 × 105 plaque forming units of HRV 14 in the phosphate buffered saline solution passed through this 2.5 cm diameter membrane (Fig. 8.19b).

8.5.3.2 Drug Delivery

Block copolymer nanotemplates with uniform pore size are desirable materials for controlled separations due to their pore size tunability, narrow size distributions of pores, and ability for selective functionalization. The template based drug delivery can minimize the denaturation of protein drugs due to its passive diffusion nature, and one can induce rate-limiting diffusion with constant release by tuning the pore sizes. Yang et al. have demonstrated the application of block copolymer nanotemplates for controlled drug release, and they were successful in achieving long-term controlled release of protein drugs by incorporating a nanoporous template into a drug-eluting device [109]. According to the hydrodynamic diameter of a target protein drug, the pore size was precisely controlled by gold deposition. By exploiting a single-file diffusion mechanism through the resulting cylindrical nanochannels, they obtained long-term controlled drug release of both bovine serum albumin (BSA) and human growth hormone (hGH) as model protein drugs with different sizes for at least 2 months in vitro. They showed that the release rate of protein drugs could also be controlled by changing the length of block copolymer nanochannels and the thickness of the Au deposition layer.

Uehara and co-workers showed controlled diffusion of BSA and glucose in a series of nanoporous polyethylene membranes [110]. These membranes were prepared by fuming nitric acid etching of polystyrene from a polyethylene-*block*-polystyrene (PE-PS) copolymer. Controlling the PS etching time resulted in controlled pore sizes and allowed selective transport of glucose over BSA. Nuxoll et al. [111] developed a composite membrane by integrating a block copolymer thin film with a 100 μm thick macroporous silicon support for the size based transport selectivity. The block copolymer layer, which contained perpendicularly oriented cylindrical channels, was prepared by spin-casting of a polystyrene-block-polyisoprene-block-polylactide (PS-PI-PLA) triblock terpolymer onto a silicon support followed by etching the PLA with a dilute base to give a mechanically robust nanoporous film. They have demonstrated that small molecules could easily pass through, while diffusion of the large macromolecule dextran blue was hindered.

8.6 Summary and Outlook

In this chapter we have attempted to cover the recent advances in the formation of nanostructures in bulk, generation of block copolymer based nanotemplates and their potential applications in nanofabrication including template nanolithography, and deposition of functional nanomaterials as well as nanoporous membranes. Most of these applications have by now been demonstrated in the laboratory, and it would now appear that real applications are under active development. For practical applications, especially in semiconductor and data storage industries, control over the long range order of the structures is very important, and it is necessary to control the long range order in block copolymer thin films via more user-friendly processes. Other applications such as a nanoporous membrane would appear highly interesting for controlled separations. These membranes are also promising materials for water purification. For this purpose, mechanical strength of the nanoporous films must be improved during fabrication processes. Use of supramolecular complexes with varieties of low molecular weight additives, for example liquid crystalline additives or metal containing additives, has not yet been explored fully and offers scope for further work. More research using more complex multiblock polymers could provide new information on how to exploit the morphologies and functionalities of these systems in thin films, and these new functional nanomaterials may open new directions.

Acknowledgements EBG is grateful to the Department of Science and Technology (Government of India) for the award of a Ramanujan Fellowship and also the Alexander von Humboldt foundation (Germany) for the research fellowship.

References

1. Bratton D, Yang D, Dai J, Ober CK. Polym Adv Technol. 2006;17:94.
2. Gates BD, Xu Q, Stewart M, Ryan D, Willson CG, Whitesides GM. Chem Rev. 2005;105:1171.
3. Ito T, Okazaki S. Nature. 2000;406:1027.
4. Chou SY, Keimel C, Gu J. Nature. 2002;417:835.
5. Kim SO, Solak HH, Stoykovich MP, Ferrier NJ, de Pablo JJ, Nealey PF. Nature. 2003;424:411.
6. Craighead HG. Science. 2003;290:1532.
7. Guo LJ. Adv Mater. 2007;19:495.
8. Bhushan B. Handbook of nanotechnology. Heidelberg: Springer; 2004.
9. Whitesides GM, Kriebel JK, Mayers BT. In: Huck WTS, editor. Nanoscale assembly. USA: Springer; 2005. Ch. 9.
10. Hamley IW. Angew Chem Int Ed. 2003;42:1692.
11. Whitesides GM, Mathias JP, Sato CT. Science. 1991;254:1312.
12. Kim H-C, Park S-M, Hinsberg WD. Chem Rev. 2010;110:146.
13. Hamley IW. The physics of block copolymers. Oxford: Oxford University Press; 1998.
14. Hamley IW. Prog Polym Sci. 2009;34:1161.
15. Stoykovich MP, Nealey PF. Mater Today. 2006;9:20.
16. Krishnamoorthy S, Hinderling C, Heinzelmann H. Mater Today. 2006;9:40.
17. Li M, Ober CK. Mater Today. 2006;9:30.
18. Xu T, Stevens J, Villa J, Goldbach JT, Guarini KW, Black CT, Hawker CJ, Russell TP. Adv Funct Mater. 2003;13:698.
19. Park S, Wang JY, Kim B, Xu J, Russell TP. ACS Nano. 2008;2:766.
20. Mansky P, Harrison CK, Chaikin PM, Register RA, Yao N. Appl Phys Lett. 1996;68:2586.
21. Thurn-Albrecht T, Steiner R, DeRouchey J, Stafford CM, Huang E, Bal M, Tuominen M, Hawker CJ, Russell TP. Adv Mater. 2000;12:787.
22. Minelli C, Hinderling C, Heinzelmann H, Pugin R, Liley M. Langmuir. 2005;21:7080.
23. Lopes WA, Jaeger HM. Nature. 2001;414:735.
24. Thurn-Albrecht T, Schotter J, Kastle GA, Emley N, Shibauchi T, Krusin-Elbaum L, Guarini K, Black CT, Tuominen MT, Russell TP. Science. 2000;290:2126.
25. Crossland EJW, Ludwigs S, Hillmyer MA, Steiner U. Soft Matter. 2007;3:94.
26. Hamley IW. Nanotechnology. 2003;14:R39.
27. Darling SB, Yufa NA, Cisse AL, Bader SD, Sibener SJ. Adv Mater. 2005;17:2446.
28. Segalman RA. Mater Sci Eng Rep. 2005;48:191.
29. Darling SB. Prog Polym Sci. 2007;32:1152.
30. Ansari IA, Hamley IW. J Mater Chem. 2003;13:2412.
31. Kim DH, Kim SH, Lavery K, Russell TP. Nano Lett. 2004;4:1841.
32. Bates FS, Fredrickson GH. Annu Rev Phys Chem. 1990;41:525.
33. Matsen MW, Bates FS. Macromolecules. 1996;29:1091.
34. Bates FS, Fredrickson GH. Phys Today. 1999;52:32.
35. Lynd NA, Meuler AJ, Hillmyer MA. Prog Polym Sci. 2008;33:875.
36. Russell TP, Coulon G, Deline VR, Miller DC. Macromolecules. 1989;22:4600.
37. Coulon G, Russell TP, Deline VR, Green PF. Macromolecules. 1989;22:2581.
38. Segalman RA, Schaefer KE, Fredrickson GH, Kramer EJ, Magonov SN. Macromolecules. 2003;36:4498.
39. Coulon G, Collin B, Ausserre D, Chatenay D, Russell TP. J Phys Fr. 1990;51:2801.

40. Fasoka MJ, Banerjee P, Mayes AM, Pickett G, Balazs AC. Macromolecules. 2000;33:5702.
41. Kim G, Libera M. Macromolecules. 1998;31:2670.
42. Kim G, Libera M. Macromolecules. 1998;31:2569.
43. Kim SH, Misner MJ, Russell TP. Adv Mater. 2004;16:2119.
44. Kim SH, Misner MJ, Xu T, Kimura M, Russell TP. Adv Mater. 2004;16:226.
45. Kimura M, Misner MJ, Xu T, Kim SH, Russell TP. Langmuir. 2003;19:9910.
46. Stoykovich MP, Mueller M, Kim SO, Solak HH, Edwards EW, de Pablo JJ, Nealey PF. Science. 2005;308:1442.
47. Segalman RA, Yokoyama H, Kramer EJ. Adv Mater. 2001;13:1152.
48. Bita I, Yang JKF, Jung YS, Ross CA, Thomas EL, Berggren KK. Science. 2008;321:939.
49. Kumar A, Whitesides GM. Appl Phys Lett. 1993;63:2002.
50. Angelescu DE, Waller JH, Adamson DH, Deshpande P, Chou SY, Register RA, Chaikin PM. Adv Mater. 2004;16:1736.
51. Villar MA, Rueda DR, Ania F, Thomas EL. Polymer. 2002;43:5139.
52. Park C, Rosa C, Thomas EL. Macromolecules. 2001;34:2602.
53. Bodycomb J, Funaki Y, Kimishima K, Hashimoto T. Macromolecules. 1999;32:2075.
54. Kellogg GT, Walton DG, Mayes AM, Lambooy P, Russell TP, Gallagher PD, Satija SK. Phys Rev Lett. 1996;76:2503.
55. Sivaniah E, Hayashi Y, Iino M, Hashimoto T. Macromolecules. 2003;36:5894.
56. Jeong U, Ryu DY, Kho DH, Kim JK, Goldbach JT, Kim DH, Russell TP. Adv Mater. 2004;16:533.
57. Sidorenko A, Tokarev I, Minko S, Stamm M. J Am Chem Soc. 2003;125:12211.
58. Tokarev I, Krenek R, Brukov Y, Schmeisser D, Sidorenko A, Minko S, Stamm M. Macromolecules. 2005;38:507.
59. Gowd EB, Nandan B, Vyas MK, Bigall NC, Eychmüller A, Schlorb H, Stamm M. Nanotechnology. 2009;20:415302.
60. Park S, Wang J-Y, Kim B, Chen W, Russell TP. Macromolecules. 2007;40:9059.
61. Li M, Coenjarts CA, Ober CK. Adv Polym Sci. 2005;190:183.
62. Kim JK, Lee JI, Lee DH. Macromol Res. 2008;16:267.
63. Marencic AP, Register RA. Annu Rev Chem Biomol Eng. 2010;1:277.
64. Angelescu DE, Waller JH, Register RA, Chaikin PM. Adv Mater. 2005;17:1878.
65. Hashimoto T, Bodycomb J, Funaki Y, Kimishima K. Macromolecules. 1999;32:952.
66. Segalman RA, Hexemer A, Hayward RC, Kramer EJ. Macromolecules. 2003;36:3272.
67. Park S, Lee DH, Xu J, Kim B, Hong SW, Jeong U, Xu T, Russell TP. Science. 2009;323:1030.
68. Park SM, Craig GSW, La YH, Solak HH, Nealey PF. Macromolecules. 2007;40:5084.
69. Tang C, Lennon EM, Fredrickson GH, Kramer EJ, Hawker CJ. Science. 2008;322:429.
70. Massey JA, Winnik MA, Manners I, Chan VZH, Ostermann JM, Enchelmaier R, Spatz JP, Moller M. J Am Chem Soc. 2001;123:3147.
71. Lin Y, Boker A, He J, Sill K, Xiang H, Abetz C, Li X, Wang J, Emrick T, Long S, Wang Q, Balazs A, Russell TP. Nature. 2005;434:55.
72. Thomas JR. J Appl Phys. 1996;37:2914.
73. Grubbs RB. J Polym Sci A Polym Chem. 2005;43:4323.
74. Chan YNC, Schrock RR, Cohen RE. J Am Chem Soc. 1992;114:7295.
75. Spatz JP, Mossmer S, Moller M. Chem Eur J. 1996;2:1552.
76. Mossmer S, Spatz JP, Moller M, Aberle T, Schmidt J, Burchard W. Macromolecules. 2000;33:4791.
77. Simon PFW, Ulrich R, Spiess HW, Wiesner U. Chem Mater. 2001;13:3464.
78. Warren SC, Disalvo JF, Wiesner U. Nat Mater. 2007;6:248.
79. Jain A, Wiesner U. Macromolecules. 2004;37:5665.
80. Mendoza C, Pietsch T, Fahmi AW, Gindy N. Adv Mater. 2008;20:1179.
81. Mendoza C, Gindy N, Gutmann JS, Fromsdorf A, Forster S, Fahmi A. Langmuir. 2009;25:9571.

82. Mendoza C, Pietsch T, Gutmann JS, Jehnichen D, Gindy N, Fahmi A. Macromolecules. 2009;42:1203.
83. Fahmi A, Pietsch T, Mendoza C, Chevel N. Mater Today. 2009;12:44.
84. Lee JS, Hirao A, Nakahama S. Macromolecules. 1998;21:274.
85. Park M, Chaikin PM, Register RA, Adamson DA. Appl Phys Lett. 2001;79:257.
86. Thurn-Albrecht T, De Rouchey J, Russell TP, Jaeger HM. Macromolecules. 2000;33:3250.
87. Chan VZH, Hoffman J, Lee VY, Iatrou H, Avgeropoulos A, Hadjichristidis N, Miller RD, Thomas EL. Science. 1999;286:1716.
88. Zalusky AS, Olayo-Valles R, Wolf JH, Hillmyer MA. J Am Chem Soc. 2002;124:12761.
89. Mäki-Ontto R, de Moel K, de Odorico W, Ruokolainen J, Stamm M, ten Brinke G, Ikkala O. Adv Mater. 2001;13:117.
90. Xu T, Goldbach JT, Misner MJ, Kim S, Gibaud A, Gang O, Ocko B, Guarini KW, Black CT, Hawker CJ, Russell TP. Macromolecules. 2004;37:2972.
91. Ikkala O, ten Brinke G. Science. 2002;295:2407.
92. Ruokolainen J, Makinen R, Torkkeli M, Makela T, Serimaa R, ten Brike G, Ikkala O. Science. 1998;280:557.
93. Valkama S, Ruotsalainen T, Nykanen A, Laiho A, Kosonen H, ten Brinke G, Ikkala O, Ruokolainen J. Macromolecules. 2006;39:9327.
94. Park M, Harrison C, Chaikin PM, Register RA, Adamson DH. Science. 1997;276:1401.
95. Cheng YJ, Ross CA, Chan VZ-H, Thomas EL, Lammerink RGH, Vancso GJ. Adv Mater. 2001;13:1174.
96. Lee JI, Cho SH, Park SM, Kim JK, Yu JW, Kim YC, Russell TP. Nano Lett. 2008;8:2315.
97. Kuila BK, Nandan B, Bohme M, Janke A, Stamm M. Chem Commun. 2009;14:5749.
98. Shin K, Leach KA, Goldbach JT, Kim DH, Jho JY, Tuominen MT, Hawker CJ, Russell TP. Nano Lett. 2002;2:933.
99. Seifarth O, Schmeißer D, Krenek R, Sydorenko A, Stamm M. Prog Solid State Chem. 2006;34:111.
100. Misner MJ, Skaff H, Emrick T, Russell TP. Adv Mater. 2003;15:221.
101. Nandan B, Gowd EB, Bigall NC, Eychmüller A, Formanek P, Simon P, Stamm M. Adv Funct Mater. 2009;19:2805.
102. Gowd EB, Nandan B, Bigall NC, Eychmüller A, Formanek P, Stamm M. Polymer. 2010;51:2661.
103. Kuila BK, Gowd EB, Stamm M. Macromolecules. 2010;43:7713.
104. Liu X, Stamm M. Nanoscale Res Lett. 2009;4:459.
105. Liu X, Stamm M. Macromol Rapid Commun. 2009;30:1345.
106. Lee DH, Shin DO, Lee WJ, Kim SO. Adv Mater. 2008;20:2480.
107. Urase T, Yamamoto K, Ohgaki S. J Membr Sci. 1996;115:21.
108. Yang SY, Ryu I, Kim HY, Kim JK, Jang SK, Russell TP. Adv Mater. 2006;18:709.
109. Yang S, Yang JA, Kim ES, Jeon G, Oh EJ, Choi KY, Hahn SK, Kim JK. ACS Nano. 2010;4:3817.
110. Uehara H, Kakiage M, Sekiya M, Sakuma D, Yamonobe T, Takano N, Barraud A, Meurville E, Ryser P. ACS Nano. 2009;3:924.
111. Nuxoll EE, Hillmyer MA, Wang R, Leighton C, Siegel RA. ACS Appl Mater Interface. 2009;4:888.

Epitaxial Growth of Metals on Semiconductors Via Electrodeposition

9

Karen L. Kavanagh

Abstract

This chapter reviews the literature on the epitaxial growth of metals on semiconductors by electrodeposition. The known examples for Si and GaAs are described with results from in-situ characterization of their surfaces prior and during metal growth in aqueous electrolytes. The application of electrodeposition to semiconductor nanowire contact formation is introduced.

9.1 Introduction

Epitaxy is the alignment of a growing crystal using the substrate as a template. Success is measured by the degree of perfection of the new crystal, and how well it mimics the crystallinity, and structure of the substrate. Epitaxial growth is easiest, therefore, when there is a match in lattice spacing of the two materials forming the new interface. Best results occur with clean surfaces.

Semiconductor surfaces free from amorphous native oxides or other contaminants are feasible in vacuum or inert gas environments. Crystal growth techniques such as molecular beam epitaxy (MBE), metal organic vapour phase epitaxy (MOVPE), or liquid phase epitaxy (LPE), carried out in ultra-high-vacuum, flowing gas, and molten solutions, respectively, are already well known for their fabrication of single crystalline, metal-semiconductor and semiconductor-semiconductor heterostructures [1]. Such interfaces are fundamental to efficient electronic device designs, as well as to our understanding of the underlying carrier transport mechanisms.

Our focus in this chapter is epitaxial growth of metals directly on semiconductors via electrodeposition. The semiconductor surface preparation and

K.L. Kavanagh (✉)
Department of Physics, Simon Fraser University, Burnaby, BC, Canada
e-mail: kavanagh@sfu.ca

M. Stepanova and S. Dew (eds.), *Nanofabrication*,
DOI 10.1007/978-3-7091-0424-8_9, © Springer-Verlag/Wien 2012

metal growth occurs within an electrically conductive solution, the electrolyte, containing metal ions. Growth is determined by electronic and ionic current through the interface and deposit, driving surface reactions. We are interested primarily in aqueous electrolytes operating at temperatures feasible with liquid water. Electrodeposition is a technique that is more than 100 years old [2, 3]. It is widely used currently to deposit polycrystalline metallic layers on conducting surfaces that are not necessarily flat including automotive and aerospace components. The capability to deposit only where electrical conduction occurs is a unique advantage. The fabrication of Cu interconnections for integrated circuits has been carried out with increasing sophistication for many years via electrodeposition [4]. With the recent growing interest in nanoscale fabrication, the advantages of electrodeposited contacts to semiconductor nanowires, for example, is being investigated [5, 6].

In general, electrodeposition offers a less complex and lower cost process than the vacuum or gas flow techniques mentioned above. Chemical preparation of semiconductor surfaces and ultra-high purity water sources are well developed. Maintenance of semiconductor surface purity in water while epitaxial nucleation and growth is occurring becomes the challenge.

We will describe the known examples of epitaxial growth of metals on conventional semiconductors using electrodeposition. Like other methods, understanding how to identify and control semiconductor surface reactions, in this case in water, is the first step. Epitaxial nucleation and growth is then a measure of the successful control of surface purity in this highly reactive environment. The application of this technique towards planar and nanostructural semiconductor contacts will be discussed.

9.2 Electrodeposition Procedures

Electrodeposition of a metal is essentially the reaction of positive metal ions with electrons (metal reduction) at the surface of the substrate, in this case a semiconductor [2, 3, 7]. This reaction is driven by a negative potential bias applied to the semiconductor (also called the cathode) with respect to a second electrode (called the anode), both immersed in an electrolyte solution (aqueous salt solution) containing the desired metal ions. A diagram of a typical experiment is shown in Fig. 9.1 [8]. The anode can be a relatively inert metal, such as Au, Pt or Pd, or a metal foil or wire of the deposit metal itself. It is important to be able to control the morphology, average rate of growth and total thickness on the cathode. For electrodeposition, the size and spacing of the electrodes must be considered to ensure uniform average deposition over the exposed surface area of the substrate. In general, the total number of atoms deposited can be obtained from measuring the current density versus time. Assuming only one reaction is occurring on the cathode, then the total number of atoms per unit area that have deposited is equivalent to the integration of current with time, or the total charge passed in the circuit, divided by the charge of the ion reacting. The average thickness then follows from the density of the deposited material.

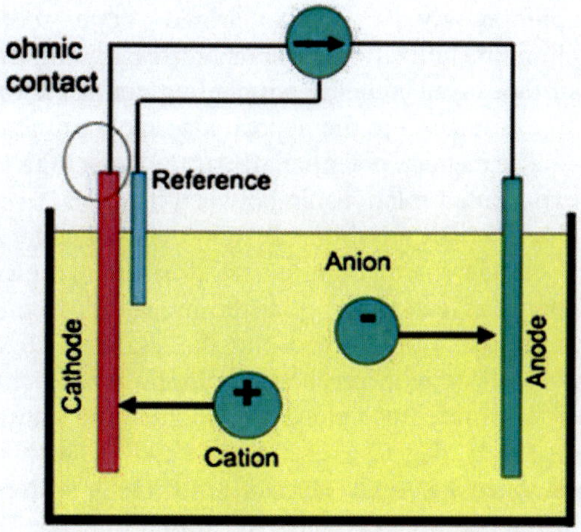

Fig. 9.1 Diagram of basic experimental procedures for electrodeposition [8]

It is common to control the electrodeposition process by maintaining either a constant total current (galvanostatic control) or a constant potential (potentiostatic control). Galvanostatic deposition would be trivial if indeed only one reaction occurred at the cathode, but this is unfortunately not the common experience. There is more likely to be parallel reactions that compete for the total current. The electrolyte is a combination of both positively and negatively-charged ions that contribute to its conductivity, and to electrode reactions. Depending on the *p*H and applied potential the reaction of water, or H^+ and OH^- ions, forming hydrogen and oxygen gas, respectively, will occur. In addition, growth of layers on the cathode or anode changes the overall resistance of the circuit. Thus, potentiostatic control is often preferred over galvanostatic control.

For potentiostatic control, the cathode potential is measured and adjusted to be constant with respect to a third, high impedance reference electrode added to the cell. Negligible current passes through the reference electrode, designed for high stability and inertness with respect to the other processes. Reactions that might be occurring at the anode, such as oxide growth, will then not interfere with the cathode potential measurement.

Measurement of the cathode potential is easier and more reproducible with a uniform, ohmic contact between the power supply and the cathode. This contact is located often on the backside of the substrate or in any convenient location not exposed to the electrolyte. An ohmic contact has a linear current-voltage characteristic meaning that the resistance is constant and not a function of the applied potential or current over the range of usage. An important difference between semiconductor and metallic cathodes is the higher resistance of semiconductors and the extra attention necessary to fabricate an ohmic contact to their surface. For n-type Si or III–V semiconductors a simple approach requiring no annealing is to use a liquid eutectic mixture of In-Ga. These metals form an adequate, low-resistance contact, as long as the semiconductor surface oxide is pre-etched appropriately [9, 10]. To provide better uniformity and lower resistivity other ohmic metallization systems are feasible for each semiconductor. A highly-doped surface layer inducing a tunneling contact is the general strategy. A larger contact area compared to the deposit also reduces the contact resistance.

The cathode potential affects the rate of the metal ion reaction on its surface in an exponential relationship between flux and "overpotential" [3]. All reactions have a characteristic equilibrium, free energy change, ΔG^o, with respect to the components at standard temperature and pressure. In electrochemistry, this change in free energy is called the equilibrium equipotential difference, ΔE^o, and is typically measured with respect to the standard hydrogen electrode (SHE) assigned $\Delta E = 0$ or with respect to a more convenient reference electrode [2]. Common types of reference electrodes include the saturated calomel electrode (SCE) based on the Hg/Hg_2Cl_2 reaction (+0.244 V with respect to a SHE electrode) or the saturated $Ag/AgCl_2$ electrode (+0.197 V with respect to SHE). An applied potential difference greater than ΔE^o upsets the equilibrium balance at the cathode and anode. The overpotential is the degree of *negative* deviation from ΔE^o. Higher overpotential increases exponentially the rate of metal film growth, as well as the initial nucleation density. As long as the flux of ions diffusing to the cathode is greater than the rate of this reaction, the growth is reaction-rate limited. Eventually, with higher overpotential the ionic flux to the surface will become the rate limiting step, and growth rates then become independent of substrate potential. Parallel reactions such as the formation of hydrogen gas, and the dissociation of water become increasingly responsible for larger fractions of the total current.

Semiconductor cathodes in an electrolytic cell behave very similarly to metal-semiconductor Schottky diodes. The surface and metal-interface potential will vary depending on the semiconductor Fermi level primarily determined by its substitutional dopant type and density. Once chemically cleaned, and adsorbates or metal atoms have reacted with the surface, native or foreign induced interfacial states are induced that also influence the position of the surface Fermi level and band bending. The conduction during electrodeposition may be carried via holes or electrons in the valence or conduction bands. These factors have been studied for many of the industrially important semiconductors via in situ analysis using cyclic voltammetry and impedance measurements combined more recently with scanning probe microscopy and x-ray scattering at synchrotron facilities [7, 10–14].

9.3 Epitaxial Metals on Si

Numerous metals and metal silicides grow epitaxially onto Si surfaces of various orientations using vacuum evaporation and annealing [15–18]. Many will begin to react at room temperature on clean Si surfaces, especially those forming metal-rich silicides, including Ni, Co, Pd, and Pt. All metals interdiffuse into Si given sufficient time and temperature. The surface reconstructions present after high temperature anneals in vacuum, and wet chemical H-passivation influence the subsequent epitaxy [19].

Many metals have been electrodeposited onto Si single crystalline substrates towards the fabrication of contacts with some of the earliest reports in the 1980s [20]. Random island nucleation is commonly observed and theories of nucleation and growth apply [21]. Photocarriers influence the reaction rates of deposition and or stripping [22–24]. Interfacial oxides are suspected based on contact properties and direct detection [14]. Rarely is epitaxial growth reported [22, 25–27].

The known exceptions include Pb [25], Cu, Co [22], and Au [26, 27] growth on H-terminated Si (111). In vacuum, Pb will deposit epitaxially at room temperature in various patterns depending on coverage [28]. For Au, interdiffusion and silicide formation is reported [29]. The lattice mismatch, *f*, defined as the difference in film and substrate lattice constants compared to the substrate ($\Delta a/a_s$) is smaller in absolute value for Pb than Au (−8.8% and 25%, respectively) on Si. Thus, Au will be happier growing on Si (001) surfaces where it can reduce the mismatch by a rotation aligning the <110> planes with the Si <100>.

Electrodeposition on H-terminated Si surfaces has been investigated in situ using cyclic voltammetry with scanning tunneling microscopy (STM), and synchrotron x-ray scattering [22, 25]. Cyclic voltammetry is a technique that cycles the voltage over a chosen range and frequency measuring current. Submonolayer reactions are routinely monitored this way. Pb was found to form facetted (111) oriented platelets indicative of interfacial ordering although continuous films were not reported [25]. The atomically-smooth NH_4F-etched *n*-Si (111) surface could be imaged prior to the deposition using a positive STM tip potential with respect to the Si electrode, consistent with a stable H termination. When a more negative potential was applied to the Si (−0.8 V SCE = −0.56 V SHE) in an acidic electrolyte (0.05 M $HC1O_4$ + 1 mM Pb^{2+}) Pb deposition occurred but only with the positive STM tip retracted. Subsequent STM imaging and *ex situ* x-ray diffraction showed preferential Pb (111) growth. The Pb layers could be stripped entirely from the surface by reversing the current indicating little Si-Pb intermixing.

Subsequent investigations by the same group using again in situ x-ray techniques including grazing incidence x-ray diffraction (GIXRD), crystal truncation rod scattering (CTRS), and x-ray standing waves (XSW) were able to confirm epitaxial ordering of Au, Cu and Co on H:Si(111) [22]. In these experiments the electrolyte composition was again small (Cu: 0.1 M H_2SO_4 + 0.1 mM $CuSO_4$; Co: 10^{-2} M H_3BO_3 + 0.2 mM H_2SO_4 + 0.1 mM $CoSO_4$) to not interfere with the surface x-ray signal, with potentiostatic control (−0.55 V to −1.05 V vs. SCE = −0.35 V to −0.85 V vs. SHE). Island morphologies developed without wetting layers. The structure at greater thicknesses towards coalescence were not reported.

Most recently, using a combination of cyclic voltammetry and ex situ atomic force microscopy (AFM), conditions where Au formed atomically smooth, continuous epitaxial layers on *n*-Si(111) and on (001) were reported [26, 27]. Epitaxial island nucleation occurred with lateral growth observed at a potential where both Au reduction and H_2 gas evolution were occurring in parallel yet before significant water dissociation. The large mismatch was relaxed within 10–20 monolayers while the Au aligned cube on cube on Si(111). On Si(001) a Au film rotation was observed retaining a (001) Au surface orientation [27]. Together with controlled *p*H and metal ion concentration (*p*H = 4, 0.1 mM $HAuCl_4$), the growth rate was constant (0.22 ML/s) and limited by diffusion to the substrate over a range of potentials (−2.0 V to 0 V Hg/Hg sulfate reference = −1.36 V to +0.64 V SHE). They concluded that the H^+ evolution reaction in parallel must maintain the H passivation of the Si surface and facilitate Au adatom diffusivity such that lateral growth of Au continues before oxidation. The exact reasons how this also improves the subsequent growth of smooth Au on Au is not clear.

The prior removal of Si native oxides, the surface passivation with H and maintenance of this passivation in the electrolyte are clearly important steps. The electrolyte *p*H influences the fraction of the total current associated with the Au reduction as well as reduces the probability of Si oxidation. Whether this particular source of Au ions and a Cl based acid are necessary are other questions. Once the Si/Au interface is complete the growth of other metals onto the Au is feasible. The magnetic properties of Au/Co, Ni, Cu and Fe multilayers have been studied with continuous layer thicknesses as small as 2–5 nm [30].

9.4 Epitaxial Growth on GaAs

There have been many more reports of epitaxial metal electrodeposition on GaAs substrates. Compared to Si the native oxides are less stable, and therefore easier to chemically or thermally remove. They regrow more slowly perhaps, but the chemically-cleaned GaAs surface is also less easy to completely passivate with H or other adsorbate. For UHV or CVD preparation, the surface is often given a wet chemical oxide etch just prior to loading in the reactor or vacuum system. Various types of chemical etch recipes including dilute HCl or NH_4OH aqueous solutions have been used [31]. Once inside the system in situ heating is also an option. The surface crystallinity, and residual surface composition after these treatments has been studied via reflected high energy electron diffraction (RHEED), x-ray photoelectron spectroscopy (XPS) and Auger spectroscopy [32]. Residual oxides and other impurities are detected as a function of the etch composition with the conclusion that As oxides predominate, with As-OH the most likely. Recent experiments with atomic layer epitaxy (ALD) for the deposition of thin layers of Hf or Al oxides onto such surfaces finds that the residual native oxides are displaced by the deposited metal. The Hf and Al reactions probably exchange with any residual Ga and As oxides in a self-cleaning mechanism [33].

The identification of the surface bonds and reactions that occur on GaAs electrodes in water as a function of applied potential have been investigated by in-situ Fourier transform infrared spectroscopy (FTIR) using attenuated-total internal-reflection prisms [34, 35], as well as with cyclic voltammetry [36]. According to FTIR investigations, the chemically-cleaned *n*-GaAs surface after removal of the native oxides (HCl 6 M) is unreconstructed and initially covered with (1 × 1) As–OH bonds consistent with the UHV investigations. In the electrolyte, these bonds are displaced by As–H bonds to monolayer coverage once under cathodic polarization. Meanwhile, the Ga^{+3} surface atoms are more likely to be reduced to

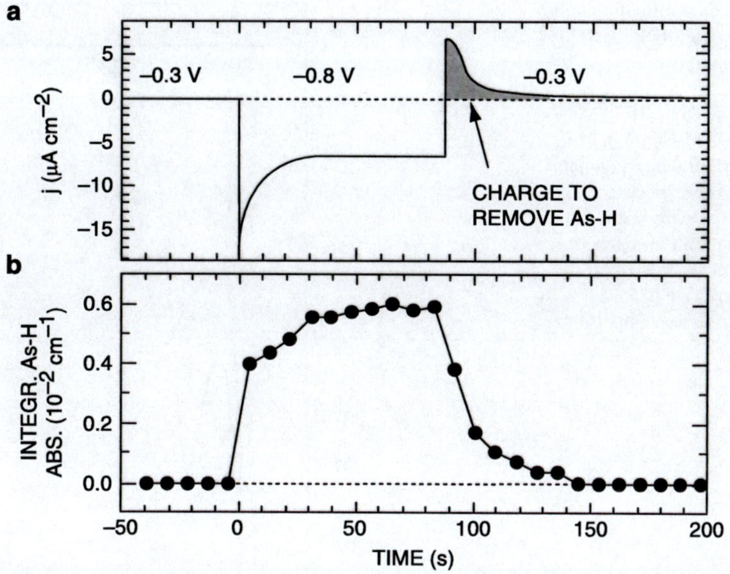

Fig. 9.2 Example of current density and FTIR absorbance as a function of applied potential and time for a *n*-GaAs surface (With permission [34])

metallic Ga rather than form H or OH bonds. Ga–OH and Ga–H are much less stable than similar As bonds. Figure 9.2 shows results from FTIR absorbance as a function of time. Only surface As–H bonds are detected as a function of the GaAs cathode potential, after a change from −0.3 V to −0.8 V with respect to a Ag/AgCl reference electrode (−0.1 V to −0.6 V SHE). The surface coverage increases exponentially until it saturates after 50 s. These measurements offer valuable insights into the presence of certain molecules and atoms on the GaAs surface and the rates of their adsorption or bonding at well defined electropotentials.

Studies of the growth of metals such as Cu, Co, and Ni onto GaAs using cyclic voltammetry detect H and OH reactions plus those generated by the metal ions [36]. One example for Cu growth on *n*-GaAs (001) is shown in Fig. 9.3. The GaAs surface was prepared by first etching to atomic smoothness by anodic oxidation followed by immersion in concentrated HCl to remove the surface oxides. Plotted is the current density versus applied cathode voltage with respect to a SCE reference for an acidic Cu sulfate electrolyte (5 mM $CuSO_4$ + 1 M H_2SO_4) [36]. The first two voltammetry scans at a frequency of 20 mV/s are shown. The peak at a negative potential of −0.65 V (−0.41 V SHE) in the first scan is due to the direct reduction of Cu^{+2} ions to Cu° forming a copper deposit on the GaAs surface in this sulfate electrolyte. The area of the peak is proportional to the total Cu thickness deposited during the scan which was less than a full monolayer. The peak shifts to a higher value (−0.61 V ≡ −0.37 V SHE) during the second scan indicating the reduction potential is different for a GaAs surface now partially covered by Cu. The shift is attributed to the formation of metal-induced interfacial states that change the

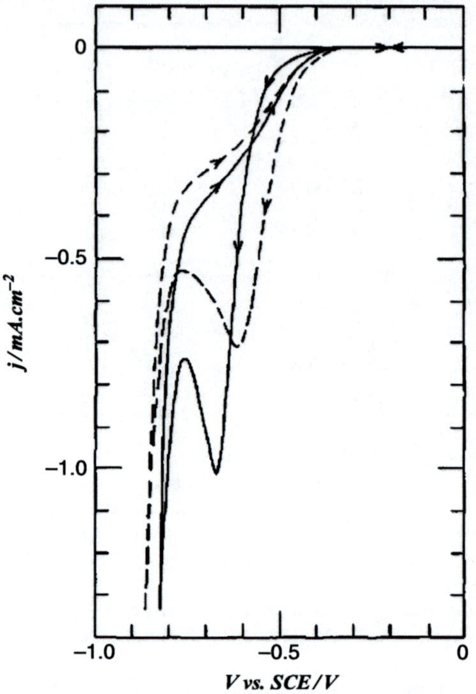

Fig. 9.3 Example of a cyclic voltammogram obtained during potentiostatic electrodeposition of Cu on *n*-GaAs (100) (5 × 10^{-3} M $CuSO_4$ + 1 M H_2SO_4). Plots of the current density versus voltage with respect to SCE from the first two scans, beginning from an open circuit potential (0.18 V) using a scan rate of 20 mV s^{-1} (With permission Fig. 10 [36])

surface potential of the new Cu surface. Linear plots of the peak current density versus square root of the cycle scan rate indicated that the growth kinetics were limited by the diffusion of Cu^{+2} ions to the cathode surface at this electrolyte concentration and GaAs potential. Studies of the process as a function of potential found that adhesive layers grew only under conditions of diffusion control. The diode barrier height was found to be a little higher than vacuum diodes, tentatively attributed to an interfacial oxide layer. They concluded that the original As–OH bonds may still be present or that other oxidation occurred during their growth. Unfortunately, no information about the crystallinity of the Cu films was provided in this study.

Others have reported the structure in the very initial stages of Cu growth at room temperature via in situ x-ray standing wave analysis [22]. Electrodeposited polycrystalline Cu-GaAs islands showed interdiffusion (2 nm) and reaction (0.5 atomic layers) similar to observations during MBE growth by the same group.

MBE with RHEED in situ found Cu on GaAs to be epitaxial, initially BCC in structure switching to FCC by 1 nm thickness [37]. Similar to Au/Si (001) the MBE Cu FCC phase grows with a 45° rotation such that the Cu <110> in-plane direction is parallel to the GaAs <001> directions. This also reduces the lattice mismatch (−10% with respect to half the GaAs lattice constant).

More recently, Cu film electrodeposition on GaAs has been investigated over a wider parameter range [38]. The process carried out galvanostatically was optimized to obtain continuous films. The crystallinity and epitaxial alignment was confirmed ex situ via observations using optical and scanning electron microscopies, and structural evaluation via x-ray and electron diffraction. The electrolyte composition was simplified towards cleaner films rather than smooth surfaces. Isolated polycrystalline islands occurred at lower or higher constant currents, while continuous and epitaxial films occurred for currents centred at 10 mA/cm^2. The $CuSO_4$ concentration (0.1 M) and electrolyte temperature (53°C) were also optimized.

Figure 9.4 shows an SEM image of the surface of a Cu film on GaAs (001) grown at 15 mA/cm^2 at 53°C. When epitaxy occurs, the Cu islands coalesce into well-oriented single crystals with aligned pyramidal surface facets, as well as strong XRD peaks. The macroscopic surface orientations are on average parallel to the substrate with local variations of up to 1° due to dislocations. The pyramidal facets develop with <100> bottom edge directions seen previously for growth on Au substrates. This is explained to be due to the influence of adsorbed impurities or to preferential reactions on surface steps. Similar phenomena are reported for Ag films [3].

Most conditions, however, produce Cu islands that do not coalescence or, at very high currents, dendrites will form. Figure 9.5 shows examples of Cu islands or dendrites that have grown in random orientations and without coalescence on *n*-GaAs (001) surfaces at room temperature at current densities of 140, 6 mA/cm^2, and 0.4 μA/cm^2 [8, 38]. Cu films also grow epitaxially onto GaAs (110) and (111) conserving the epitaxial preference for alignment of Cu <110> with GaAs <100> directions [8]. The variability in the GaAs doping density, sample area, and the ohmic contact meant that potentiostatic control was less reproducible than that of galvanostatic. Unfortunately, there has not yet been an accurate potentiostatic measurement of the cathode potential to enable comparisons to the earlier results described above.

Continuous films of single crystalline $Cu_{0.3}Ni_{0.7}$ alloys have been reported on (001) GaAs by pulsed electrodeposition (−2.5 V SCE ≡ −2.3 V SHE) although pure Ni did not grow well [39]. The final preparation of the GaAs surface and the electrolyte for these experiments also avoided any chloride chemistry.

The metal with the smallest lattice mismatch to GaAs is BCC Co ($a_o = 0.2827$ nm, $f = +0.1\%$) a phase that is metastable, first grown by vacuum deposition [40].

Fig. 9.4 SEM image of Cu electrodeposited onto GaAs (001) at 53°C, 15 mA/cm^2 (With permission [38])

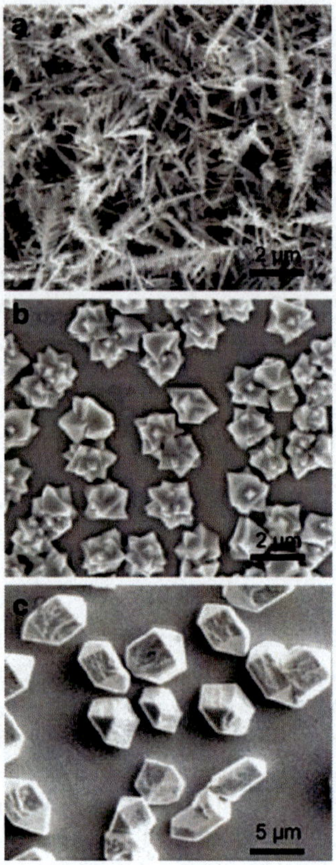

Fig. 9.5 SEM images of polycrystalline Cu/GaAs (001) electrodeposits as a function of current density (0.1 M $CuSO_4$, room temperature): (**a**) 140 mA/cm^2, (**b**) 6 mA/cm^2, and (**c**) 0.4 μA/cm^2 (With permission [8, 38])

At thicknesses greater than a critical value (36 nm for vacuum deposits) the Co films relax to the equilibrium HCP phase (lattice constant: $c = 0.40695$ nm). The in-plane cubic [200] and [110] diffraction peaks (RHEED) split towards the HCP structure. The transformation to HCP Co is such that the *c*-axis direction becomes aligned with a GaAs <110> in-plane direction with a <1120> surface orientation finally with a larger mismatch (1.77% with respect to half the GaAs lattice constant).

The same epitaxial arrangement dominates during Co electrodeposition on GaAs. Electrodeposited in acidic electrolytes ($CoSO_4$ (1.5 M) + H_3BO_3 (0.5 M), $p\text{H} = 3.45$), Co/GaAs (001) has structural and magnetic properties comparable to those of vacuum-deposited films, including the initial growth of the BCC phase [41]. In a electrolyte that is less acidic (*p*H 5–6), buffered using ammonium sulfate, and using a lower Co sulfate concentration (0.1 M) without boric acid, Co films grown galvanostatically (10 mA/cm^2), develop vertical (0001) facets on top of the BCC interfacial layer [42]. Growth is isotropic on the (0001) facets resulting in circular discs. Figure 9.6 shows SEM images of Co/GaAs as a function of GaAs

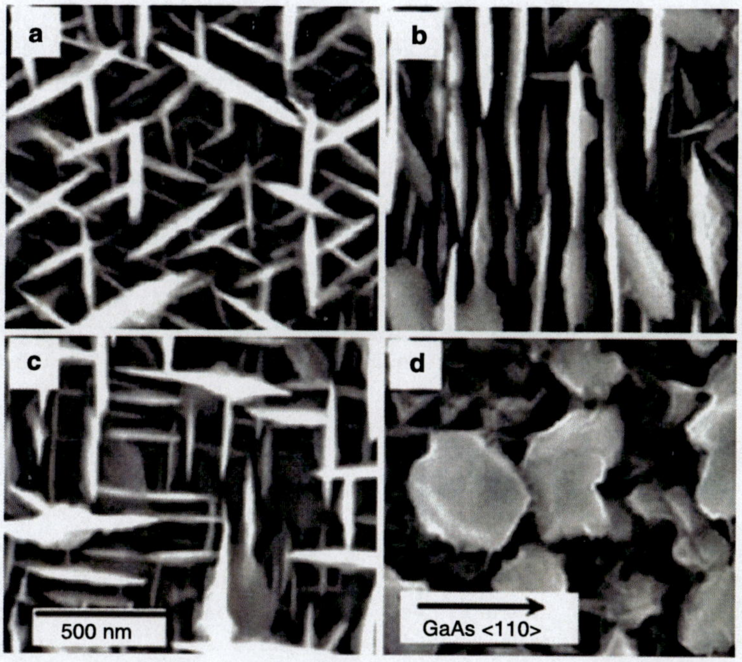

Fig. 9.6 SEM images of Co nanodiscs electrodeposited on GaAs: (**a**) (111)B, (**b**) (110), and (**c**) (001) substrates from $CoSO_4$ (0.1 M) plus $(NH_4)_2SO_4$ (1.0 M) electrolyte. The perpendicular to the plane of the discs is always aligned with a GaAs <110> direction. (**d**) An SEM image of a Co film grown on GaAs (001) without $(NH_4)_2SO_4$ (With permission [42])

substrate orientation. There are discs arranged parallel to each in-plane GaAs <110> direction. Their aspect ratio, thickness to diameter, is 0.05–0.2. With the same current density, but lower electrolyte concentration (0.01 M $CoSO_4$), a similar shaped disc is obtained, but only 25% the size of these. At lower electrolyte temperature (4°C) the discs are also smaller and they do not occur for electrolyte temperatures above 22°C [42]. Random alignment without facets results for deposition in pure $CoSO_4$ electrolyte (no $(NH_4)_2SO_4$), acidic (H_2SO_4), or basic $(NH_4)OH$ $CoSO_4$ solutions. These experiments were carried out with heavily-doped, ($2 \times 10^{18}/cm^3$) *n*-type GaAs single crystals using a Pt wire anode with a cathode-anode potential difference of −4 V. The total current density increased rapidly with voltage in this range due to the parallel reduction of H^+ to hydrogen gas or dissociation of water (depending on the substrate resistivity and *p*H). The fraction of current attributable to Co film growth was as low as 10%. The formation of facets is likely occurring due to differences in the rate of reaction at Co surface orientations or to preferential adsorption of other ions in the electrolyte such as sulfate.

Co films also grew epitaxially on lower doped *n*-type substrates ($2 \times 10^{17}/cm^3$) for diode preparation [8]. The electrical barrier heights were measured ex-situ from current-voltage (*I-V*) and capacitance-voltage characteristics for (001), (011) and (111) *B* GaAs substrates. The Co films prepared in the same manner at room temperature gave identical values of 0.76 ± 0.02 eV independent of the substrate orientation with the highest *I-V* ideality factor occurring for the most polarizable (111) *B* orientation (1.07). These results are curious in that they are independent of orientation and method. Often a small difference between *I-V* and *C-V* is blamed on image force lowering (0.03 eV for a dopant density of $2 \times 10^{17}/cm^3$) which occurs during an *I-V* measurement but not during *C-V* [9]. The lack of any effect of the orientation indicates that any changes in the interface state density or semiconductor work function are small or self compensating. An interfacial dielectric layer such as an oxide, or deviation in the surface substrate carrier concentration from impurities, e.g. H, would cause a deviation in the measured barrier height between the two methods.

The metal with the second closest lattice match to GaAs is Fe (BCC 1.4%). Fe films grow epitaxial onto GaAs surfaces by UHV or electrodeposition [43]. For MBE growth the GaAs surfaces once in the UHV system are cleaned of native oxides by heating or ion beam sputtering followed by growth of GaAs epitaxial layers or further annealing to remove damage. Growth on substrate orientations of (001) and (110) have been reported with MBE. The Fe films are grown very slowly (monolayers per hour) at temperatures at or near room temperature. The Fe nucleates in islands that coalescence after three atomic layers. The epitaxy is cube on cube with the strain initially accommodated elastically. At a thickness of 20 atomic layers (2.8 nm) strain relaxation begins to be detected in situ via changes in the magnetic properties and splitting of reflection high energy electron diffraction (RHEED) spots [44]. Relaxation continues presumably by misfit dislocations with increasing thicknesses exponentially reducing the residual strain 90% by 15 nm thickness. These thin Fe layers have seen much attention due to their interesting magnetic anisotropies attributed to the interfacial structure. Spin injection and detection through thin MBE Fe/GaAs tunnel barriers have been studied using lateral Hall devices or optically via emission from buried quantum wells [45, 46].

Electrodeposited epitaxial Fe/GaAs has been reported for (100), and (110), and, in addition, (111)*B* orientations of bulk substrates [47, 48]. Like MBE, the films in all cases are oriented cube on cube with the substrate, consistent with the low mismatch. Epitaxial Fe was first reported via galvanostatic control ($2.5 mA/cm^2$) using a Fe sulfate (0.1 M) plus boric acid electrolyte adjusted to a *p*H of 2.5 (sulphuric acid) at room temperature [47]. The native oxides were etched in ammonia and it was noticed that $FeCl_2$ electrolytes produced inferior Fe layers based on poorer magnetic properties.

Subsequent work was carried out at a higher *p*H via buffered ammonium sulfate electrolytes and at higher electrolyte temperature, using Fe sulfate, ammonia for native oxide etching, and galvanostatic control ($10 mA/cm^2$). These conditions resulted in better structural films based on x-ray diffraction peak widths [48]. Higher acidity with additions of sulphuric or chloride-based acids reduced or destroyed the epitaxy [8]. Optimal conditions result in growth rates of approximately 100 nm/min, much faster than is typical of MBE. Islands form with diameters on the order of 50 nm, but their coalescence may not occur until thicknesses between 20 and 50 nm. Island growth is epitaxial but strain relaxation occurs prior to their coalescence, probably at a similar thickness as does MBE, 3 nm [44]. A third investigation confirmed the detrimental effects of chloride electrolytes and identified the nature of observed faults seen in all films to be (221) oriented twinning defects [49].

The interface is found to be atomically abrupt to within a few monolayers without evidence of oxides or reaction based on cross-sectional TEM [49]. However, fabrication of TEM cross-sectional samples without heating the interface during thinning is difficult and this may influence these results. Figure 9.7 shows an example of a high resolution TEM cross-sectional view of a electrodeposited Fe/GaAs(001) interface produced with minimal heating during polishing, and ion milling at liquid nitrogen temperatures. The FCC GaAs fringes are (111) planes while those from the BCC Fe are (110) planes (0.20 nm). It is difficult to exactly locate the position of the interface. There is a less ordered region with a thickness of about five monolayers. There may be a misorientation in the surface of the substrate or roughness from etching the oxide. These factors would increase the apparent interfacial roughness.

Relaxation of the lattice mismatch strain seen from x-ray diffraction measurements is confirmed by the edge dislocations visible. Two regions with an extra GaAs plane are circled in the image. The spacing between them is 17.5 nm consistent with the expected average spacing for such dislocations (14.3 nm) if they are to relax completely the 1.4% mismatch. Like Co, Fe diodes fabricated for the three major GaAs surfaces resulted in similar barrier heights of 0.88 ± 0.02 eV comparable to those fabricated in vacuum systems [48, 50].

The purity of the Fe films can be evaluated by its lattice constant. High resolution x-ray diffraction has measured the average perpendicular Fe lattice constant with respect to the GaAs substrate with an accuracy of ±0.0005 Å. Fe films grown at room temperature are found to have a lattice constant of 2.8605 ± 0.0005 Å, increasing to value of 2.8650 ± 0.0005 Å at 64 C, values slightly smaller than bulk Fe (2.8665 Å) [48]. Since the concentration of dissolved oxygen in the electrolyte decreases with increasing electrolyte temperature, this small contraction of the lattice is likely due to dissolved oxygen reacting in parallel. Oxygen impurities

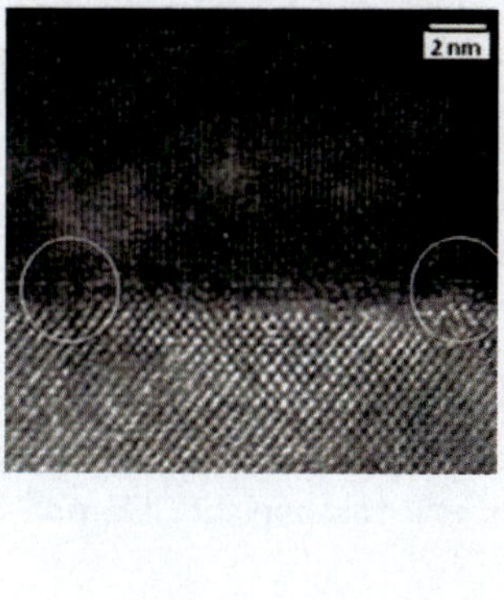

Fig. 9.7 High magnification lattice image of an electrodeposited Fe/GaAs (001) interface looking along a <110> direction. The circled regions have a misfit dislocation with an extra GaAs plane

forming Fe oxides within the Fe film are also consistent with inhomogeneities found in the magnetic properties [51].

Alloys of Fe including Fe_xCo_{1-x} and Fe_xNi_{1-x} have also been epitaxially grown onto GaAs using sulfate metal salts and ammonium sulfate buffering similar to Fe/GaAs. Thick BCC Fe_xCo_{1-x} films grow on GaAs over a range of compositions up to 64 at.% Co [52]. At greater Co concentrations a cubic structure was not detected by XRD and was assumed to be unstable for the same growth conditions or too thin to detect. Similar to BCC Fe films, twinning with a (221) orientation is observed to occur in BCC Fe_xCo_{1-x}. The proportion of twinned material varied with Co composition. An abrupt interface is observed via TEM. Magnetic properties are comparable to those of MBE Fe_xCo_{1-x} films.

Fe_xNi_{1-x} alloys grow in an FCC structure on GaAs (001) from x = 0 to 0.3, and BCC from x = 0.4 to 1 at room temperature (ammonia sulfate electrolyte). The composition in between is a two phase region where both BCC and FCC phases co-exist [50]. The FCC Ni-rich phases grew with compositions close to that of the electrolyte consistent perhaps with a mass-transport-limited process. If the reaction was rate limiting then the reaction of Ni is expected to be faster than Fe considering its higher equilibrium potential. Both types of metal ions have similar aqueous diffusion coefficients. Nevertheless, the Ni deposition was inhibited for the higher BCC Fe_xNi_{1-x} alloys. In these alloys the ratio of Fe in the film exceeded that of the electrolyte by a value proportional to the composition increasing up to 50% at 80 at.% Fe, despite otherwise similar electrodeposition conditions. This type of preferential deposition of the less noble metal has been carefully studied and can be explained by preferential adsorption of Fe ions or reactants compared to Ni at higher cell current densities or overpotentials [53]. The growth of Ni using similar electrodeposition conditions resulted in polycrystalline films. Earlier reports using more acidic electrolyte were more successful in aligning Ni on GaAs (001) and (011). The reasons for this difference is not understood [54].

The final metal reported to grow epitaxially on GaAs is Bi a trigonal crystal structure ($a = 0.4546$ nm, $c = 1.1862$ nm) [55, 56]. Initial experiments reported results from a Bi deposition carried out in a two step process involving a short nucleation pulse at a higher overpotential (−0.275 V vs. Ag/AgCl = −0.078 V SHE) followed by growth at a lower value (−0.02 V = +0.18 V SHE) to the final film thickness. The electrolyte was 20 mM BiO + and 2 M $HClO_4$ at room temperature. XRD measurements showed that the film had a strong (018) preferred orientation on the GaAs (011) surface [55].

A less acidic electrolyte consisting of $(NH_4)_2SO_4$ with a saturated solution of bismuth (III) nitrate pentahydrate ($Bi(NO_3)_3{\cdot}5\,H_2O$), resulted in better epitaxial Bi films on three major GaAs orientations. Bi grows *c*-axis oriented on (111)*B* and (100) GaAs and again (018) oriented on (011) GaAs [56]. The crystallinity and degree of lateral coalescence was best for an elevated growth temperature of 70°C [56]. No evidence of interfacial layers were detected via cross-sectional TEM investigations. There was a larger variation in the barrier heights than for Co or Fe diodes, particularly for (111)*B* diodes. A comparison between this As-polarized surface and the opposite Ga-rich (111)*A* side would be interesting. Experiments

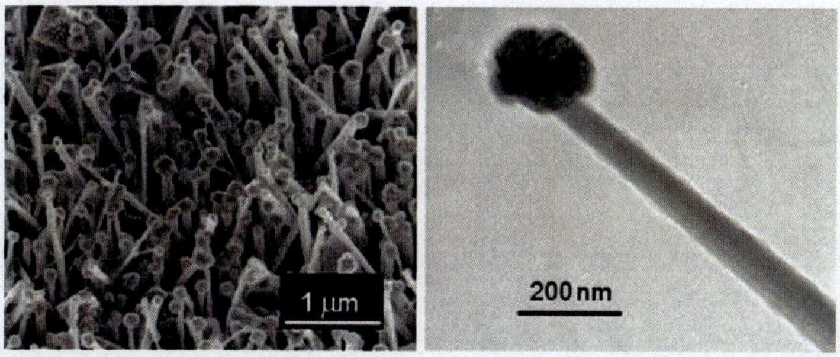

Fig. 9.8 SEM (*left*) and TEM (*right*) images of Cu electrodeposited onto Au/GaAs nanowires [6]

with Au electrodeposition have found significant differences in the nucleation and deposition potential for the two (111) orientations consistent with their different surface chemistry [57].

9.5 Semiconductor Nanowires

Electrodeposition has been applied to the deposition of metals on Si and GaAs nanowires grown by vapour liquid solid mechanisms [5, 6]. The technique can be used for the fabrication of contacts to the ends or sides of the wire probing their conductivity and composition. If the wire is poorly conducting or is covered in an insulating oxide, then deposition occurs only onto the end of the wire where there is the Au catalyst. Figure 9.8 shows SEM and TEM images of GaAs wires after the electrodeposition of Cu. These wires were not intentionally doped so their conductivity was poor. Nevertheless, there was sufficient conduction from the substrate for Cu to deposit around the Au catalyst at its tip but not on the sidewalls. The Au catalyst is single crystalline as is the Cu layer grown onto it. Intentional doping to increase the wire conductivity results in sidewall deposition. In the case of conductive Si nanowires, once the surface oxide is removed, deposition of Ni occurs onto the sides as well as the Au tip [5]. Similar to flat surfaces, the control of epitaxial nucleation and orientation should be feasible via appropriate surface preparation and cell conditions.

9.6 Comments and Remarks

A summary of metals, reported to have been epitaxially grown on Si or GaAs via aqueous electrodeposition are listed in Table 9.1. Included is their lattice constant, a_o, lattice mismatch, epitaxial alignment with the substrate, film orientation, the native oxide etchant, electrolyte, and references to the literature. The removal of native oxides is certainly an important requirement. The formation of a passivation layer of H bonds on Si or OH bonds on GaAs after oxide etching is observed in situ

Table 9.1 A summary of the properties of metals reported to grow epitaxially by electrodeposition on Si or GaAs substrates

Metal	Lattice const. (Å)	f (%)	Film/subst. in-plane	Film orient.	Surface oxide etchant	Electrolyte/pH	Cell control (mA/cm^2 (V SHE))	References
Si (111) 5.4321								
Pb	4.9500	−6.6	[112]‖[112]	(111)	NH_4F	$HClO_4$ + $PbHClO_4$	(−0.56)	[25]
Cu	3.6100		"	(111)	NH_4F	H_2SO_4 + $CuSO_4$	(−0.85)	[22]
Co	4.0695	−25	"	(0001)	NH_4F	H_2SO_4+ $CoSO_4$	(−0.85)	[22]
Au	4.0800	−25	"	(111)	NH_4F + $(NH_4)_2SO_3$	$HAuCl_4$ + K_2SO_4 + KCl + H_2SO_4/4	(−0.6)	[26, 27]
GaAs 5.6535								
Co	2.827	+0.01	[100]‖[100]	(001)	HCl + NH_3	$CoSO_4$/3.45	10(−2.1)	[41]
BCC								
HCP	c = 4.0695	+1.77	[001]‖[110]	(1102)	NH_3	$CoSO_4$ + $(NH_4)_2SO_4$/6.5	10	[8, 42]
Co_xFe_{1-x}		+0.5 to +1.4	[100]	(001)	NH_3	$CoSO_4$ + $FeSO_4$ $(NH_4)_2SO_4$	10	[52]
BCC ($0 \leq x < 0.7$)								
Fe	2.8665	+1.4	[100]‖[100]	(001)	NH_3	$FeSO_4$ + ascorbic acid + $B(OH)_3$/2.5	2.5	[47]
BCC			"	(011)				
			"	(001)	NH_3	$FeSO_4$ + $(NH_4)_2SO_4$/5–6	10	[48]
			"	(011)				
			"	(111)				
Fe_xNi_{1-x} BCC and FCC		+1.4 to −13.5	[100]‖[100] [001]‖[110]	(001)	NH_3	$FeSO_4$ + $NiSO_2$ $(NH_4)_2SO_4$/5–6	10	[50]
Cu	3.610	−9.5	[110]‖[100]	(001)	H_2SO_4	H_2SO_4 + $CuSO_4$	(−0.3)	[12]
FCC								
			[110]‖[100]	(001)	NH_3	$CuSO_4$	10	[8, 38]
				(111)				
$Ni_{0.7}Cu_{0.3}$	3.59	−12	[110]‖[100]	(001)	NH_3	$CuSO_4$ + $NiSO_4$ + H_2SO_4/2	(−2.3)	[39]
FCC								
Ni	3.5238	−13.5	[110]‖[100]	(001)	NH_3	$NiSO_4$/2.3	4	[54]
FCC			–	poly	"	$NiSO_4$ + $(NH_4)_2SO_4$/5–6	10	[8, 50]
Bi	a = 4.55	−12.1	(110)	(018)	HCl	BiO^+ + $HClO_4$	(−0.06)	[55]
Trigonal	c = 11.86		(001) (111)B	(111)	NH_3	$(NH_4)_2SO_4$+ $(Bi(NO_3)_3 \cdot 5H_2O)$	10	[56]
			(011)	(018)				

Listed are the metals, their lattice constant, the mismatch with the substrate, f, the resulting film/substrate alignment, resulting film orientation, the substrate surface preparation, electrolyte composition and pH, constant current or potential used during deposition, and references via FTIR and cyclic voltammetry, and ex-situ by XPS. The subsequent displacement of these bonds by the metal, once cathodically polarized, is supported by epitaxial growth and the lack of detectable interfacial oxide layers. Native-oxide etching using HCl solutions is less effective compared to NH_4OH, although the formation of As–OH bonds is detected nonetheless using this etchant. The presence of Cl ions may contribute directly to the passivation and or interfere with the oxidation control. This ion is not a necessary component to native oxide removal and does not aid in epitaxial growth.

The importance of other factors for GaAs including using a higher *p*H, and the presence of sulfate, are apparently important but also not understood. In most cases, accurate potentiostatic investigations in combination with structural investigations have not yet been carried out. Except for Cu all of these metals have negative equilibrium reduction potentials, meaning that H^+ reduction reaction will always be occurring in parallel to some extent. The addition of ammonium sulfate for Cu deposition on GaAs did not improve the epitaxy although the formation of CuO at the apex of pyramidal surface features was likely delayed. Higher electrolyte temperature was beneficial in more than one metal likely related to residual oxygen in the electrolyte. A certain fraction of the total current being a H^+ reaction may be a necessary requirement for uniform, lateral growth on GaAs, as already suggested for Si formation. While Cu, Co, Fe, and Bi form epitaxial layers on GaAs, it is not clear why other metals including Ni and Cr are less successful.

Acknowledgments The author thanks SFU student collaborators Bao, Majumder, Chao, Ahktari-Zavareh, Grist, Spiga, Shaw, Berring, Radich, Bratvold, Cheng, Jensen, and Abbet. She is also grateful for many useful discussions with attendees of the Gordan Conferences on Electrodeposition, and funding support from NSERC.

References

1. Mathews JW. Epitaxial growth. New York: Academic; 1975.
2. Lobo VMM, Quaresma JR. Handbook of electrolyte solutions, Physical science data, vol. 41. Amsterdam: Elsevier; 1989. Bradford SA. Corrosion control. 2nd ed. Edmonton: CASTI Publishing Inc.; 2002; Switzer JA, Hodes G. Mat Res Soc Bull. 2010;10:743–49.
3. Budevski E, Staikov G, Lorenz WJ. Electrochemical phase formation and growth. Weinheim: VCH Verlagsgesellschaft mbH; 1996.
4. Reid J, McKerrow A, Varadarajan S, Kozlowski G. Solid State Technol. 2010;53:14–6.
5. Ingole S, Manandhar P, Wright JA, Nazaretski E, Thompson JD, Picraux ST. Appl Phys Lett. 2009;92:223118–3.
6. Liu C, Einabad OS, Watkins S, Kavanagh KL. 217th ECS meeting, Abstract #1577. Electrodeposition of metal on GaAs nanowires. Master of Science Thesis, Department of Physics, Simon Fraser University, 2011.
7. Gerischer H. J Vac Sci Technol. 1978;15:1422–8.
8. Bao ZL. Epitaxial metal-GaAs contacts by electrodeposition. PhD thesis, Simon Fraser University; 2006.
9. Rhoderick EH, Williams RH. Metal-semiconductor contacts. Oxford: Clarendon Press; 1988.
10. Woodall JM, Freeouf JL, Pettit GD, Kirchner P. J Vac Sci Technol. 1981;19:626–7.
11. Magnussen OM, Hotlos J, Nichols RJ, Kolb DM, Behm RJ. Phys Rev Lett. 1990;64:2929–32.

12. Scherb G, Kazimirov A, Zegenhagen J, Lee TL, Bedzyk MJ, Noguchi H, Uosaki K. Phys Rev B. 1998;58:10800–5.
13. Memming R. Semiconductor electrochemistry. Weinheim: Wiley-VCH; 2001.
14. Muñoz AG, Lewerenz HJ. J Electrochem Soc. 2009;156:D184–7.
15. Tseng WF, Liau ZL, Lau SS, Nicolet M-A, Mayer JW. Thin Solid Films. 1977;46:99–107.
16. Murarka SP. Silicides for VLSI applications. New York: Academic; 1983.
17. Tu KN, Mayer JW, Feldman LC. Electronic thin film science: for electrical engineers and materials scientists. New York: McMillan; 1992.
18. Poate JM, Tu KN, Mayer JW. Thin films: interdiffusion and reactions. New York: Wiley; 1978.
19. Grupp C, Taleb-Ibrahimi A. Phys Rev B. 1998;57:6258–61.
20. Ghosh K, Chowdhury NKD. Int J Electron. 1983;54:615–23.
21. Hoffmann PM, Radisic A, Searson PC. J Electrochem Soc. 2000;147:2576–80. Radisic A, Ross FM, Searson PC. J Phys Chem B. 2006; 110: 7862–7868; Guo L, Searson PC. Electrochim Acta. 2010;55:4086–91.
22. Zegenhagen J, Renner FU, Reitzle A, Lee TL, Warren S, Stierle A, Dosch H, Scherb G, Fimland BO, Kolb DM. Surf Sci. 2004;573:67–79.
23. Forment S, Van Meirhaeghe RL, De Vrieze A, Strubbe K, Gomes WP. Semicond Sci Technol. 2001;16:975–81.
24. Oskam G, Long JG, Nikolova M, Searson PC. Mater Res Soc Symp. 1997;451:257–66.
25. Ziegler JC, Reitzle A, Bunk O, Zegenhagen J, Kolb DM. Electrochim Acta. 2000;45:4599–605. Kolb DM, Randler RJ, Wielgosz RI, Ziegler JC. Mater Res Soc Symp. 1997;451:19–30.
26. Prod'homme P, Maroun F, Cortès R, Allongue P. Appl Phys Lett. 2008;93:171901–3.
27. Prod'homme P, Warren S, Cortès R, Jurca HF, Maroun F, Allongue P. ChemPhysChem. 2010;11:2992–8.
28. Ganz E, Hwang I-S, Xiong F, Theiss SK, Golovehenko J. Surf Sci. 1991;257:259–73.
29. Yeh J-J, Hwang J, Bertness K, Friedman DJ, Cao R, Lindau I. Phys Rev Lett. 1993;70:3768–71.
30. Allongue P, Maroun F. Mater Res Soc Bull. 2010;35:761–70.
31. Clawson A. Mater Sci Eng R Rep. 2001;R31:1–438.
32. Yoon HJ, Choi MH, Park IS. J Electrochem Soc. 1992;139:3229–34.
33. Hinkle CL, Sonnet AM, Vogel EM, McDonnell S, Hughes GL, Milojevic M, Lee B, Aguirre-Tostado FA, Choi KJ, Kim HC, Kim J, Wallace RM. Appl Phys Lett. 2008;92:071901–3.
34. Erné BH, Ozanam F, Chazalviel J-N. J Phys Chem B. 1999;103:2948–62.
35. Erné BH, Stchakovsky M, Ozanam F, Chazalviel JN. J Electrochem Soc. 1998;145:447–56. Erné BH, Ozanam F, Chazalviel JN. Phys Rev Lett 1998;80:4337–40.
36. Vereeken PM, Vanden Kerchove F, Gomes WP. Electrochim Acta. 1996;41:95–107. Strubbe K, Vereecken PM, Gomes WP. J Electrochem Soc. 1999;146:1412–20.
37. Tian Z, Tian CS, Yin LF, Wu D, Dong GS, Jin X, Qiu ZQ. Phys Rev B. 2004;70:012301-1-4.
38. Bao ZL, Grist S, Majumder S, Xu LB, Jensen E, Kavanagh KL. J Electrochem Soc. 2009;156: D138–45.
39. Hart R, Midgley PA, Wilkinson A, Schwarzacher W. Appl Phys Lett. 1995;67:1316–8.
40. Prinz GA. Phys Rev Lett. 1985;54:1051–4.
41. Ford A, Bonevich JE, McMichael RD, Vaudin M, Moffat TP. J Electrochem Soc. 2003;150: C753–9.
42. Bao ZL, Kavanagh KL. J Cryst Growth. 2005;287:514–7.
43. Prinz A, Krebs JJ. Appl Phys Lett. 1981;39:397–9.
44. Kebe T. SQUID-magnetometry on Fe monolayers on GaAs(001) in UHV. Ph.D. Thesis, University of Duisberg-Essen, 2006.
45. Isakovic A, Carr DM, Strand J, Schultz BD, Palmstrøm CJ, Crowell PA. Phys Rev B. 2001;64:161304–4.
46. Hanbicki AT, Jonker BT, Itskos G, Kioseoglou G, Petrou A. Appl Phys Lett. 2002;80:1240–2.

47. Liu YK, Scheck C, Schad R, Zangari G. Electrochem Solid-State Lett. 2004;7:D11–3. Scheck C, Evans P, Schad R, Zangari G. J Appl Phys 2003;93:7634–36.
48. Bao ZL, Kavanagh KL. J Appl Phys. 2005;98:066103–3.
49. Svedberg EB, Mallett JJ, Bendersky LA, Roy AG, Egelhoff WF, Moffat TP. J Electrochem Soc. 2006;153:C807–13.
50. Bao ZL, Majumder S, Talin AA, Arrott AS, Kavanagh KL. J Electrochem Soc. 2008;155: H841–8.
51. Majumder S, Arrott AS, Kavanagh KL. J Appl Phys. 2009;105:07D543-3.
52. Mallett JJ, Svedberg EB, Vaudin MD, Bendersky LA, Shapiro AJ, Egelhoff WF, Moffat TP. Phys Rev B. 2007;75:85304-1-7.
53. Matlosz M. J Electrochem Soc. 1993;140:2272–9.
54. Evans P, Scheck C, Schad R, Zangari G. J Mag Magn Mater. 2003;260:467–72.
55. Yang FY, Liu K, Chien CL, Searson PC. Phys Rev Lett. 1999;82:3328–31. Vereeken PM, Rodbell K, Ji C, Searson PC. Appl Phys Lett 2005;86:121916–3.
56. Bao ZL, Kavanagh KL. Appl Phys Lett. 2006;88:022102–3. J Vac Sci Technol B. 2006;24:2138–43.
57. Depestel LM, Strubbe K. J Electroanal Chem. 2004;572:195–201.

Part IV

Nanoscale Etching and Patterning

Chemical Mechanical Polish for Nanotechnology

10

L. Nolan and K. Cadien

Abstract

Chemical mechanical polish (CMP) is a process technology that was adapted from wafer polishing to IC fabrication and thereby enabled the semiconductor industry to extend optical lithography and invent novel approaches such as damascene interconnects. In this chapter, we discuss the fundamentals of CMP as well as applications of the technique to nanotechnology. These applications include established IC techniques such as damascene and dual damascene fabrication, and shallow trench isolation, and emerging processes such as gate-last technology. CMP has recently found broader applications within the nano-technology community, and has been adapted to produce extremely smooth surfaces for materials, such as sapphire and gallium nitride, that are far beyond the usual scope of IC manufacture. It is also being explored as a fabrication technique for novel interconnect and memory materials, such as phase-change memory. This chapter examines the features of CMP that make it a fundamental top-down nanofabrication technique for such a wide range of applications.

10.1 Introduction

Polishing was developed centuries ago to produce copper and bronze mirrors, and to shape glass optics. All mirrors and lenses used in astronomy have been shaped and smoothed using polishing. Samples prepared for metallography are polished to produce a mirror-like surface finish for optical inspection. All semiconductor substrates are polished as part of the final step of substrate manufacturing to produce a flat, smooth, and defect free surface.

L. Nolan (✉) • K. Cadien
Department of Chemical and Materials Engineering, University of Alberta, Edmonton, Alberta, Canada
e-mail: kcadien@ualberta.ca

M. Stepanova and S. Dew (eds.), *Nanofabrication*,
DOI 10.1007/978-3-7091-0424-8_10, © Springer-Verlag/Wien 2012

CMP was developed from substrate manufacturing technology in the 1980s [1] by IBM to address topography issues associated with adding a second metal layer. CMP was used to planarize SiO_2 interlayer dielectrics (ILD), reducing topography and thus extending the usefulness of photolithographic techniques by allowing a smaller depth of focus to be used. Planarity was further enhanced at the transistor level by the development of shallow trench isolation (STI) technology that required the polishing of SiO_2 and stopping on the silicon nitride (Si_3N_4) masking layer. Since then, CMP has been adapted for use with metals such as tungsten plugs, where tungsten polish replaced the tungsten plasma etch back process with dramatic improvement in defects and reduction in plug recess.

Copper has a significantly lower resistivity than aluminum. It also has a higher melting point, and thus better resistance to atom mobility due to high current density (electromigration). However, Cu did not replace Al until the 130 nm technology generation due to significant issues with integration of copper interconnects into integrated circuit process flow. Due to pattern density restraints, copper could not be wet etched, and there was not a viable technology for plasma etching Cu. The breakthrough for Cu implementation came from the invention of copper damascene technology [2]. In Cu damascene, a copper interconnect is fabricated by plasma etching a trench in the ILD, then blanket coating the trench with a diffusion barrier and a copper seed layer. The wafer is then coated with copper using electroplating, filling the trench. The excess copper, seed layer, and diffusion barrier are then removed using Cu CMP. There are several strategies for damascene Cu, including dual and single damascene, which involve the fabrication strategy for the interconnect and the via (vertical metal connection that connects different metal layers). CMP has also enabled novel integration schemes for transistors, such as the gate last approach for high-k metal gate transistors [3]. Today, Cu CMP comprises approximately 50% of the CMP process steps in logic manufacturing.

It is clear from the above discussion that CMP has been a key enabling technology for the semiconductor industry and this is one of the main attributes of CMP. Another attribute of CMP is that it creates new surfaces and eliminates defects on incoming wafers. Finally, CMP planarizes surfaces on many different length scales from the nano to meso and macro scales. Achieving such planarity is important in the continued use of optical lithography, and is finding fresh application in fields such as opto-electronics. In the following chapter, we will discuss the fundamentals of CMP, and three applications of CMP to nanotechnology, namely, nano-interconnects, nanodevices, and nano-smooth surfaces.

10.1.1 CMP Fundamentals

As the name suggests, CMP combines both mechanical and chemical polishing to achieve a very high quality surface finish. However, this process cannot be understood as a simple addition of chemical and mechanical effects; the underlying mechanisms by which polishing occurs during CMP are not fully understood. The outcome of a CMP process is affected by a wide range of inputs relating to both the chemical and mechanical aspects of the technique, such as the speed and pressure of polishing, the roughness and hardness of the polishing pad and the pH of the polishing slurry. Although CMP is used to manipulate the surface of a wafer at a very fine scale, the process itself is controlled almost exclusively by these macroscopic parameters.

In CMP, a wafer is pressed against a rotating polishing pad in the presence of a polishing slurry. The slurry contains both chemically active ingredients, such as oxidizers and surfactants, and mechanically active abrasive particles. A schematic diagram of the polishing process is shown in Fig. 10.1. During polishing, there are also many other process and equipment variables such as pad conditioning and various rotation rates and directions, as well as sweep directions shown on the

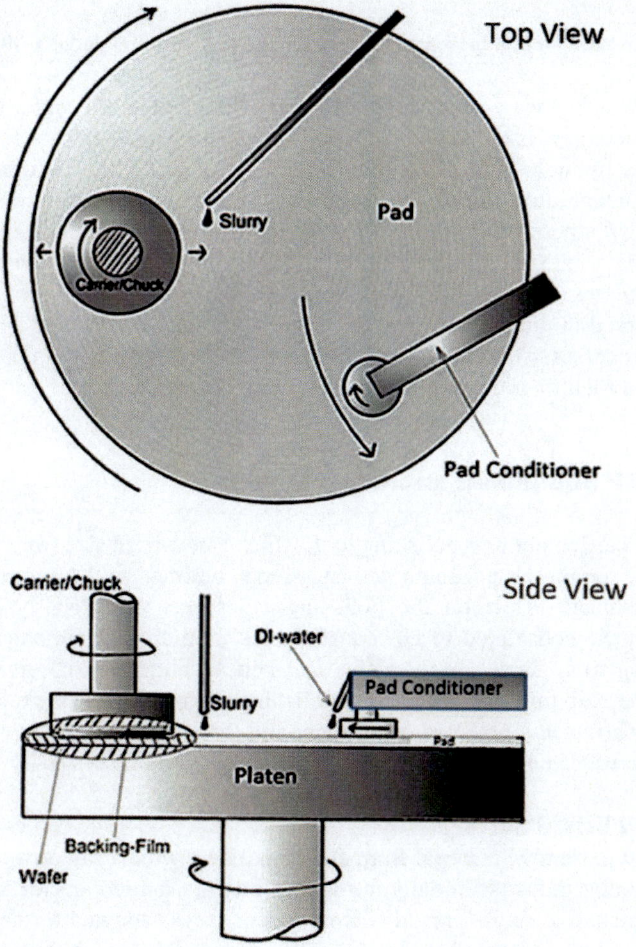

Fig. 10.1 Schematic diagram of CMP process showing the top and side views [4]

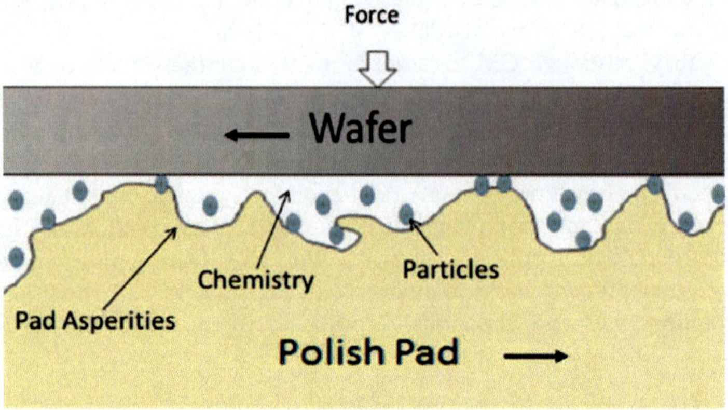

Fig. 10.2 Expanded view of the region between the pad and the wafer during CMP

figure. A magnified view of the region between the pad and the wafer is shown in Fig. 10.2. This figure is not to scale, but during polishing, the gap between the wafer and the pad is thought to be 60–80 μm [5] under typical CMP conditions, which is of a similar magnitude as the pad roughness. The abrasive particles are typically smaller, with diameters of 10 nm to 5 μm. In addition to removing material from the wafer surface at rates of several hundred nanometers per minute, planarity lengths of 20–30 mm are achieved [2].

As discussed in the previous section, CMP combines mechanical and chemical polishing processes. These processes will be described separately in the following sections, and then their combined effects will be discussed.

10.1.2 CMP Equipment and Consumables

Polishing is carried out on a polishing tool, which can vary in size from table-top to the size of a room. The polishing pad and slurry, outlined in the previous section, have a significant effect on the polishing outcomes. Additionally, both these components are considered to be 'consumables' and are not permanent parts of the polishing tool. They are expensive and hence contribute significantly to the cost-of-ownership for CMP processors. Polishing uniformity and the rate of defects are strongly affected by the pad conditioner and the wafer cleaning equipment and procedure used. The role of these items will be described individually here.

10.1.2.1 Polished Surfaces

CMP is used to remove material from and planarize a wide array of surfaces in the microelectronics industry. Initially introduced into the semiconductor industry as a method to planarize single-crystal silicon wafers, it was adapted for device manufacture in 1983 when it was used to planarize reflown glass in isolation trenches [6] Even though this process was never used commercially, it led to the development of

Table 10.1 Surfaces processed using CMP

Metal, Metalloids	Ceramics	Emerging
Silicon	Silicon dioxide	Carbon nanotubes [7]
Tungsten	Silicon nitride	Zinc oxide [8]
Aluminum	Low-k dielectrics	Bi_2Te_3 [9]
Copper	Tantalum nitride	
Tantalum	Gallium nitride	
Germanium		

CMP processes to planarize other insulating materials, such as SiO_2 and Si_3N_4 for shallow trench isolation. It is also used in back-end-of-line (BEOL) processing to form vias made of tungsten, aluminum and copper. CMP enables other fabrication techniques such as damascene and dual-damascene copper technology, where polishing of both metals and barriers such as Ta and TaN is required. Current research topics in CMP include the planarization and removal of low-k dielectrics, and planarization of surfaces such as germanium and gallium nitride for optical devices.

A broad but not exhaustive list of surfaces processed using CMP is shown in Table 10.1.

10.1.2.2 Polishing Tools

All CMP tools are based on just two essential components; the wafer carrier and the polish platen. The wafer carrier must apply load and rotate. The polishing platen must resist the load applied by the carrier and also rotate. The simplest CMP tools consist of little more than this, with the slurry feed and conditioning systems typically added by the user. Improvised tools using a conventional metallographic polishing wheel as a platen, and a drill press or other electric motor as a carrier, have been used extensively in research (e.g. [10–12]). These tools typically have a pad diameter of 0.3 m, or 12″, and can be used to polish wafers up to 0.1 m, or 4″, in diameter.

Several of the CMP consumables, such as the polishing pad and the conditioner head, must be kept damp in order to avoid damaging them. Integrated, research-scale polishers such as the CETR Tribopol and the Logitech Tribo are still small enough to be housed on a laboratory bench top (hence their common designation as "benchtop polishers") but contain integrated conditioning and fluid delivery systems. These incorporate automatic pad-wetting functions. They may also have a specially designed carrier to distribute the pressure across the wafer surface more evenly. Additionally, as these are intended primarily for research, they usually incorporate a number of metrology tools, such as pad temperature monitoring and real-time friction and acoustic emission analysis. Like the improvised polishers, benchtop tools typically can be used for polishing wafers up to 0.1 m or 4″ in diameter, but may have platen diameters of up to 0.5 m or 20″.

Polishing for production in a fabrication facility has several additional challenges that have resulted in the development of larger and more sophisticated tools, though the basic components remain the same. Due to the large volumes of wafers polished, modern tools accept cassettes of wafers and automatically load, polish and unload them. CMP involves the extensive use of small particles which are typically an anathema to cleanroom facilities. As a result, "dry-in, dry-out" loading and unloading, with wafer cleaning incorporated into the process, is also becoming the industry standard, so that all the polishing occurs in a contained environment isolated from the rest of the fabrication facility. In order to increase throughput, these tools typically have two or more polishing platens. For example, copper CMP for damascene manufacturing is typically carried out on three platens, each with its own specialized pad and slurry. On the first platen, the bulk of the copper is removed with a high rate polish. The second platen is used to remove the last of the copper at a much lower rate while stopping on the diffusion barrier. Some dishing or other topography may develop at this stage as some areas are overpolished to ensure complete removal of the copper from other regions. At the third and final platen, the barrier is removed and the wafer is planarized to remove the topography developed in the previous stage. The wafer is then cleaned by a wafer scrubber before being dried and reloaded into a cassette. A schematic of this polishing sequence is shown in Fig. 10.3.

Each process is optimized to take approximately the same length of time to maximize throughput and manufacturing efficiency. Industrial CMP tools typically can polish wafers up to 0.3 m (12″) in diameter, with wafer diameter increasing to

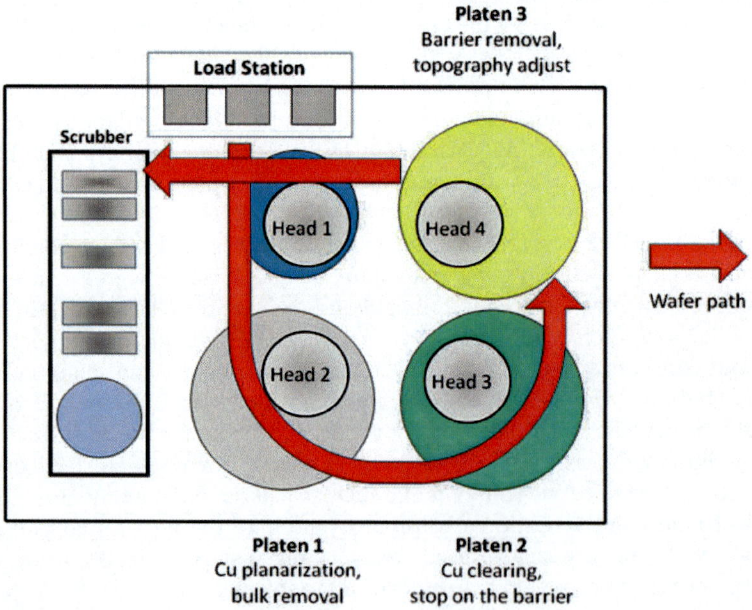

Fig. 10.3 Schematic diagram of a production CMP machine

0.45 m in the future. The platen size is correspondingly larger and may be up to 0.75 m or 30″ in diameter.

10.1.2.3 Slurry

Polishing slurries vary in composition according to their intended use. However, as suggested by the process name, they all contain both chemically and mechanically active components. The mechanically active role is performed by abrasives in the slurry, while the chemically active portion of the slurry consists of substances chosen to either soften or oxidize the polished surface. Both these components will be discussed here, as well as other additives that may be used but do not contribute directly to polishing.

In all slurries, the active components are diluted in water, usually de-ionized. In many cases, these active components only make up a few percent of the total slurry volume, with the balance consisting of water.

Mechanical Components: Abrasives

Abrasives used for CMP are typically hard, small oxide particles suspended in the slurry. For polishing silicon oxide and nitride, such as in shallow trench isolation (STI), the abrasives are usually fumed or colloidal silica, alumina or ceria. These typically range in size from several nanometers to several microns in diameter. The size distribution is generally narrow, especially for colloidal particles formed by precipitation techniques. Oversize particles can cause scratches on the wafer surface and hence are removed by filtration prior to use. Aggregates of the particles can also lead to scratching due to their size and so they are also eliminated from the slurry, typically by stirring and using chemical dispersants. These substances will also be briefly discussed in the description of chemical additives. The most common particle morphology is a sphere, although the effect of using particles that are plate-like [13] or roughly cubic [14] has also been investigated.

A number of composite particles have also been fabricated and used in oxide CMP. These particles consist of a softer core coated with smaller, harder and occasionally more reactive particles. The aim of such particles is to prevent gross scratching defects on the wafer by making the particles soft, while still maintaining high polish rates with the encrusted oxide. Polymeric cores with yttria crusts [15] and silica cores with ceria crusts [16] have both been examined; these particles are shown schematically below. A TEM image of Fig. 10.4b is shown in Fig. 10.5. The particles are usually present in the slurry in concentrations of between 1 and 10 wt%, though this can vary for specialized applications.

The effect of the particle composition, size, shape and concentration on the polishing process is complex and not well understood. Some research suggests that the rate of polishing decreases with increasing particle size and increases with increasing particle concentration [17, 18]. Other studies on different substrates suggest that increasing the particle size increases the polishing rate [19], while still others suggest that the particle size has very little effect on polish rate [20]. Similarly, many researchers do not differentiate between particle compositions in their models of the CMP process, (for example the work of Fu, Chandra et al. [21]

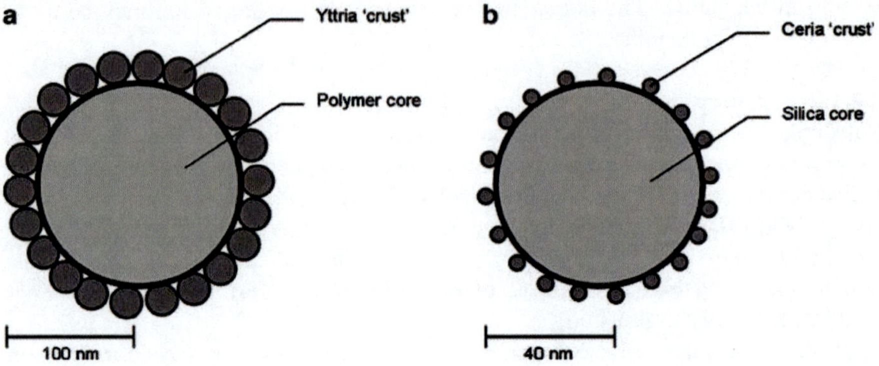

Fig. 10.4 Schematic diagram of (**a**) polymer and (**b**) silica primary particles with yttria and ceria nanoparticle surface layers or 'crusts'

Fig. 10.5 TEM image showing ceria nanoparticles attached to the surface of silica particles

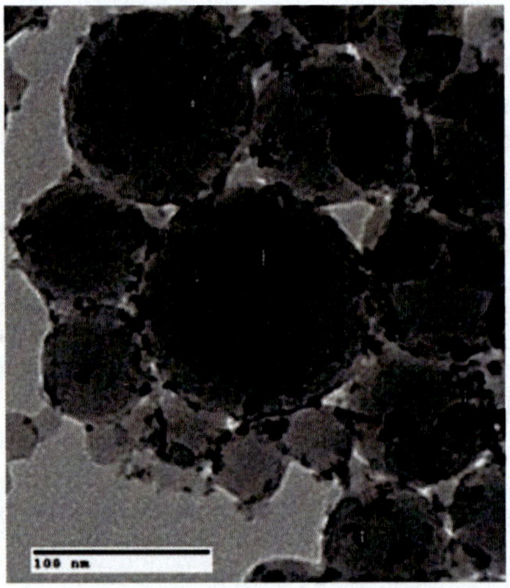

and Che, Guo et al. [22]), while others suggest that chemical interactions between the particle and substrate are key to polishing [23]. This concept will be discussed further in the next section.

Chemical Components

The roles of the chemicals in the polishing slurry are diverse; the most commonly used classes of additives, based on their function in the slurry, will be discussed here.

pH Adjusters and Buffers

The pH of the polishing slurry is of high importance for virtually all types of wafers and abrasives. One of the primary methods for preventing particle aggregation in the polishing slurry is to maintain a strong surface charge of the same polarity on all the particles, so that they repel each other. This is demonstrated in the schematic zeta potential – pH plot shown in Fig. 10.6.

From this plot, it is evident that the surface charge on a material can be manipulated to achieve a desired outcome using the solution pH. As demonstrated in the schematic, two different surfaces (such as an abrasive and a wafer) may have different surface charge characteristics. In the case shown here, this leads to a pH region ($6 < pH < 10$) where the surfaces have charges of opposite polarities and hence attract each other. This may be useful during polishing to maintain high polish rates. Similarly, at a pH above or below this range, the particles and wafer will have the same polarity and hence will repel each other, a property that is often utilized for cleaning the wafer.

The slurry pH also has a significant effect on the polishing process by modifying the wafer surface. The dissolution rate of silicon dioxide (SiO_2) and other silica-based materials increases with pH. This is illustrated in Fig. 10.7 for amorphous silica at around ambient temperature. This accelerates removal of the SiO_2 surface such that polishing rates increase significantly at pH above 10. The mechanisms behind this increase will be discussed in detail in Sect. 2.1.

The pH also affects the oxidation behavior of metallic surfaces, which can in turn alter the polishing rate. Pourbaix diagrams depict regions of thermodynamic oxidation behavior over a range of pH and potential, and are a useful tool in designing slurry pH. The Pourbaix diagram for copper in water is shown in Fig. 10.8 as an example; from this plot, it is evident that copper tends to oxidize at $6.5 < pH < 13.5$, and may dissolve at lower pH values. For many metal surfaces, polishing is achieved by oxidizing the surface and then removing it by abrasion, so controlling the pH is important in controlling this mechanism. This will be discussed further in the following section.

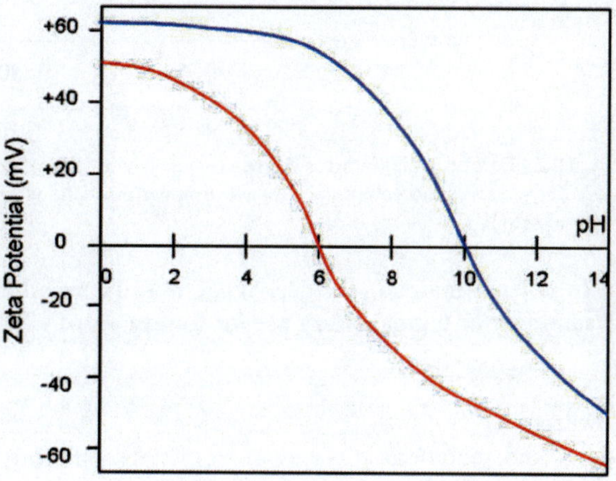

Fig. 10.6 Schematic plot of zeta potential vs. pH for two different substances

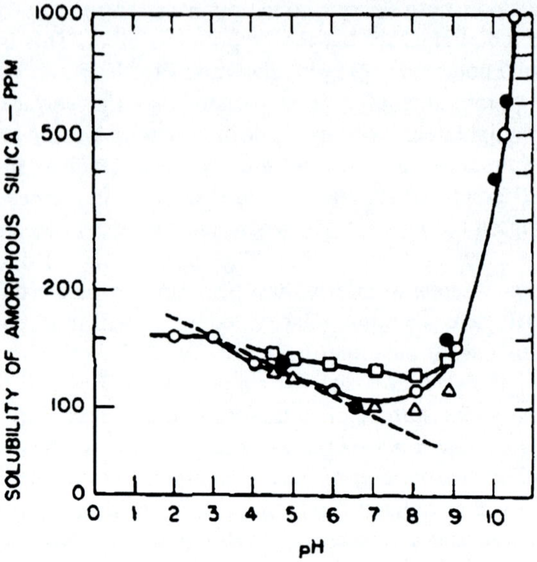

Fig. 10.7 Solubility of amorphous silica vs. pH [24]

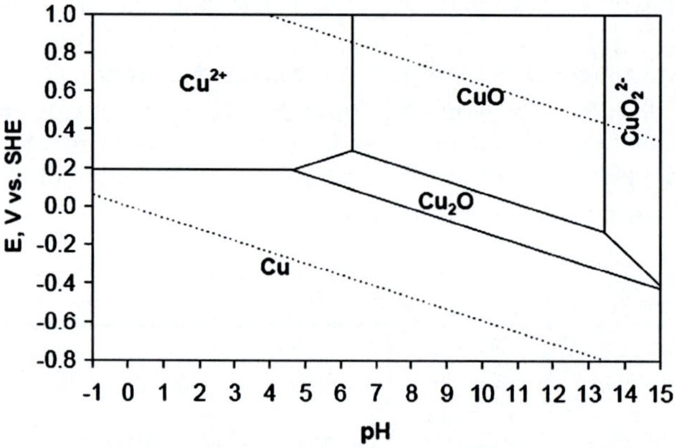

Fig. 10.8 Pourbaix diagram for copper in water, with a total copper activity in solution of 10^{-5} [25]. The y-axis is the potential of the environment in volts with respect to the standard hydrogen electrode (SHE)

In order to maintain the design pH, buffers are often added to polishing slurries. A summary of typical slurry pH for four types of CMP is shown in Fig. 10.9.

Oxidizers

Metal and metalloid surfaces are typically oxidized during polishing to enable rapid, even, defect-free polishing. The mechanisms by which this occurs are

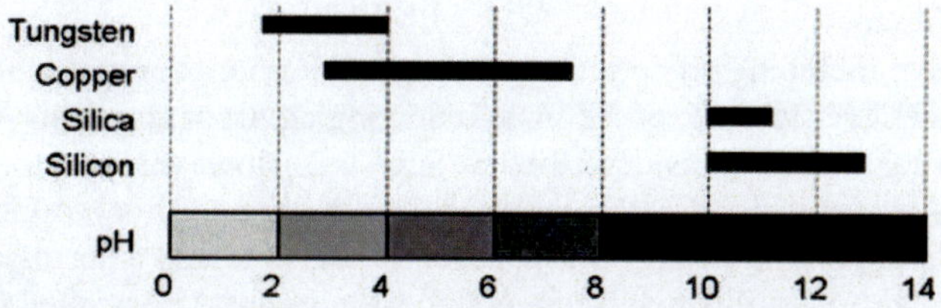

Fig. 10.9 Typical slurry pH values for various CMP processes

discussed in Sect. 2.2; however, regardless of mechanism, oxidizers are required to induce this process and will be discussed here.

A successful CMP oxidant should rapidly induce oxidation of the wafer surface, forming an adherent, continuous and stable oxide layer. It should also do so uniformly, without generating pits or preferentially etching grain boundaries. Where no oxidizer is known that meets all these criteria, a corrosion inhibitor (such as those described in the next subsection) can be used in conjunction with the oxidizer to help stabilize the film formed. The emphasis on oxidizer selection is then on rapid reaction, and pitting and etching resistance.

Because the wafer surface essentially undergoes controlled corrosion during polishing, electrochemical techniques are very often used to evaluate the performance of oxidizers. A description of electrochemical characterization methods is not within the scope of this work, but several references are available (e.g. [26]).

Oxidizing acids, specifically nitric acid, were the first oxidizers adopted in CMP. Nitric acid was found to give poor results when used with copper as the oxide layer formed was unstable and provided no passivation of the underlying metal [27].

To achieve oxidation in basic conditions, ammonium hydroxide is often used as the oxidizing agent. A 1 M solution of NH^{3+} in water has a pH of 11.6, which is sufficiently high for hydration and softening of the oxide layer in STI-type processes. Ammonium hydroxide may also be used to oxidize metals, especially where the Pourbaix diagram suggests that a high pH is desirable for polishing. However, it also acts as a chelating agent for a number of metals, including copper, silver, cobalt and zinc [28], and their oxides, including CuO [2]. This destabilizes the passive film formed and can lead to preferential etching of grain boundaries in copper, excising grains and roughening the metal surface [29].

Generally, hydrogen peroxide has been found to be the most suitable oxidizer in metal CMP. It avoids the issues apparent with HNO_3 in that it passivates the underlying surface, and unlike NH_4OH it does so without dissolving the oxide film or metal at the same time. The use of H_2O_2 with copper in particular has the additional advantage that Cu^{2+} ions catalyze the decomposition of the H_2O_2 into OH^* free radicals, which are stronger oxidizers than H_2O_2 alone and further increase the slurry's ability to quickly and evenly passivate the metal surface [30]. This process is known as the Fenton cycle and it also occurs with other transition metals, most notably iron.

Corrosion Inhibitors

Many surfaces, including silicon, tungsten and aluminum, form a continuous and adherent oxide layer when exposed to an oxidizing environment. This layer is then removed during polishing, as described in Sect. 2.2. However this does not occur with all surfaces for which CMP is required, leading to poor final results in several areas, notably copper polishing. One solution to this is to add a corrosion inhibitor to the slurry. Although this may seem counter-intuitive, as controlled corrosion is desirable and necessary for successful polishing, the role of the corrosion inhibitor in these situations is to stabilize any surface oxide that does form and prevent it detaching from the surface, rather than prevent the corrosion reaction from happening, for example by preferentially reacting with the cathodic processes in play. This stabilization is usually achieved by chemisorption of the corrosion inhibitor onto the existing film as a chelate, bonding it tightly. As most other wafer surfaces processed using CMP are passivating, the discussion in this section will be restricted to copper.

The most common corrosion inhibitor used in copper CMP is benzotriazole, or BTAH. This molecule has been used extensively and for many years as a corrosion inhibitor for copper in applications such as water cooling towers. The thermodynamic suitability of BTAH is demonstrated by the expanded passivity region of the Pourbaix diagram, shown shaded in Fig. 10.10.

The slow reaction kinetics and potential environmental harms associated with BTA [31] have led to considerable research on alternatives, especially amongst

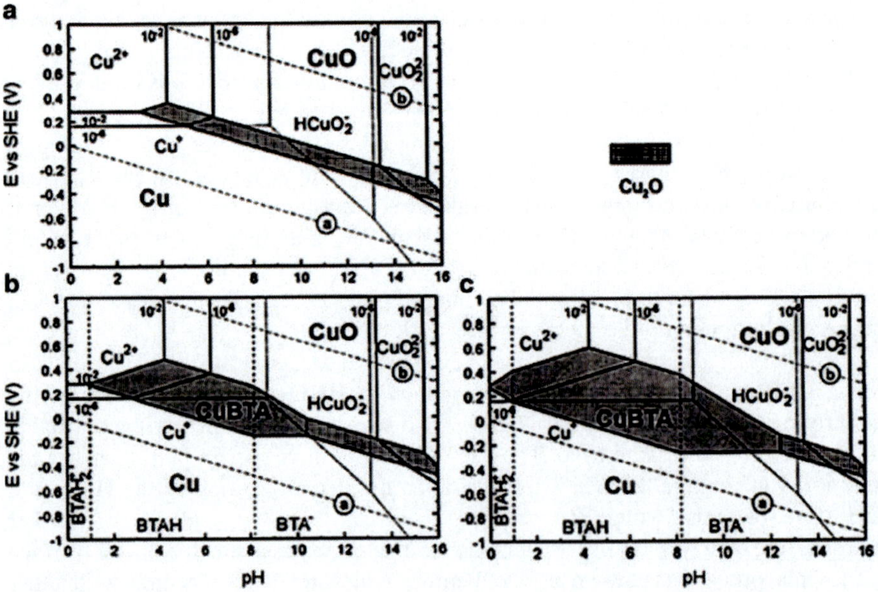

Fig. 10.10 Pourbaix diagrams for the copper-water-BTA system, with (**a**) no BTA, (**b**) $\{a_{BTA}\} = 10^{-4}$ and (**c**) $\{a_{BTA}\} = 10^{-2}$ [32]. Reprinted with permission from J. Electrochem. Soc., 145, L42 (1998). Copyright 1998, The Electrochemical Society

surfactants as many of these are known to chemisorb onto surfaces in a manner similar to that of BTA. At the forefront of these alternative corrosion inhibitors are ammonium dodecyl sulphate (ADS), Triton X-100 and D-TAB [2, 33–35].

Chelating Agents

Chelating agents are added to CMP slurries to dissolve any polishing debris removed from the wafer surface and avoid scratching. In metal polishing processes, the chelating agents may also prevent the redeposition of removed material on to the polished surface. A number of chelating agents are typically used. For copper CMP, these include citric acid and glycine. Chelating agents and corrosion inhibitors have a complex impact on polishing rates. The addition of either can lower or raise the polishing rate, depending on their relative concentration and the oxidizer concentration present.

Other Additives

Several other substances are typically added to commercial slurries to prolong the slurry pot life and stability. These may include stabilizers for volatile compounds such as hydrogen peroxide and colloid stabilizers. Biocides to prevent bacteria growth and other proprietary substances to modify the slurry viscosity [36] may also be added.

10.1.3 Pads

Polishing pads are typically made from cross-linked polyurethane, and are a few millimeters in thickness. They are fixed to the platen with pressure-sensitive adhesive on one side and are grooved on the uppermost, polishing face. Pads for different applications have different hardness and microstructures, and may be grooved in different patterns.

The pad has multiple roles in the CMP process, and both its microstructure and macrostructure are important in achieving those roles. As well as holding the abrasives and slurry against the wafer surface to achieve polishing, the pad controls the distribution of the slurry from the centre of the pad to the edge through the use of grooves. Common grooving patterns are shown in Fig. 10.11 and a single pad may combine two or more of these, with simultaneous radial and concentric grooving frequently used.

It is interesting to note that the actual utilization of slurry in CMP processes, while dependent on a number of process variables, has been shown to be as low as 5% [37]. Because of their role in slurry transport and in establishing the lubrication regime under the wafer, grooving may also have a significant effect on the polishing temperature (Fig. 10.11).

Away from the grooves, pads generally have either a porous or fibrous microstructure, with the porous type being more common. Representative SEM images of both types are shown in Fig. 10.12. These microstructures result in a rough polishing surface, with average roughness on the order of 2 μm and peak-to-valley heights as high as 20 μm.

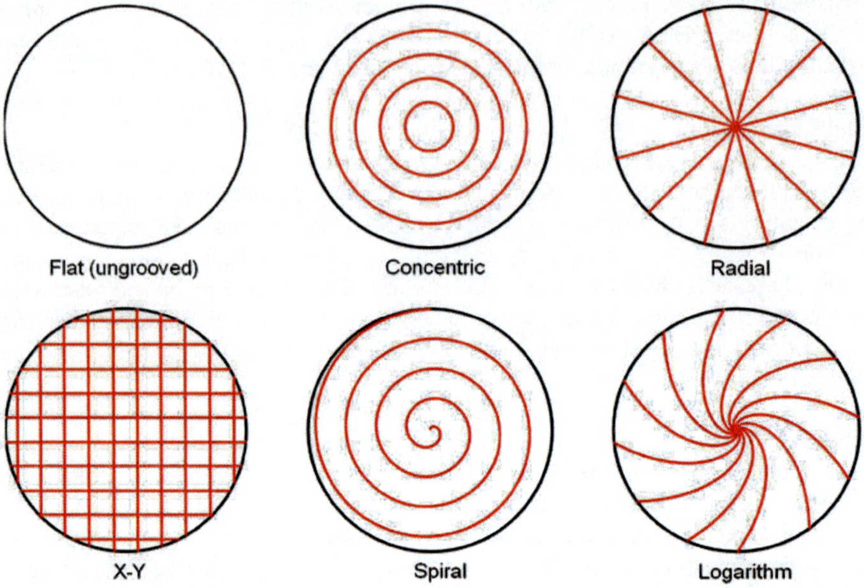

Fig. 10.11 Pad grooving patterns

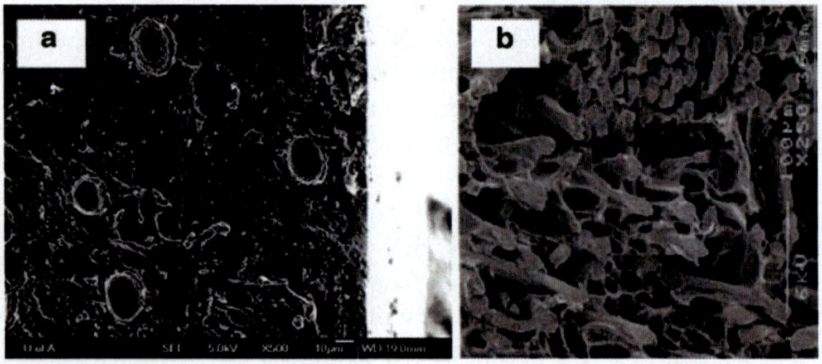

Fig. 10.12 SEM micrographs for pads with (**a**) porous microstructure [38], and (**b**) fibrous microstructure [39]

The mechanical properties of the pads vary significantly between pad types, within pads and over time. Pads are designed with different hardness for different applications, with harder pads being used for silicon dioxide and other hard wafers, and soft pads for metals, primarily copper. The hardness of the pad is controlled by the chemical make-up of the polyurethane, especially the degree of crosslinking, but also by the size and number of pores present. This can lead to significant variation in hardness within each pad, as shown in the spatial plot of hardness recorded for a typical pad in Fig. 10.13. In this figure, the detected pad hardness

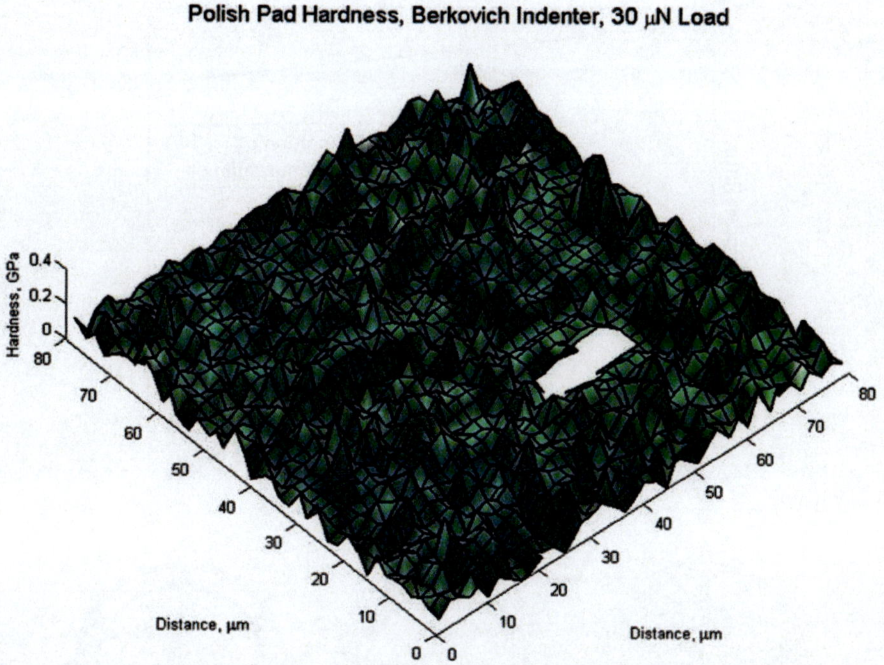

Fig. 10.13 Map of pad hardness, demonstrating its variability [38]

varies from 0.004 to 0.4 GPa over an 80 μm × 80 μm area. The pad mechanical properties also vary significantly over time due to the combined effects of exposure to slurry chemistry, constant immersion in water, and conditioning. Some studies suggest that the pad storage modulus can decrease by as much as 20% after prolonged exposure to liquid [40].

Conditioning consists of dragging a rotating diamond grit disk back and forth across the pad surface. Conditioning is necessary to refresh the pad surface, which otherwise becomes 'glazed'; without conditioning, the polish rate quickly drops away, as shown schematically in Fig. 10.14 for polishing materials such as silicon dioxide. Additionally, the pad is conditioned before use to cut the 'skin' off the pad and open up the pores. The process of constant renewal does however wear the pad, requiring it to be replaced periodically. This is why the pad is also classed as a consumable in CMP technology.

10.1.4 Pad Conditioners

As described above, conditioning is the process of renewing the pad surface with a diamond grit disk to ensure consistent polishing rates. This is essential to

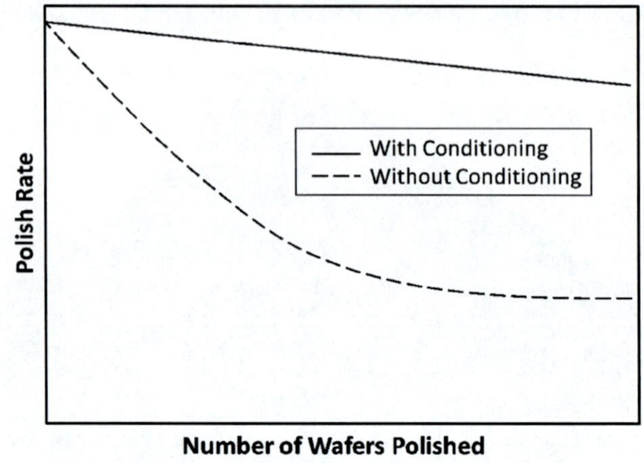

Fig. 10.14 Schematic diagram comparing the decay of polish rate for polishing with and without conditioning

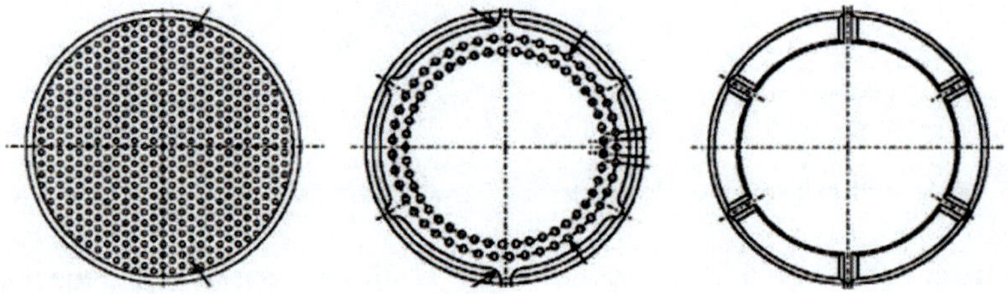

Fig. 10.15 Conditioner configurations (Adapted from Ref. [5])

maintaining consistent polishing conditions, although pad wear debris generated by conditioning also correlates with higher levels of wafer scratching [41].

Conditioning plays a role in distributing slurry across the pad by dragging it from the centre, where it is poured onto the platen, to the edge. The style of conditioner used, its speed and the conditioning pressure can therefore influence fluid distribution in polishing [5]. Some examples of conditioner styles produced by Mitsubishi Materials Corporation are shown in Fig. 10.15; however many other designs are available.

Like the polishing pad, the conditioning disk must be kept wet at all times to avoid damage. The disk can also wear, causing diamond particles to detach and scratch the wafer. The diamond disks are therefore also classed as 'consumables' for CMP and are periodically replaced.

10.1.5 Cleaning

Following polishing, wafers must be cleaned to remove residual chemicals and adherent particles that could contaminate future fabrication stages. Particle

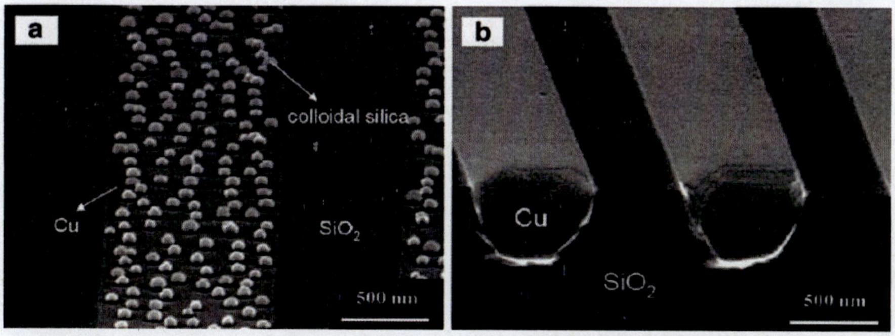

Fig. 10.16 Copper and silica surfaces after polishing and rinsing with DI water (**a**), and following cleaning (**b**) [42]

adhesion in particular is problematic as the adhesion force between particles and substrates can be high, approaching 500 nN for alumina particles on various substrates in some studies [44]. An SEM image of a silica substrate with copper features fabricated by a damascene process is shown in Fig. 10.16. The device on the left has been polished using silica particles, then rinsed thoroughly in water. Many particles still adhere to the copper surfaces. In contrast, the same device is shown on the right after undergoing a brush-scrubbing cleaning technique.

Particles can be prevented from adhering to the wafer using electrostatic repulsion, or removed after adsorption by mechanical or fluid shearing, or both. Electrostatic repulsion between the wafer surface and the particles is achieved by varying the pH to point where both particles have the same polarity. This is demonstrated in the zeta potential – pH diagram shown in Fig. 10.6. The image above shows this in a practical sense; after polishing, the colloidal silica particles adhere to the copper but not the silica substrate. This is because, in the same solution, the silica particles and silica substrate must necessarily have the same charge and repel each other. Particles can also be removed by etching away the surface to which they are attached, or 'underetching'. Studies have shown that approximately 3 nm of the surface must be removed to reliably detach the particles [44]. As device size decreases, this removed material forms a larger proportion of the device itself and may not be acceptable, depending on the dimensions and applications of the device.

Where preventing contamination in this manner is either impractical or impossible, particles are removed from the wafer by mechanically shearing with either a brush or a fluid. Brush scrubbing uses rolling brushes to gently buff the polished surface in conjunction with a cleaning fluid. The brushes generate a sufficient level of fluid shear to roll or lift adherent particles from the wafer surface [44]. These are prevented from redepositing on the wafer by using the cleaning solution chemistry to develop electrostatic repulsion with or high wettability of the wafer surface.

Brush scrubbing, though effective, is a contact technique that applies a pressure to the surface, and the particles on that surface. It can therefore lead to scratching of the wafer. Megasonic streaming, by contrast, is a non-contact technique for cleaning. In this technique, the wafer is submerged in a fluid that is agitated by sound waves at a very high frequency (up to around 700 kHz). The sound waves generate fluid streams in the tank and moreover induce very thin boundary layers on the submerged surfaces. The combination of high fluid velocities and thin boundary layers induces very high viscous shear forces along the wafer surfaces, removing any adherent particles. As in brush cleaning, the cleaning fluid can be optimized to prevent redeposition of any particles through electrostatic repulsion or wettability [45].

10.2 Polishing Mechanisms

From the previous section, it is evident that polishing is a complex, multi-parameter process that combines many macro-scale inputs to control a surface at the nano-scale. While the individual roles of many polishing components are known, the mechanisms behind polishing are not as well understood. In this section, proposed polishing mechanisms for three classes of process will be discussed. These are CMP of hard materials such as oxide, metals, and mixtures of materials (achieving selectivity). Attempts to mathematically model the polishing process, and the difficulty in doing so, are then discussed.

10.2.1 Polishing Silicon Dioxide: The "Chemical Tooth"

The process of polishing hard materials is thought to be largely mechanical, with chemical contributions from the slurry pH and the activity of the polishing particle.

"Chemical Components" in Sect. 1.2.3 briefly refers to dissolution of silica as fundamental to polishing. In this context, dissolution is a process of siloxane bond-breaking and hydration at the silica surface. In this sense, it is analogous to the depolymerization of silica. This process can be represented by:

$$\equiv Si - O - Si \equiv + H_2O \Leftrightarrow 2 \equiv Si - OH$$

In this expression, water penetrates the siloxane bonds on the surface and forms ≡Si–OH species. The importance of this process on polishing has been demonstrated in a number of experiments; polishing has been carried out in liquids with varying hydroxyl concentrations, such as simple alcohols in series from methanol to n-dodecanol, and also in liquids without hydroxyl groups, such as oils and paraffin. These experiments confirm the importance of hydroxyl groups; polishing rates are close to zero where no hydroxyl groups are present and increase with increasing hydroxyl concentration [23].

Why does a hydroxylated surface contribute so greatly to polishing for surfaces of this nature? When contacted by a polishing particle, ≡Si–OH species allow one

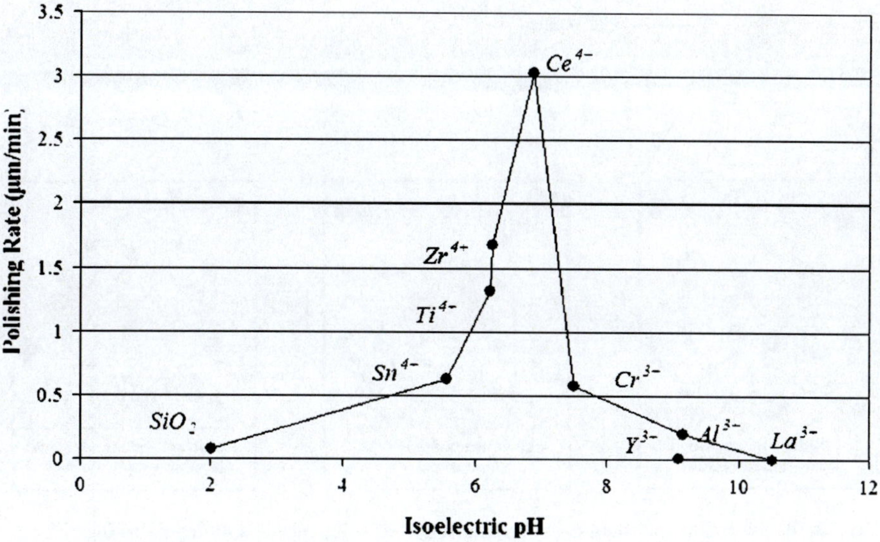

Fig. 10.17 Variation of material removal rate with oxide ion valence (Adapted from Ref. [23])

or more silica tetrahedra to chemisorb on to that particle and lift away from the surface. In contrast, fully networked silica that has not been hydroxylated is not able to chemisorb on to the surface and is not removed.

One side of this chemisorption couple is obviously the surface ≡Si–OH species developed on the wafer surface. The other side is the particle; thus, the chemical activity of the polishing particle and its ability to chemisorb ≡Si–OH also affects polishing. This has been borne out by polishing studies that examine the activity of polishing particles at their near-neutral surface charge. These studies indicate that increasing particle surface activity, represented by valence, corresponds with a higher polishing rate, as shown in Fig. 10.17.

This suggests that silica is slow to polish silica surfaces. While this can be true, using a silica-silica system has the advantage that no foreign oxides are introduced into the system. Additionally, as noted in section "Mechanical Components: Abrasives", silica particles are readily available, inexpensive and generally have good size control.

Following chemisorption of the wafer material onto the particle, the particle moves away from the wafer surface. At this point, the bonded material may adhere to the particle and be removed or may remain on the wafer, depending on the relative bond strengths. If it is removed, it may either remain on the particle, leading to a gradual build-up of material there, or enter the polishing fluid where redeposition onto the silica surface can occur.

This process of material removal, known as "chemical tooth" and described by Cook in his seminal 1990 paper [23], is shown schematically in Fig. 10.18 with silica as the abrasive material.

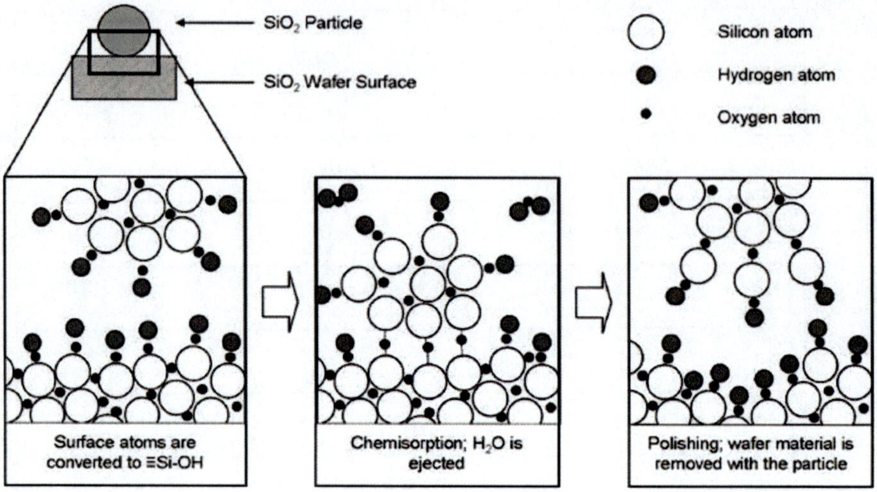

Fig. 10.18 Schematic outlining the chemical 'tooth' of silica on silica during polishing

Oxide polish is optimized primarily through the manipulation of the slurry pH and the particle type. These are altered to maximize the adhesion of the wafer surface molecules to the abrasives, and minimize redeposition of material back onto the wafer. Polishing pressure and velocity are used to control the number of particle-wafer contacts and the shear on the particles at the surface.

This same mechanism is used to polish silicon surfaces. In this case, an oxidizer is added to the slurry to convert the silicon surface to SiO_2. Polishing then proceeds as outlined above.

10.2.2 Polishing Metals

As with silicon, metal surfaces are polished first by converting them to an oxide. This is achieved by adding an oxidizer, selected by the criteria outlined in "Chemical Components" in Sect. 1.2.3, to the polishing slurry. Both oxidation and abrasion are required to effect polishing. Numerous studies ([46] for example) have demonstrated that polishing rates with abrasives only, or oxidizer only, are near negligible and that polishing is achieved only by the combination of oxidizers and abrasives.

The advantages of oxidizing a metal surface prior to polishing it are two-fold. Firstly, the oxide layer is hydrated due to the aqueous chemistry of polishing slurries and is very often softer than the underlying metal. This makes removal by mechanical abrasion easier and increases polishing rates. Secondly, the oxide layer, if continuous and adherent, passivates the surface it covers. This protects that surface from damage and dissolution when it is not directly in contact with an abrasive and/or the pad. As described in section "Chemical Components",

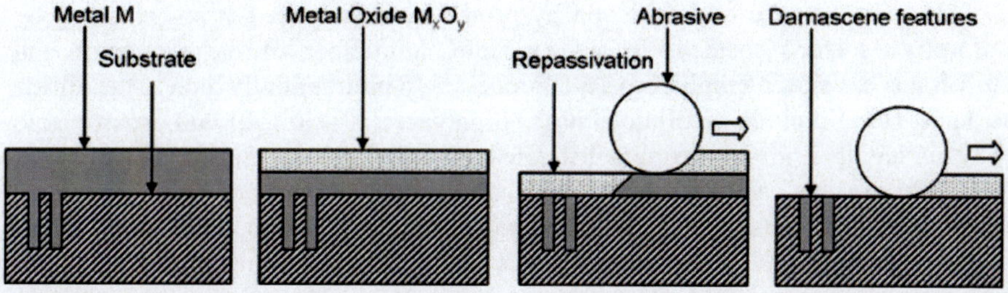

Fig. 10.19 The process of passivation, mechanical removal, and repassivation as used in damascene fabrication

a corrosion inhibitor can be added to the slurry to enhance this effect in metals that do not form such an oxide naturally.

Once the oxide layer is removed by abrasion, the underlying metal is quickly repassivated. This cycle of passivation, removal and repassivation is repeated until the desired amount of metal is removed. This is shown schematically in Fig. 10.19 for the formation of a damascene feature.

Metals which are known to form a passivating film during CMP include tungsten and aluminum. Copper, in contrast, does not form such a layer and considerable research has been dedicated to identifying slurry additives that promote passivation.

The mechanical action of the abrasives on this passive film is not well understood. As well as the "chemical tooth" described in the previous section, researchers have postulated that the abrasives remove the oxide by mechanical scratching, with a particular emphasis on sliding rather than rolling wear [47]. This may be enhanced by particle adhesion [43, 48], or by the intersection of particle paths [22]. Fatigue of wafer asperities by fluctuating hydrodynamic pressures, leading to fracture, has also been proposed as a removal mechanism [49]. As in silica polishing, redeposition of material removed from the wafer surface can occur [50] and is controlled by chelating agents added to the slurry.

This lack of understanding makes CMP of metals more difficult to predict and control than oxide CMP. At the same time, while oxide surfaces are often simply planarized using CMP, the technique is used with metal for fabrication processes such as damascene and dual-damascene that require greater control.

10.2.3 Polishing Mixed Surfaces (Selectivity)

There are many situations in which CMP of a mixture of materials is required. One common example is shallow trench isolation (STI), in which CMP is used to remove silicon dioxide until an underlying silicon nitride layer is reached. Another example is damascene fabrication where copper is polished until an underlying barrier layer is exposed. The barrier layer is then also removed. In both these situations, the ability to polish one surface but not another is critical to achieving a planar end product.

Selectivity is generally achieved by protecting some areas from polishing by adsorptive surface species. As an example, a number of organic acids and surfactants have been employed in STI processes to preferentially bind to the nitride surface. These include picolinic acid [51], polyacrylic acid [52] and aspartic acid [53], achieving oxide-to-nitride polish rate ratios of 32, 77 and 80 respectively. The size of the particle can also play a role in selectivity. A number of researchers have noted that ceria abrasive size can affect selectivity, even when all the other slurry components are held constant [52]. This is thought to be because the anionic adsorbing species are able to interact with both the nitride and the ceria surfaces. Silicon dioxide has a point of zero charge (PZC) of pH ~ 2, while the PZCs for ceria and silicon nitride are ~7 and ~6.5. Thus, at an intermediate pH, the surfaces of both the nitride and ceria are positively charged and can interact with negative ions. While these ions may initially suppress removal of the nitride surface, they will adsorb onto the ceria particles as they polish. The perceived size effect is thought to stem from this phenomenon; the specific surface area of small particles is considerably greater than that of large particles and they are able to absorb more surfactant from the nitride surface at a given weight or volume concentration.

Selectivity in copper CMP is more complex, in that selectivity between two metals (namely copper and tantalum) is typically required. This is often achieved in practice by manipulating the pH of the slurry [34]. As noted in "Chemical Components" in Sect. 1.2.3, the Pourbaix diagram for copper shows regions of passivation, dissolution and immunity, with the passive regime preferred for polishing. In contrast, the Pourbaix diagram for tantalum indicates passivity across most of the pH range (Pourbaix). Consequently, copper displays a strong dependence of polishing rate on pH, while polish rate is nearly independent of pH for tantalum [54]. A range of copper-to-tantalum polish rate ratios are therefore achievable, based on the slurry pH.

A second form of selectivity required in Cu CMP is selectivity between high and low regions of the device being polished. In this case, selectivity is used to fabricate planar surfaces from grossly non-planar precursor surfaces. Polishing of low-lying areas is suppressed by forming a strongly passivating layer over these surfaces, protecting them from removal until they are mechanically abraded. This highlights the importance of effective passivation in achieving good polishing outcomes. This mechanism as it pertains to smoothing is discussed further in Sect. 3.1.

10.2.4 Modeling the Polishing Process

The film thickness on a wafer cannot be measured in-situ during polishing, and so the user must have some other method of deciding when sufficient material has been removed. From this need has sprung many polishing models that attempt to link the material removal rate (MRR) with process parameters such as pressure and velocity. These have been employed with varying degrees of success and will be discussed here.

The first quantitative description relating polishing rate to process parameters came from the glass industry. Preston proposed his now widely-known equation in 1927, which stated that the rate of material removed from a surface, per unit time, was proportional to the product of the polishing velocity and pressure [55]. Expressed mathematically, this is

$$MRR = K_p \times PV \tag{10.1}$$

where MRR is the material removal rate, P is the pressure applied to the polishing couple, V is the relative velocity between the pad and wafer and K_p is an empirically-determined constant, often referred to as the Preston coefficient. This relationship has been successfully used for silicon dioxide polishing, and is moderately successful at describing metal polishing. An example of its application to copper polishing, and the Preston coefficients obtained, are shown in Fig. 10.20.

It is clear that the relationship breaks down at zero pressure or zero velocity. In either of these situations, both P × V and MRR should be zero; however this is not borne out by the data shown. To account for this, some workers have suggested that the non-zero MRR at P × V = 0 represents the chemical reaction rate of the polished surface in the slurry [56]. Preston's equation is modified to include the removal rate at zero pressure and velocity, R_c, and the experimentally determined constants, K and B, as follows:

$$MRR = (KP + B)V + R_c \tag{10.2}$$

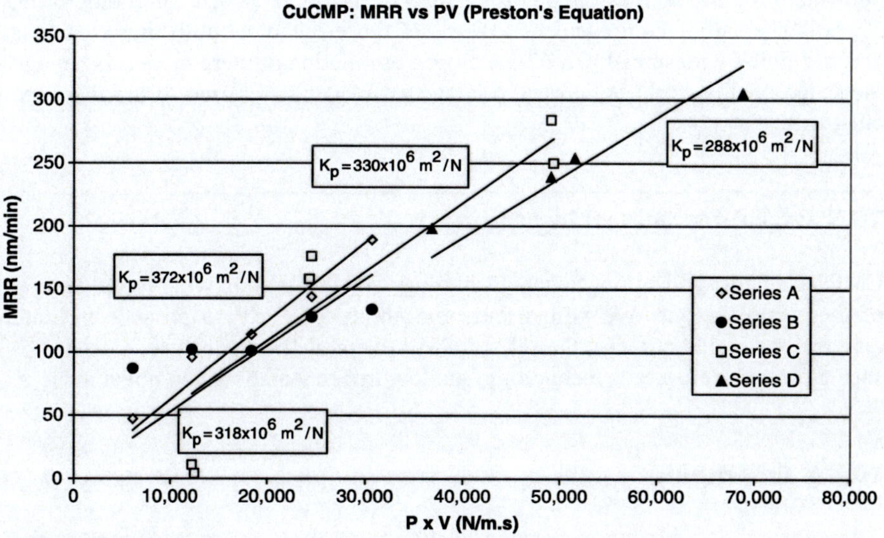

Fig. 10.20 Polishing rate of copper vs. the product of the down force and polishing velocity, demonstrating the application of Preston's equation [38]

Others have suggested that in fact the MRR remains zero until a minimum pressure, known as the threshold pressure, is reached. The threshold pressure represents the abrasives' transition from rolling to sliding motion and is accompanied by a significant increase in MRR [47]. As well as a threshold, they also suggest that the MRR varies sublinearly with pressure, representing the increase in actual contact area between the pad and wafer as pressure increases. This is expressed as

$$MRR = \begin{cases} K \times (P^{\frac{2}{3}} - P_{th}^{\frac{2}{3}})V, P \geq P_{th} \\ 0, P < P_{th} \end{cases} \tag{10.3}$$

where P is the applied pressure and P_{th} is the threshold pressure.

All three of these expressions are simple in form, and make no explicit reference to any other polishing parameter, such as the concentration of the abrasive particles or roughness of the pad. Instead, these effects are bound up in the constant K_p or K which must be empirically determined. This excludes these models from use in *a priori* process design and optimization.

A second class of models have thus been developed which attempt to calculate MRR from first principles, for instance by the summation of material removed per particle contact per unit time. These are numerous, and no one model has found wide acceptance in the polishing community. Several are semi-empirical in nature and include fitting parameters or other process parameters that may only be obtained experimentally, such as effective hardness [21, 48, 57]. It is notable that most of these include a statistical treatment of at least one component of the polishing process, such as the pad roughness or the abrasive size distribution [58, 59, 60]. The use of finite element analysis of the mechanical polishing aspects is also becoming widespread [61, 62]. A close examination of these models is beyond the scope of this work; however the interested reader is referred to the literature cited here.

10.3 CMP for Nanofabrication

CMP has many applications to nanotechnology and this section will discuss several of these applications. We will discuss the ability of CMP to produce flat and atomically smooth surfaces, the role of CMP to enable novel integration schemes such as copper damascene technology, shallow trench isolation, and novel devices.

10.3.1 Smoothing

"Smoothness" is, in many respects, a relative term; micrometer-scale topography may be considered "smooth" in comparison to a marble-sized object, but not in comparison to an ant-sized object. Similarly, as devices become smaller, the roughness of their surfaces must meet smaller and smaller absolute limits in order to ensure they are 'smooth' in relation to the object of which they are a part. An illustration of this phenomena and the role of CMP in achieving smoothness is in the manufacture of molecular devices. Because the components of such devices are very small, even minor surface roughness can lead to very low yields. CMP has been shown to produce roughness less than an angstrom; in devices composed of Langmuir-Blodgett monolayers of eicosanoic acid on platinum substrates, smoothing of the platinum by CMP increased yield from less than 50% to 100% [63]. CMP therefore has applications in virtually all aspects of top-down nanofabrication, including microfluidics and optoelectronics, as well as traditional applications in IC manufacture.

In microfluidic devices, excessive wall roughness can lead to unacceptably high fluid friction, and interrupt the flow characteristics of the device. This becomes increasingly problematic as channel dimensions decrease [64]. CMP has been shown to produce good quality polymer surfaces for MEMS and microfluidic applications [65]. Optoelectronic devices have stringent requirements both for short-range surface roughness and for longer range undulations in the surface. Both roughness and undulations, or 'waviness', can induce signal losses in optical waveguides. According to the Marcuse bending loss equation, the logarithmic degree of loss is inversely proportional to the radius of the waviness; thus, as waviness becomes more pronounced, the signal loss increases exponentially. CMP has been shown to reduce both short- and long-range surface undulations in optoelectronic integrated circuits to the extent that these bending losses are negligibly small [66].

Smoothing is important during integrated circuit (IC) manufacture because residual topography can generate defects on the device during subsequent fabrication steps, and can ultimately lead to failure of the device. Reactive ion etching (RIE), a commonly used IC manufacturing technique, removes material from the chip surface in a highly directional fashion. Because of this, topography on the device can lead to incomplete etching and the formation of 'stringers' which may lead to short-circuits in operation. A comparison of devices made with and without CMP is shown schematically in Fig. 10.21.

Roughness can also lead to increases in the electric field experienced by components of the IC device. This increased field can lead to accelerated breakdown of the dielectric layer, rendering the device non-functional and decreasing yield [67].

CMP is an ideal technique for the purpose of smoothing as it has inherent selectivity between high and low areas, allowing for the fabrication of extremely smooth surfaces on both metals and other materials. This high/low selectivity is a unique consequence of the anti-lubricating characteristics of the CMP process. Several researchers have found that hydrodynamic lubrication, as indicated by the Stribeck curve, is not achieved under normal CMP conditions [68, 69]. Hydrodynamic lubrication occurs when two surfaces are completely separated by a fluid, due to the pressure developed in that fluid. This occurs at low applied pressures

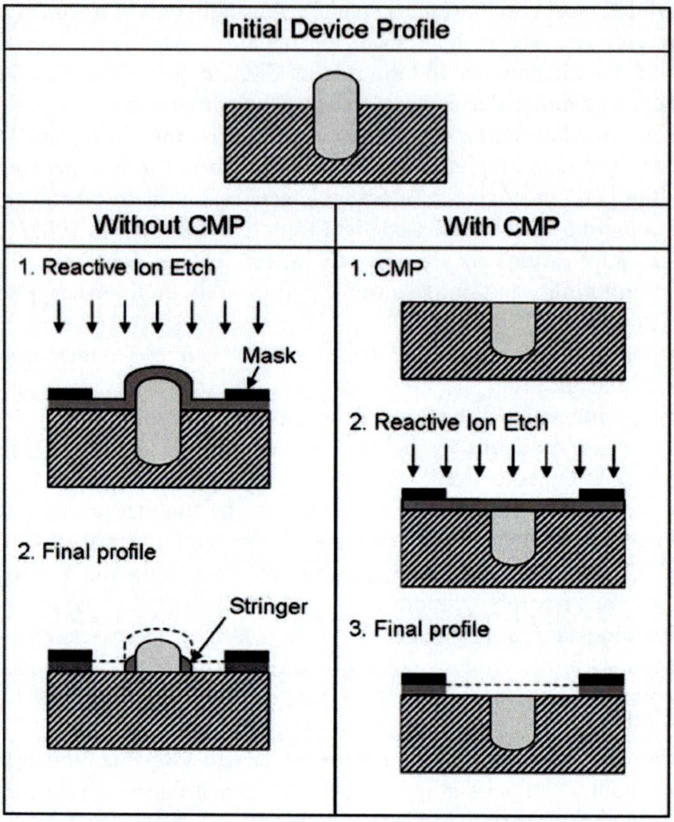

Fig. 10.21 Schematic demonstrating the formation of topography-related defects; a non-planar surface leads to incomplete RIE etching, which in turn leads to the formation of a 'stringer'. This is avoided if the surface is initially planarized with CMP

and/or high relative velocities, signified by high values of the Sommerfield number, *So*, a dimensionless number defined as

$$So = \frac{\eta V}{\delta P} \tag{10.4}$$

where η is the fluid viscosity, V is the relative velocity between the two surfaces, δ is the separation between the surfaces and P is the pressure applied to the polishing couple. Plotting the Stribeck curve (the coefficient of friction against the Sommerfield number) identifies three zones of contact, shown on the schematic curve in Fig. 10.22.

In many situations where wear is undesirable, for instance in bearings, the friction couple operates in the hydrodynamic region. CMP, however, usually occurs in the mixed lubrication regime, indicating that there is significant but not complete

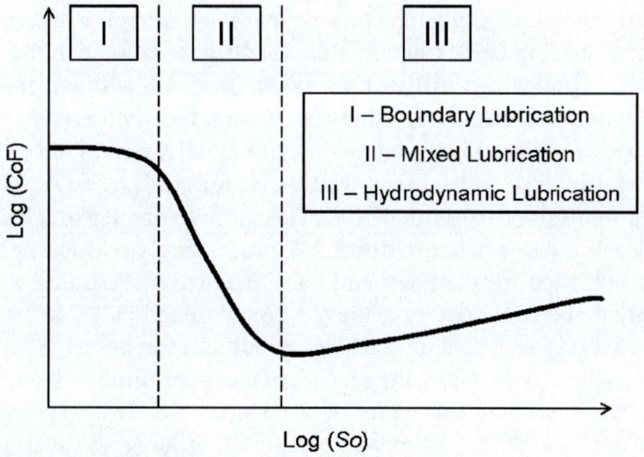

Fig. 10.22 Theoretical Stribeck curve

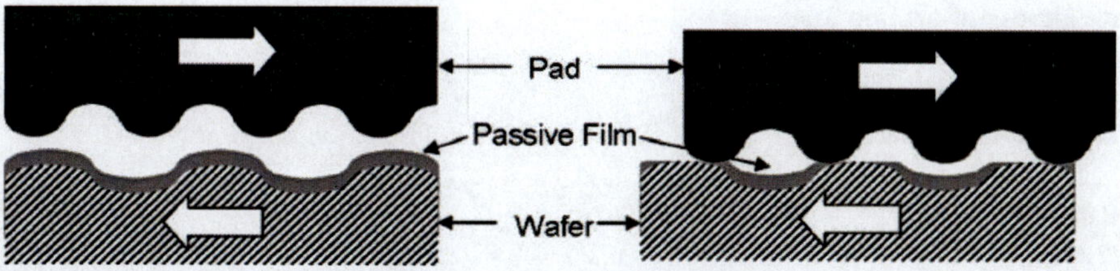

Fig. 10.23 Schematic of the high/low selectivity mechanism in CMP

contact between the polishing pad and the wafer [10]. This is thought to be a consequence of the pad's rough texture, and of the presence of grooves and porosity that drain fluid from the polishing interface before pressure can develop.

The contact between the pad and wafer that does occur is located at the asperities of both surfaces. It is at these points of contact that wear occurs, so the high points on the wafer are preferentially removed. Passivation of the low lying areas ensures that they do not dissolve until they are also relatively high enough to come in to contact with the pad. This process is shown schematically in Fig. 10.23.

The focus of CMP for ultrasmooth surfaces is therefore passivation of the polishing surface. As a consequence, the removal rates during this kind of polishing can be very low [70]. The polishing parameters are typically chosen to be mild, with very low pressures in particular. Additionally, the CMP tool may be isolated from external vibration, and a lubricant such as carboxylic acid (with an alkyl chain length of at least 10) may be added [68].

Difficulties in CMP for smoothing arise with the wide variety of surfaces that are processed, including noble metals and ceramics which are not easily oxidized. Solutions have been found for individual cases by a process of trial-and-error, or by using a slurry originally intended for a different material; for instance, silver has been processed using slurry developed for copper [71], though with limited success. Another issue with using CMP for smoothing is the generation of defects. Scratches and embedded particles are difficult to avoid in CMP without careful process design, and again this generally is achieved through trial-and-error.

CMP has been used on metals such as platinum, silver and gold [71], hard disk drive substrates [63, 72], and semiconductors such as CdZnTe [73, 74], GaN [75] and InSb [76]. With the exception of CdZnTe, sub-nanometer roughness has been recorded for all of these substrates after CMP processing, and sub-angstrom roughness has been achieved for platinum and hard disk drive substrates. A selection of surfaces polished and roughnesses achieved are tabulated in Table 10.2.

The dramatic improvement in surface smoothness achievable using CMP is demonstrated in Fig. 10.24. Here, the surface of a copper film is shown as-deposited on the left, then following polishing for 90 s in slurry of hydrogen peroxide, BTAH, glycine, citric acid and 85 nm silica particles. CMP reduced the average roughness of the surface from 61 Å to 4 Å, a 15-fold decrease, during this process. For comparison, a surface scan of a commercial single crystal silicon wafer is shown in Fig. 10.25. This surface was also finished using CMP and has an average roughness of around 1 Å.

Table 10.2 Comparison of roughness values obtained for different substrates using CMP

Material	Roughness R_a (Å)	References
Hard disk substrate	<0.4–0.9	[68, 72]
Platinum[a]	0.8	[63]
GaN	1	[75]
InSb	2–5	[76]
Sapphire[a]	6.83	[77]
CdZnTe	14.78–18.56	[74, 73]

[a]Value quoted is RMS roughness

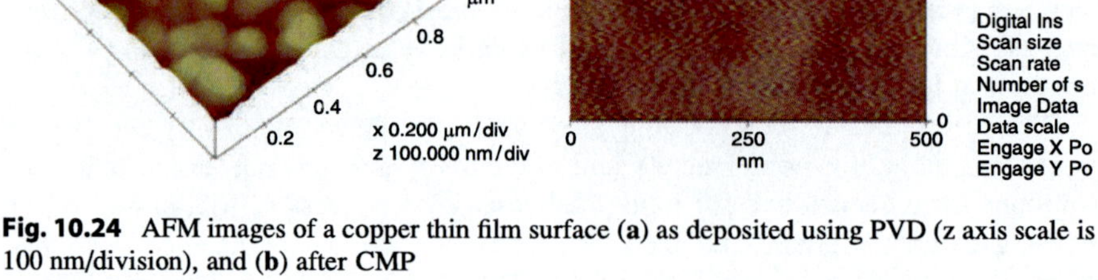

Fig. 10.24 AFM images of a copper thin film surface (**a**) as deposited using PVD (z axis scale is 100 nm/division), and (**b**) after CMP

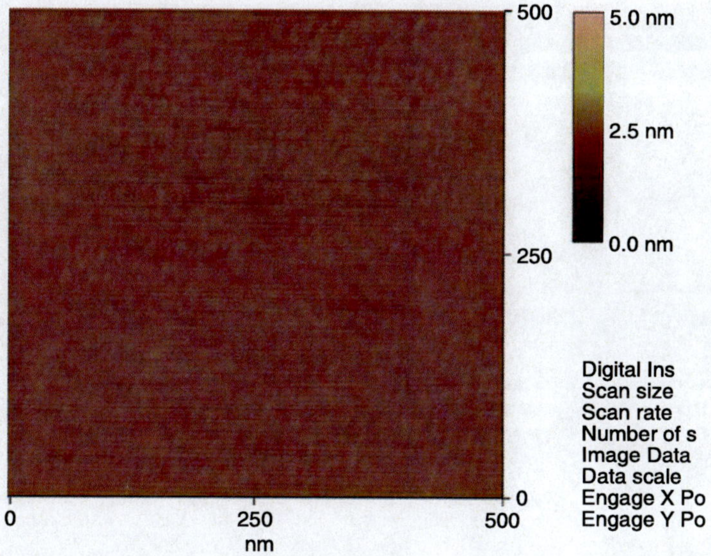

Fig. 10.25 AFM scan of a silicon single crystal wafer

10.3.2 Novel Integration Processes

CMP removes material by chemical and mechanical means, as discussed earlier. This means that while it removes material it also planarizes it on many length scales. This ability to planarize material while it is being removed enables novel process integration schemes. In this section, we will review the use of CMP to enable damascene nano-interconnects and electrical isolation between nano devices, and we will look at the application of CMP to phase change memory and molecular devices.

10.3.2.1 Copper Damascene Process

The semiconductor industry historically manufactured interconnects by depositing a blanket layer of aluminum (Al), then using lithography and etching technology, excess Al was removed, leaving behind an interconnect structure. In this subtractive approach, tungsten (W) vias were used to connect different Al interconnect layers together, as shown in Fig. 10.26. In this figure, ILD 1 and 2, and the white W plugs have been polished. The first ILD layer is borophosphosilicate glass, BPSG.

The interconnect line width control was determined by the etch bias of the metal and the lithographic process. As dimensions shrank, control of line width became exceedingly difficult, because the metal sidewall profile needed to be vertical to allow for higher packing densities. This required the use of reactive ion etching with large substrate biases to make ions impinge orthogonally to the surface. However, ion bombardment eroded the edge of the photoresist profile leading to degradation

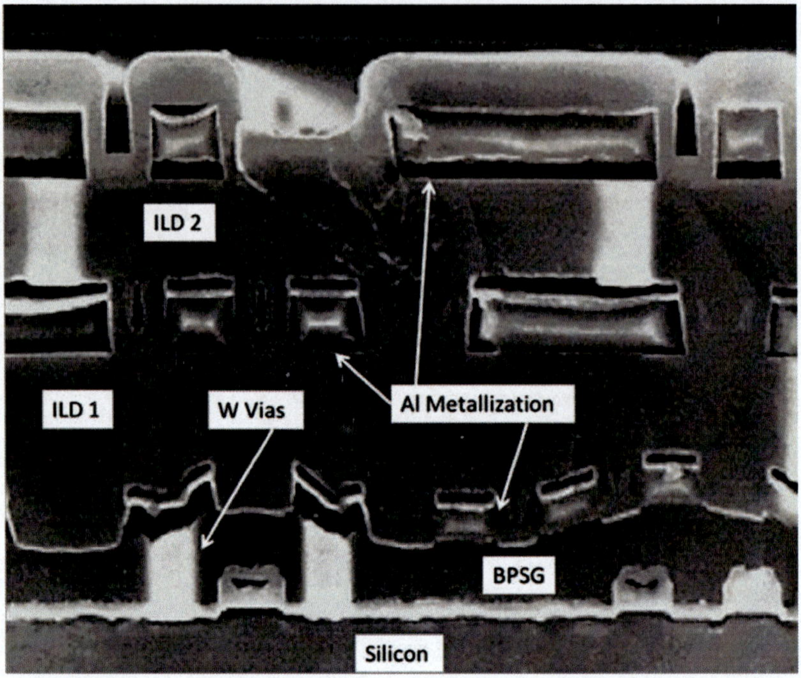

Fig. 10.26 Cross-section of a 0.5 μm technology devices showing aluminum interconnect fabricated using plasma etch process. Note that ILD 1 and 2 were planarized using SiO_2 CMP

Table 10.3 Properties of interconnect materials

Property	Aluminum	Silver	Copper	Gold
Resistivity, μΩ cm	2.67	1.59	1.67	2.35
Melting point, °C	659	961	1083	1063
Adhesion to SiO_2	Good	Poor	Poor	Poor
Diffusion barrier	No	Yes	Yes	Yes
Wet etching	Yes	Yes	Yes	Yes
Dry etching	Yes	No	No	No

of the Al edge profile and larger etch bias. Concurrent with these fabrication issues with Al, there was ongoing research on replacing Al with copper (Cu).

As shown in Table 10.3, Cu has a much lower resistivity than Al, and only a marginally higher value than silver (Ag). The issue with copper is that it is not easily etched by dry etching techniques and it also requires an adhesion layer and a diffusion barrier. Copper is a fast diffuser in SiO_2 and is a deep trap in Si. Despite these issues, Cu was selected to replace Al for two reasons. Firstly, Cu has 40% lower resistivity than Al. This is important because signals propagated along interconnects are delayed due to the resistance (R) of the interconnect and the capacitance (C) with respect to other interconnects and the substrate. This gives rise to RC delay. As devices shrink, the distance between interconnects decreases and capacitance goes up. Decreasing the resistivity of the interconnect reduces the RC delay. Secondly, Cu has the highest melting point of all of the potential interconnect materials. This is important because a primary failure mechanism of interconnects is electromigration, that is, the movement of lattice atoms in the direction of electron flow due to transfer of momentum from the electrons to the atoms. At the high current densities seen in interconnects, ~1×10^6 A cm^{-2}, material is transported in the direction of the electrons creating voids and hillocks at conductor discontinuities. Electromigration is a diffusion type process that occurs more readily along grain boundaries and interfaces that are aligned in the direction of current flow. The mean time to failure for an interconnect is given by Black's [78] equation

$$t_{50} = CJ^{-n}e^{\frac{E_a}{kT}} \tag{10.5}$$

where t_{50} is the median time to failure, C is a constant, J is the current density, n is an integer between 1 and 7, but is believed to be 2, T is temperature in kelvins, k is Boltzmann's constant, and E_a is the activation for electromigration. Typical electronic devices operate between room temperature and 373 K. Electromigration at device operating temperatures occurs more readily in low melting point materials.

While copper has desirable properties, it cannot be processed using the subtractive approach used to fabricate Al lines. Copper interconnects were enabled by CMP using the damascene process as shown in Fig. 10.27 where the right hand panel shows the damascene process and the left hand panel shows the shallow trench isolation process (STI) that will be discussed later. In the damascene process, the ILD is patterned with trench and vias. There are many ways to arrange the order of sequence of the trench and via fabrication [79]. In the damascene process, Cu CMP is used to remove the excess copper leaving the surface flat. Non-optimal polishing of patterned surfaces can lead to dishing or erosion of these features. Both of these are strongly affected by the size and density of the features. Dishing tends to occur across features that are large compared to the abrasives, while erosion occurs in areas of high pattern density [2]. These phenomena are illustrated in Fig. 10.28.

In order to prevent erosion and dishing, different areas of the wafer are required to have different polishing rates. For instance, the erosion and dishing shown in the schematic above could be reduced or eliminated if the metal polish rate was very low, while the barrier layer (substrate) rate was high. This selectivity principle is used to improve polishing outcomes on mixed surfaces.

The damascene process for copper has enabled the ability to add metal layers. An example of a multilevel metal interconnect structure fabricated by Cu damascene is shown in Fig. 10.29. Here, an eight layer copper interconnect structure is shown and it is evident that all of the layers are flat, and there appears to be no impediment, apart from cost, to adding many more metal layers.

As CMOS scaling dimensions approach the 20 nm milestone, the width of the copper interconnects at the critical metal layers (those layers with critical

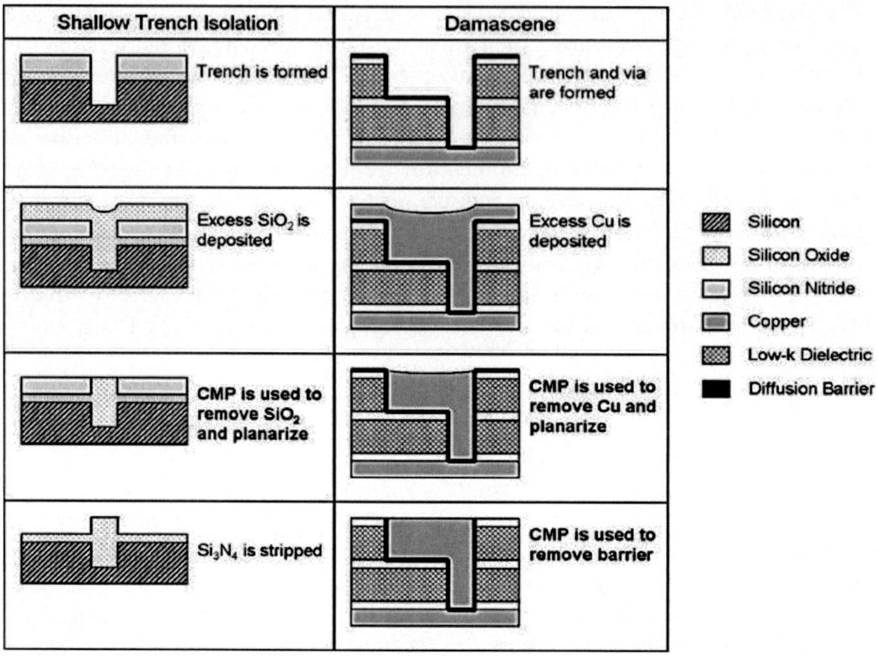

Fig. 10.27 Schematic diagram showing the process for shallow trench isolation (STI) and Damascene Cu interconnects

Fig. 10.28 Schematic diagram comparing ideal and suboptimal copper polish

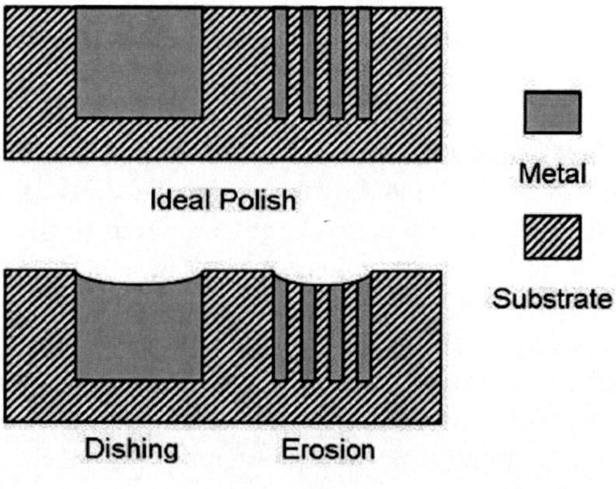

dimensions) are also approaching 20 nm. At these dimensions, control of the CMP process is critical.

10.3.2.2 Shallow Trench Isolation

It is necessary to electrically isolate nanodevices in order to prevent leakage and interaction between adjacent devices. Isolating devices must be a very efficient

Fig. 10.29 65 nm multilevel device. Each level has been created by depositing an excess of copper and then polishing it down to create a planar interface (Courtesy of Intel Corporation)

process so as to not add significantly to chip area. Shallow trench isolation was invented in 1977 [80] but not fully implemented in IC manufacturing until the 1990s. It was the application of CMP to STI that made this approach viable.

The STI process is shown schematically in left hand panel in Fig. 10.27. The substrate is covered with a thin thermal oxide, and then a deposited silicon nitride layer. The nitride layer is patterned to form a mask for the silicon trench reactive ion etch. This etch tends to erode photoresist, so a hard, chemically resistant, mask must be used. The trench is then filled with CVD oxide. Excess oxide is removed by CMP and the process stops on the nitride layer. Finally the silicon nitride masking layer is removed. There are many subtle and important features about STI. For example, the bottom and top corners of the trench must be rounded to prevent the concentration of electric fields. There are also requirements on the amount of silicon nitride left after polishing, which means that the polish rate selectivity of STI slurry must be high. STI polish has the most stringent requirements of any polish process.

10.3.2.3 Novel Devices

CMP has been used to fabricate phase change memory (PCM) devices with a nanoscale damascene structure [81]. The damascene structure confined the Joule heating, and the smoothness of the surface of the phase change material, $Ge_2Sb_2Te_5$ (0.8 nm RMS) improved the contact resistance with the net result that the PCM had improved switching reliability.

The yield of molecular scale devices has been improved by CMP. This was achieved by polishing sputtered platinum electrodes and improving the smoothness from 2 to 0.08 nm. This improvement increased the water contact angle and led to better packing of self-assembled alkanethiolate molecules on the polished surface [63]. Cross bar molecular devices fabricated with the polished Pt electrodes gave dramatic improvement in yield for self-assembled (2X) and Langmuir-Blodgett (7X) films.

10.3.2.4 Issues

As with any technology, CMP has many virtues but also some shortcomings. CMP does require down force for polishing to occur. A polishing pressure of 34.5 kPa (5 psi) on a 0.1 m (4″) diameter wafer has a down force of 280 N. This force also results in shear stress between the pad and the wafer surface. These stresses and forces may be too high for delicate nanostructures. CMP can also cause nano-scratches in soft materials. These can be circumvented to a certain degree by use of a soft pad and much lower polishing pressures. In addition, novel integration schemes can also be devised. For example, instead of building an interconnect structure, as shown in Fig. 10.29, on top of an array of nanodevices, one could build the interconnect structure first and then build the devices on this structure. This is shown schematically in Fig. 10.30. In this concept, the interconnects are fabricated on one side of the wafer, then through-substrate vias are formed to the other side of the wafer. Finally, nanodevices are fabricated on the other side of the substrate. This is a hypothetical diagram that may be too complicated to build; however, it does indicate one possible integration approach to separate the high stress and temperature steps from the sensitive nanodevice steps.

Conclusions

CMP is an extremely versatile technology that uses mechanical force and chemical action to planarize surfaces and to make them smooth at the atomic level. Furthermore, CMP is an enabling technology as discussed in this chapter, and we believe that CMP will an even greater role to play in the future of nanofabrication and nanotechnology.

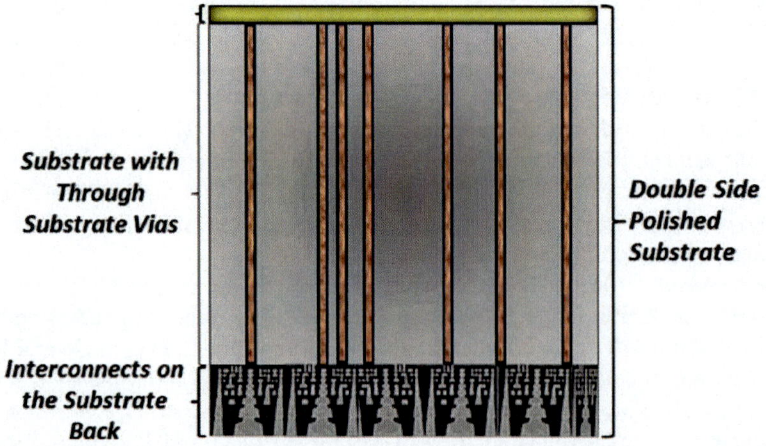

Fig. 10.30 Schematic of possible integration scheme where the interconnects are manufactured first, and the top surface planarized to allow nanodevice arrays to be built on top

References

1. "Interconnect" in The International Technology Roadmap for Semiconductors 2007 Edition and 2008 Update. http://www.itrs.net/. Accessed 28 Jan 2009.
2. Li Y. Microelectronic applications of chemical mechanical planarization. Hoboken: Wiley; 2008. ISBN 9780471719199.
3. Bohr MT, Chau RS, Ghani T, Mistry K. IEEE Spectrum. 2007;44:29.
4. After Horiba Group. http://www.horiba.com/us/en/scientific/products/particle-characterization/applications/cmp/. Accessed June 2011.
5. Li Z, Lee H, Borucki L, Rogers C, Kikuma R, Rikita N, Nagasawa K, Philipossian A. J Electrochem Soc. 2006;153:G399.
6. Beyer K. IBM MicroNews. 1999;5:4.
7. Chiodarelli N, Li Y, Cott D. Microelectron Eng. 2011;88:837.
8. Gupta S, Kumar P, Chakkaravathi A. Appl Surf Sci. 2011;257:5837.
9. Feng H-P, Yu B, Chen S. Electrochim Acta. 2011;56:3079.
10. Philipossian A, Olsen S. Jpn J Appl Phys 1. 2003;42:6371.
11. Higgs III CF, Ng SH, Borucki L, Yoon I, Danyluk S. J Electrochem Soc. 2005;152:G193.
12. Mueller N, Rogers C, Manno VP, White R, Moinpour M. J Electrochem Soc. 2009;156:H908.
13. Li Y. ICSICT Proc. 2004;1:508.
14. Lu Z, Lee S-H, Gorantla VRK, Babu SV, Matijevic E. J Mater Res. 2003;18:2323.
15. Kawahashi N, Matijević E. J Colloid Interf Sci. 1991;143:103.
16. Lin F, Cadien KC. Unpublished work, 2011.
17. Bielmann M, Mahajan U, Singh RK. Mater Res Soc Symp Proc. 2000;566:103.
18. Luo J, Dornfeld DA. IEEE Trans Semicond Manuf. 2003;16:469.
19. Jung S-H, Singh RK. Mater Res Soc Symp Proc. 2004;816:49.
20. Li Z, Ina K, Lefevre P, Koshiyama I, Philipossian A. J Electrochem Soc. 2005;152:G299.
21. Fu G, Chandra A, Guha S, Subhash G. IEEE Trans Semicond Manuf. 2001;14:406.
22. Che W, Guo Y, Chandra A, Bastawros A. J Manuf Sci Eng Trans ASME. 2005;127:545.
23. Cook LM. J Non-Cryst Solids. 1990;120:152.
24. Iler RK. Chemistry of silica – solubility, polymerization, colloid and surface properties and biochemistry. New York: Wiley; 1979. ISBN 047102404X.
25. Aksu S. Mater Res Soc Symp Proc. 2005;867:15.
26. Perez N. Electrochemistry and corrosion science. Boston: Kluwer Academic Publishers; 2004. ISBN 1402077440.
27. Carpio R, Farkas J, Jairath R. Thin Solid Films. 1995;266:238.
28. Washburn EW. International critical tables of numerical data, physics, chemistry and technology. 1 Electronicth ed. Norwich: Knovel; 2003. ISBN 9781591244912.
29. Ein-Eli Y, Abelev E, Rabkin E, Starosvetsky D. J Electrochem Soc. 2003;150:C646.
30. Wardman P, Candeias LP. Radiat Res. 1996;145:523.
31. Janna H, Scrimshaw MD, Williams RJ, Churchley J, Sumpter JP. Environ Sci Technol. 2011;45:3858.
32. Tromans D. J Electrochem Soc. 1998;145:L42.
33. Zheng JP, Roy D. Thin Solid Films. 2009;517:4587.
34. Hong Y, Devarapalli VK, Roy D, Babu SV. J Electrochem Soc. 2007;154:H444.
35. Hong Y, Patri UB, Ramakrishnan S, Roy D, Babu SV. J Mater Res. 2005;20:3413.
36. Grover GS, Liang H, Ganeshkumar S, Fortino W. Wear. 1998;214:10.
37. Philipossian A, Mitchell E. Jpn J Appl Phys 1: Reg Pap Short Notes Rev Pap. 2003;42:7259.
38. Nolan L, Cadien KC. Unpublished work, 2009.
39. Lu H, Fookes B, Obeng Y, Machinski S, Richardson KA. Mater Charact. 2002;49:35.
40. Moinpour M, Tregub A, Oehler A, Cadien K. MRS Bull. 2002;27:766.
41. Prasad YN, Kwon T-Y, Kim I-K, Kim I-G, Park J-G. J Electrochem Soc. 2011;158:394.
42. Chen P-L, Chen J-H, Tsai M-S, Dai B-T, Yeh C-F. Microelectron Eng. 2004;75:352.
43. Cooper K, Gupta A, Beaudoin S. J Electrochem Soc. 2001;148:G662.

44. Xu K, Vos R, Vereecke G, Doumen G, Fyen W, Mertens PW, Heyns MM, Vinckier C, Fransaer J, Kovacs F. J Vac Sci Technol B. 2005;23:2160.
45. Busnaina AA, Elsawy TM. J Electron Mater. 1998;27:1095.
46. Renteln P, Ninh T. Mater Res Soc Symp Proc. 2000;566:155.
47. Zhao B, Shi FG. Electrochem Solid-State Lett. 1999;2:145.
48. Bastaninejad M, Ahmadi G. J Electrochem Soc. 2005;152:G720.
49. Haosheng C, Jiang L, Darong C, Jiadao W. Tribol Lett. 2006;24:179.
50. Xu G, Liang H, Zhao J, Li Y. J Electrochem Soc. 2004;151:G688.
51. Liang-Yong W, Bo L, Zhi-Tang S, Wei-Li L, Song-Lin F, Huang D, Babu SV. Chin Phys B. 2011;20:038102.
52. Oh M-H, Nho J-S, Cho S-B, Lee J-S, Singh RK. Powder Technol. 2011;206:239.
53. Manivannan R, Victoria SN, Ramanathan S. Thin Solid Films. 2010;518:5737.
54. Janjam SVS, Surisetty CVVS, Pandija S, Roy D, Babu SV. Electrochem Solid-State Lett. 2008;11:H66.
55. Preston FW. J Soc Glass Technol. 1927;11:214.
56. Luo Q, Ramarajan S, Babu SV. Thin Solid Films. 1998;335:160.
57. Wang C, Sherman P, Chandra A. IEEE Trans Semicond Manuf. 2005;18:695.
58. Luo J, Dornfeld DA. IEEE Trans Semicond Manuf. 2001;14:112.
59. Borucki L. J Eng Math. 2002;43:105.
60. Yu T-K, Yu CC, Orlowski M. Technical Digest – International Electron Devices Meeting, 1993, p. 865.
61. Bastawros A, Chandra A, Guo Y, Yan B. J Electron Mater. 2002;31:1022.
62. Seok J, Sukam CP, Kim AT, Tichy JA, Cale TS. Wear. 2003;254:307.
63. Islam MS, Li Z, Chang S-C, Ohlberg DAA, Stewart DR, Wang SY, Williams RS. 5th IEEE international conference on nanotechnology 2005;1:80.
64. Xiong R, Chung JN. Microfluid Nanofluid. 2010;8:11.
65. Zhong ZW, Wang ZF, Tan YH. Microelectron J. 2006;37:295.
66. Barkley E, Fonstad Jr CG. IEEE J Quantum Electron. 2004;40:1709.
67. Lee S-C, Oates AS, Chang K-M. IEEE Trans Dev Mater Reliab. 2010;10:307.
68. Lei H, Luo J. Wear. 2004;257:461.
69. Higgs CF, Terrell EJ, Kuo M, Bonivel J, Biltz S. Mater Res Soc Symp Proc. 2007;991:333.
70. Saif Islam M, Jung GY, Ha T, Stewart DR, Chen Y, Wang SY, Williams RS. Appl Phys A. 2005;A80:1385–9.
71. Logeeswaran VJ, Chan M-L, Bayam Y, Saif Islam M, Horsley DA, Li X, Wu W, Wang SY, Williams RS. Appl Phys A. 2007;A87:187.
72. Lee W, Qi Z, Lu W. 12th IFToMM world congress, Besançon; 2007.
73. Zhang Z, Gao H, Jie W, Guo D, Kang R, Li Y. Semicond Sci Technol. 2008;23:105023.
74. Zhou H. Mater Manuf Process. 2010;25:418.
75. Xu X, Vaudo RP, Brandes GR. Opt Mater. 2003;23:1.
76. Vangala SR, Qian X, Grzesik M, Santeufemio C, Goodhue WD, Allen LP, Dallas G, Dauplaise H, Vaccaro K, Wang SQ, Bliss D. J Vac Sci Technol B. 2006;24:1634.
77. Zhang Z, Liu W, Song Z, Hu X. J Electrochem Soc. 2010;157:H688.
78. Black JR. IEEE Int Reliab Phys Symp Proc. 2005;43:1.
79. Wolf S. Silicon processing for the VLSI era, vol. 4. Sunset Beach: Lattice Press; 2004. ISBN 096167217X.
80. Bondur JA, Pogge HB. United States Patent number 4,104,086, Method for forming isolated regions of silicon utilizing reactive ion etching, 1 Aug 1978.
81. Zhong M, Song ZT, Liu B, Wang LY, Feng SL. Electron Lett. 2008;44:322.

11 Deposition, Milling, and Etching with a Focused Helium Ion Beam

P. F. A. Alkemade and E. van Veldhoven

Abstract

The recent successful development of the helium ion microscope has produced both a new type of microscopy and a new tool for nanoscale manufacturing. This chapter reviews the first explorations in this new field in nanofabrication. The studies that utilize the Orion helium ion microscope to grow or remove material are described, concentrating on helium ion beam deposition, milling, and etching. Helium ion beam induced deposition combines the advantage of electron beam deposition, namely high spatial resolution, with that of heavy-ion beam induced deposition, namely high efficiency. Helium milling is much slower than gallium milling, but ideal for structuring thin slabs of material with high precision. A handful of studies has demonstrated the possibility of helium ion beam etching. Experimental and theoretical studies suggest that secondary electron emission is the dominant mechanism in helium ion beam induced processing.

11.1 Introduction

When a beam of energetic ions crashes into a material, the composition and structure of the material change. This phenomenon lies at the heart of many techniques for manufacturing structures on the nanometer scale. In *lithography*, the material is covered with a resist, the solubility of which changes with exposure to the beam. In *ion beam induced processing* (IBIP), the physical and chemical

P.F.A. Alkemade (✉)
Kavli Institute of Nanoscience, Delft University of Technology, Delft, The Netherlands
e-mail: p.f.a.alkemade@tudelft.nl

E. van Veldhoven
TNO Science and Industry, Delft, The Netherlands
e-mail: emile.vanveldhoven@tno.nl

M. Stepanova and S. Dew (eds.), *Nanofabrication*,
DOI 10.1007/978-3-7091-0424-8_11, © Springer-Verlag/Wien 2012

conversion of an adsorbed layer is sufficient to deposit or etch the desired nanostructure. Until recently, ion beam induced processing was carried out with Ga^+ ions (Ga-IBIP). In general, Ga-IBIP is much more efficient in terms of processed volume per incident particle than the related technique of electron beam induced processing (EBIP). Furthermore, purer materials can be obtained. However, the minimum feature sizes are larger than for EBIP. Finally, the inevitable ion implantation and ion beam sputtering might lead to process complications. For these reasons, an alternative ion species was highly desirable. However, the use of light ions was not initially considered to be effective. In an early study of gold deposition with a broad helium beam, low deposition yields were measured [1].

Nevertheless, the recent launch by Alis Corporation (now part of Carl Zeiss) of the Orion-Plus helium ion microscope with a sub-nanometer ion beam led scientists to be optimistic that an alternative beam was available for nanometer scale materials processing [2, 3]. Indeed, the first He-IBIP growth experiments performed by the manufacturer [4] and follow-ups at Zeiss [5], at TNO in Delft [6–8], and at the University of Southampton [9] proved the feasibility of this new materials processing technique. Moreover, the first experiments displayed favorable qualities. In particular, it was found that He-IBIP combines the efficiency of Ga-IBIP with the spatial resolution of EBIP.

This chapter presents an overview of the recent work on helium ion beam induced processing (He-IBIP), particularly direct milling, beam induced deposition (He-IBID), and beam induced etching (He-IBIE). As this field is still very young, only a limited number of studies has been published. Almost all of them are explorative and only a few actual devices have been made so far. Nevertheless, we believe these studies cover all relevant aspects of He-IBIP. We discuss in this chapter the published studies and some unpublished ones. In addition, we summarize the recent studies on nanofabrication by direct helium ion beam milling [8, 10–13]. Wherever possible, we will explain the observed differences with the related techniques of Ga-IBIP and EBIP.

Extensive descriptions of the interaction of helium ions with matter and of the only instrument currently available commercially – the Orion helium ion microscope – are given in Chap. 4. Therefore, only a brief summary of the fundamentals and the equipment are given here. Furthermore, there are excellent recent reviews of ion and electron beam induced processing, notably by Randolph et al. [14], by Utke et al. [15], and by van Dorp and Hagen [16], and therefore we will not discuss general aspects of IBIP and EBIP here.

11.2 Interaction of Helium Ions with Materials

11.2.1 Penetration of Ions in Matter

During imaging, the Orion helium ion microscope (HIM) irradiates the material under study with a highly focused beam between 0.1 and 10 pA of helium ions [3]. The beam scans a rectangular area of the surface. The smallest spot size is typically between 0.4 and 2 nm, depending on the beam current. The acceleration voltage is between 10 and 30 kV. For most imaging applications, the ion induced secondary electrons (iSE's) generated by the primary ions (PIs) are the active particles, *i.e.* they generate the signal for imaging. Recent studies on nanofabrication by helium ion beams suggest that they are also the actinic particles, *i.e.* they initiate the chemical reactions for nanofabrication [4, 6–9].

Several relevant interaction events can take place when an ion impinges on a surface. The most important ones are sketched in Fig. 11.1. We note that the typical penetration depth of 30 keV He ions in matter is several hundreds of nanometers.

The incoming ion (1) can be scattered back into the vacuum by single (2) or multiple (3) scattering events in the target material. During these scattering events, the ion may be neutralized (4). Interactions inside the material may lead to the emission of secondary electrons (5). The momentum imparted to the atoms by the incoming ion can cause a collision cascade, resulting in the ejection of atoms, ions, or molecules, *i.e.* sputtering (6). Instead of being scattered back into the vacuum, the projectile may be adsorbed on the surface (7) or implanted in the interior of the target material (8). All these processes are stochastic and the associated probabilities depend strongly on the energy or velocity of the ions. A schematic diagram of the various phenomena is shown in Fig. 11.2.

Figure 11.2 shows that implantation, sputtering, charge transfer, dissociative scattering, and collision-induced dissociation are the most relevant processes for 30 keV helium ions. These processes may be desirable, irrelevant, or harmful depending on the application in mind. Nevertheless, it is clear they should be known and considered when He-ion beam processing is being applied for nanofabrication. For instance, the growth of a 1 nm thick deposit requires a dose of typically 10^{15} ions/cm^2. In that case, substrate damage is unavoidable, see *e.g.* Fig. 9 in Chap. 4.

Figure 11.3 from Ref. [18] illustrates the interaction volume of gallium ions, helium ions, and electrons in silicon. It shows how changing the beam energy changes the size and shape of the interaction volumes. Compared to gallium ions, helium ions penetrate deeper and remain laterally more confined. Electrons

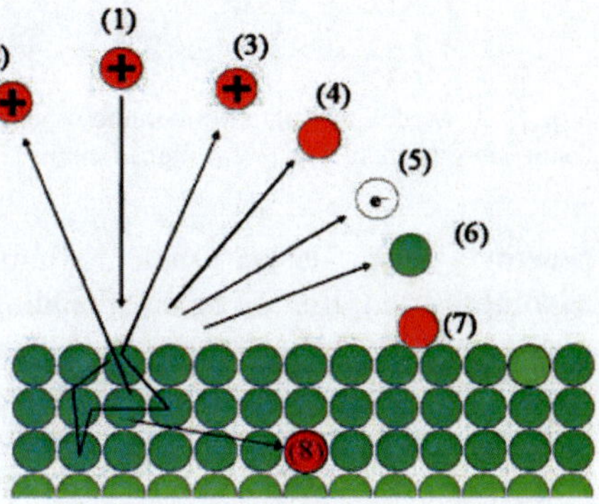

Fig. 11.1 Ion-solid interactions (see text for explanation)

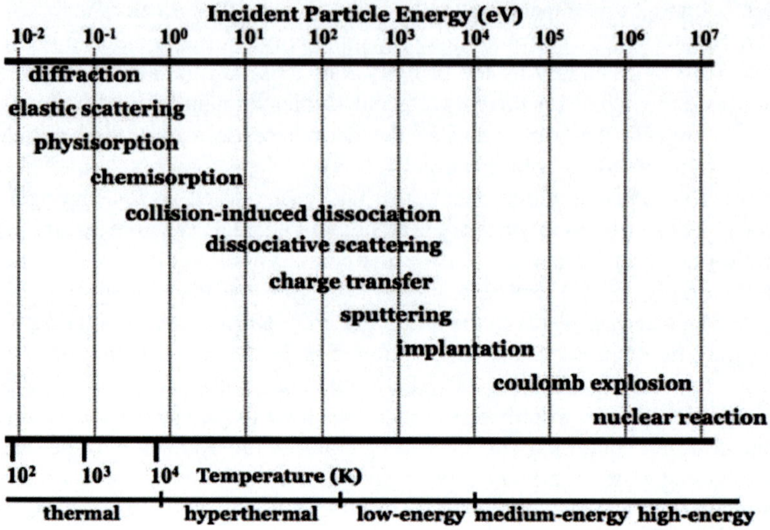

Fig. 11.2 Selected particle-surface processes as a function of ion kinetic energy (Reproduced from J. Rabalais [17])

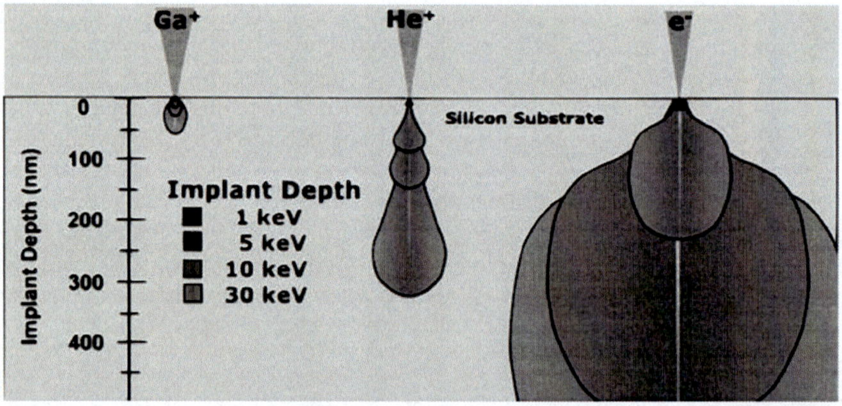

Fig. 11.3 Modeled interaction volume of gallium ions, helium ions, and electrons for equivalent beam energies in silicon (Reproduced from R. Livengood and S. Tan, Intel Corporation [18])

penetrate even deeper than helium, yet are laterally less confined. For nanofabrication, limited beam spreading is vital.

Figure 2 in Chap. 4 compares calculated trajectories of 30 keV electrons and 30 keV helium ions in silicon. The top panel shows that very few helium ions are scattered back to the surface. The bottom panel shows the primary particle trajectories in the outermost 20 nm, which are two to four times the escape depth or mean free path of the secondary electrons (SE). At a depth of 20 nm, the lateral spreading of the helium beam is less than 1 nm, whereas for 30 keV electrons it is several nanometers. These two observations demonstrate two beneficial attributes of nanofabrication with helium ions compared to electrons: more localized near-surface interactions, implying high resolution, and weak if any proximity effects.

11.2.2 Helium Ion Beam Milling

Removal or addition of material are the fundamental physical processes of nanofabrication. Removal of material by direct helium ion sputtering (*milling*) is possible, but slow. Nevertheless, milling limits the minimal detectable object size in helium-ion microscopy, which for instance can be several nanometers for tin balls [19]. At normal incidence and at 1–50 keV beam energies, the typical sputtering rate is between 0.01 and 0.1 atoms per incident ion, which is about two orders of magnitude lower than for Ga^+ ions, see for instance Fig. 11.4a. At oblique incidence, the helium sputtering yield rises strongly, see Fig. 11.4b, and values above 1 atom/ion can be reached at incidence near 90°. For a sputtering rate of 0.1 atom/ion, removal of a 1 nm thick layer requires a dose of typically 10^{17} ions/cm^2. At this dose, substantial subsurface damage can occur [18]. In the experiment shown in Fig. 11.5, a dose above 10^{18} helium ions/cm^2 has been used, which should have caused a recess of the bombarded copper surface of about 10 nm. Owing to swelling by the formation of dislocations and nanobubbles, the actual surface has moved outward. Due to the low removal rate and possible substrate damage, helium ion beam milling will probably be used in special cases only.

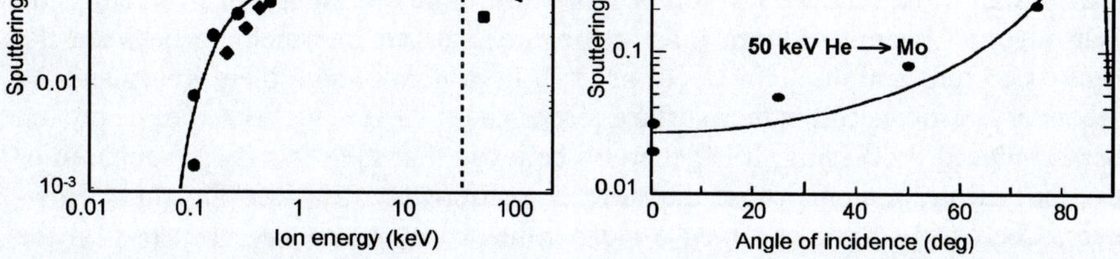

Fig. 11.4 (**a**) Au sputtering yield as a function of He beam energy at normal incidence. At 30 keV, the yield is 0.1 atoms/ion, whereas for Ga, it is 10 atoms/ion (*open circle*). Note that gold is one of the fastest eroding materials under ion bombardment. (**b**) Mo sputtering yields for Ga and He, as a function of angle of incidence (Data from Ref. [20])

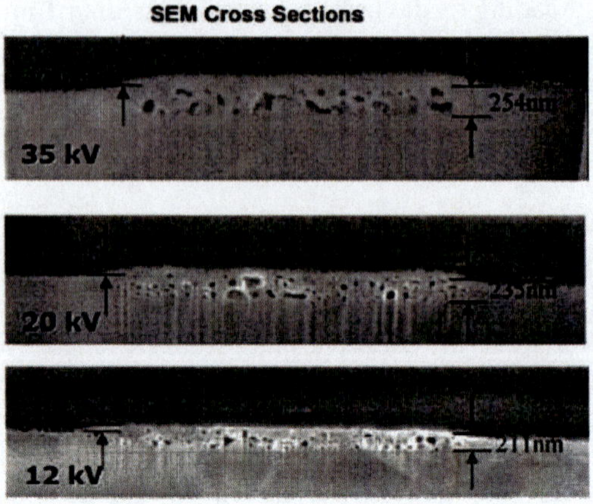

Fig. 11.5 SEM cross sections of a copper sample, implanted with a helium dose of $1.3 \times 10^{18}/\text{cm}^2$ with varying beam energies. Nanobubbles have formed at the end of range of the ions (Reproduced from R. Livengood and S. Tan, Intel Corporation [18])

11.2.3 Ion and Electron Induced Processing

Particle induced physical and chemical reactions at the surface depend of course on the type and state of the solid. Solids covered with a thin layer of adsorbates might be subjected to a chemical reaction with the adsorbates or their fragments. Coverage of a solid surface by a thin layer of adsorbates can be maintained during beam exposure by injecting a precursor gas into the vacuum chamber. Beam-induced chemical reactions between the adsorbed precursor molecules and the atoms at the solid surface constitute the mechanism of EBIP and IBIP. With a judicious choice of adsorbate, one can locally either deposit or remove material. One can select the sites of conversion by directing the charged particle beam. For keV-energy electron and ion beams, this can be done routinely with a spatial and temporal precision better than 10 nm and 100 ns, respectively. As the adsorbed precursor molecules are consumed during the conversion, a proper match between the fluxes of precursor molecules and of ions or electrons is needed for optimal processing.

The emitted secondary particles can be electrons, atoms, molecules, or ions – see Fig. 11.1. Owing to the large mass difference, the emission of secondary atoms, ions, and molecules are rare in the case of electron beams. Furthermore, at beam energies above several keV's, ion beams cause more secondary electron emission than electron beams. Helium ions are regarded as an intermediate between the heavy Ga^+ ions and the light electrons [15]. In principle, both the primary and the secondary particles can induce surface reactions [15] – see Fig. 11.6. For heavy ion beam-induced processing, it is generally believed that the secondary (sputtered or excited) substrate atoms cause the surface reactions [1, 15] – see Fig. 11.6. However, Chen et al. have observed a clear influence of secondary electrons on the growth by Ga-IBID [21].

The primary electron beam in most electron microscopes has an energy of between a few hundreds of eV's and several tens of keV's in scanning electron microscopes (SEM) and up to several hundreds of keV's in transmission electron

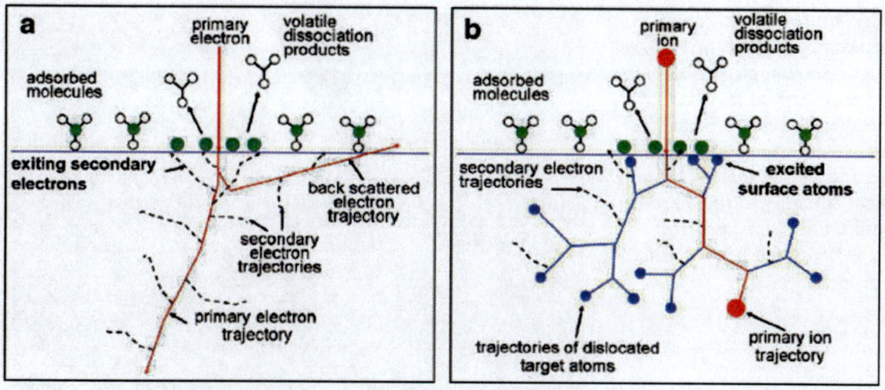

Fig. 11.6 (**a**) Interactions of primary electrons generating a flux of secondary electrons and backscattered electrons. All electrons can dissociate adsorbed precursor molecules via electronic excitation. (**b**) Interactions of primary ions generating secondary electrons and a collision cascade in the substrate. The non-sputtered target atoms remain as excited surface atoms contributing to molecule dissociation (Reproduced from I. Utke [15])

microscopes (TEM). Their associated secondary electrons have energies between a few eV's and several hundreds of eV's. It is still not known whether the primary or the secondary electrons contribute most to the electron induced reactions.

For deliberate processing, a needle of a gas injection system (GIS) is brought to within of a few hundred micrometers to the impact point of the focused particle beam, see Figs. 11.7 and 11.8. The needle can deliver a precursor gas from an external reservoir to the area accessible by the beam. Precursor decomposition results in growth of a deposit (Fig. 11.8a) or etching of the substrate (Fig. 11.8b). The composition of the deposit is related to the chemical nature of the precursor molecule, but many applications need a sufficiently pure deposit. For instance, tungsten wires can be grown by beam induced decomposition of $W(CO)_6$, but the deposited material contains large quantities of carbon and oxygen as well. Unfortunately, the additional organic or sometimes non-organic subgroups are indispensable for the delivery of the molecules from the gas-phase to the sites of interest. If a heavy-ion beam is used for beam induced processing, growth of the deposit is slowed down substantially by the inevitable sputter erosion or ion beam milling. *A fortiori*, if the supply of precursor molecules is too low, sputter erosion might dominate, leading to a net removal of material instead of growth. Obviously, the use of a helium ion beam would reduce concomitant sputter erosion by at least an order of magnitude down to a few percent. Consequently, a low flux of precursor molecules retards the growth but does not lead to net removal of material. Hence, balancing the fluxes of ions and precursors is less critical for He-IBID than for Ga-IBID.

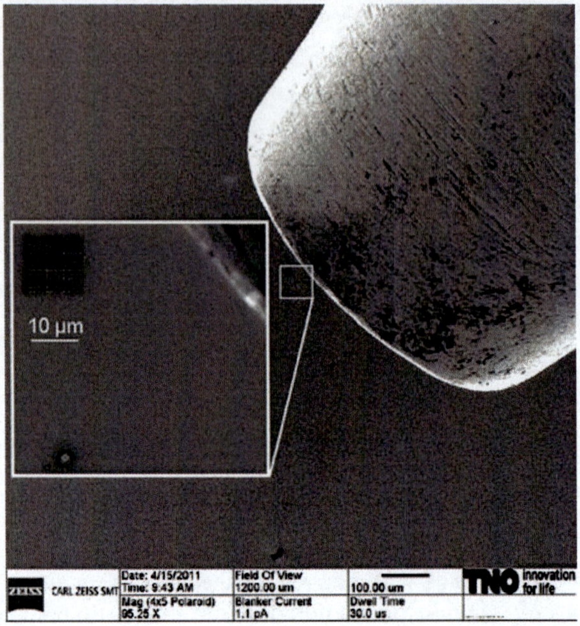

Fig. 11.7 Helium-ion-microscope (HIM) image of a processed sample surface plus a part of the outlet of a GIS needle. The center of the image corresponds to the central axis of the ion beam line. An array of fine needles and a 1 × 1-µm^2 box had been deposited (Unpublished)

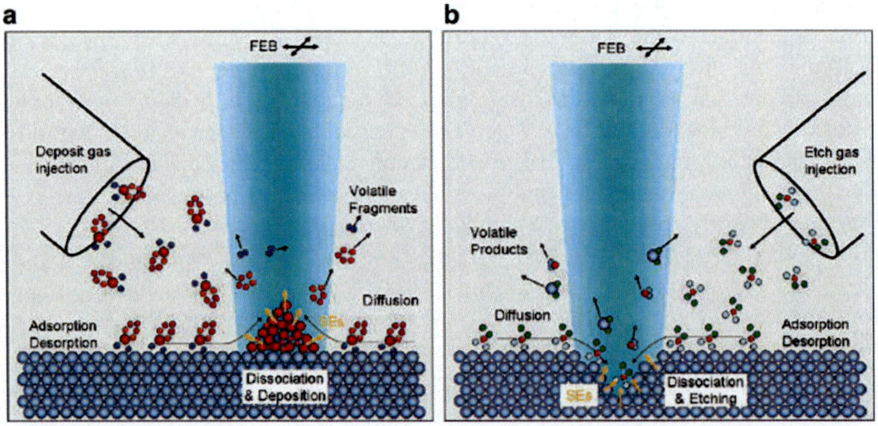

Fig. 11.8 Precursor depletion and replenishment in focused electron beam (FEB) induced processing (EBIP): molecules adsorb, desorb, and diffuse at the surface and are dissociated under electron impact. (**a**) FEB-induced deposition: the nonvolatile dissociation products form the deposit, growing coaxially into the beam. Volatile fragments are pumped away. (**b**) FEB-induced etching: the adsorbed molecules dissociate under electron impact into reactive species, the latter react to form volatile compounds with the substrate material (Reproduced from I. Utke [15])

11.2.4 Theoretical Aspects of Beam Induced Processing

The processes that determine beam induced growth and etching are related to the physical and chemical attributes of the precursor molecules, of the actinic primary particles, and of the material surface. These attributes determine the rate of adsorption, spontaneous or beam-induced desorption, surface diffusion, and beam-induced decomposition. Furthermore, they determine on the behavior of the decomposition products: desorption or incorporation into the surrounding material. If an ion beam is used, there is additional atom removal by milling. The deposition or etch rate $R(r)$, expressed in volume per unit of time, at a distance r to the center of the beam spot on the surface, is [15]

$$R(r) = \sigma V f(r) n(r), \tag{11.1}$$

where σ is the cross section for the deposition or etching process, V is the deposited or removed volume, $f(r)$ is the flux of primary or secondary particles (particles per unit of area per unit of time), and $n(r)$ is the surface density of adsorbed precursor molecules.

The entire process of beam induced deposition or etching is described by the second-order differential equation [15]

$$\frac{\partial n}{\partial t} = sJ\left(1 - \frac{n}{n_0}\right) + D\left(\frac{\partial^2 n}{\partial r^2} + \frac{1}{r}\frac{\partial n}{\partial r}\right) - \frac{n}{\tau} - \sigma f n \tag{11.2}$$

The four terms on the right-hand side represent adsorption, diffusion, desorption and decomposition, respectively. J is the incoming precursor flux, n_0 is the saturation surface density of the precursors, D is the surface diffusion constant, and τ is the mean surface residence time of the precursor. An extended discussion of the role of primary and secondary particles can be found in Ref. [15].

Solving Eq. (11.2) is rarely simple, especially when a three-dimensional object of deviating composition develops. Sometimes steady-state solutions for a broad beam exist or the initial growth behavior can be calculated [15]. For that reason, several investigators used Monte Carlo methods to calculate the time dependent growth or etching [22–24].

Interestingly, helium ion beams take a central position between heavy-ion beams and electron beams. For instance, the concomitant occurrence of ion beam milling during deposition with helium is much lower than with gallium. Moreover, the role of excited or sputtered secondary atoms in precursor decomposition is assumed to be very small for He-IBIP. On the other hand, both helium and gallium ion beams induce secondary electrons in relatively high quantities. Therefore, one expects in the case of helium and gallium ion beams similar contributions to the deposition by the emitted secondary electrons. In any case, the availability of a third beam type of a different nature will provide additional and independent tests of theoretical models for beam induced processing.

Fig. 11.9 The Orion helium ion microscope of TNO Delft, equipped with an OmniGIS gas injection unit. The inset shows the entire GIS unit

11.2.5 Equipment

Only a limited number of studies on deposition and etching with a helium ion beam have been conducted so far, and even fewer have been published [3–9]. All authors have used the same or similar equipment: the Orion helium ion microscope from Zeiss [2, 3], equipped with an OmniGIS gas injection system from Omniprobe. Precursor gases were methylcyclo-pentadientyl trimethyl platinum ($C_9H_{16}Pt$) for platinum deposition [3, 5–9], tungsten hexacarbonyl $W(CO)_6$ for tungsten deposition [4, 9], tetra-ethoxy-silane (TEOS) for silicon oxide deposition, and XeF_2 for etching.

The OmniGIS system (Fig. 11.9) houses three gas containers and a nitrogen purge or carrier line, all connected via a gas reservoir to a long needle that can be brought close to the beam impact point. The typical distance between the lowest point of the needle and the sample surface measures 200–300 μm – see Fig. 11.8. The diameter of the needle outlet is 500 μm and the angle between the needle and the beam axis is 65°. In this system, gases are mixed with a carrier gas, usually nitrogen, for more stable flows. This gas is also used for purging.

11.3 Review of Recent He-IBIP Work

11.3.1 Deposition of Boxes

Probably the simplest structure that can be made by beam induced deposition is a box. It is obtained when the ion beam is scanned continuously within a rectangular area of the specimen surface. An example from the University of Southampton is shown in Fig. 11.10 [9]. The precursor was $C_9H_{16}Pt$ and the substrate Si. The He^+ ion current was varied between 0.5 and 4.0 pA, and the helium beam energy was

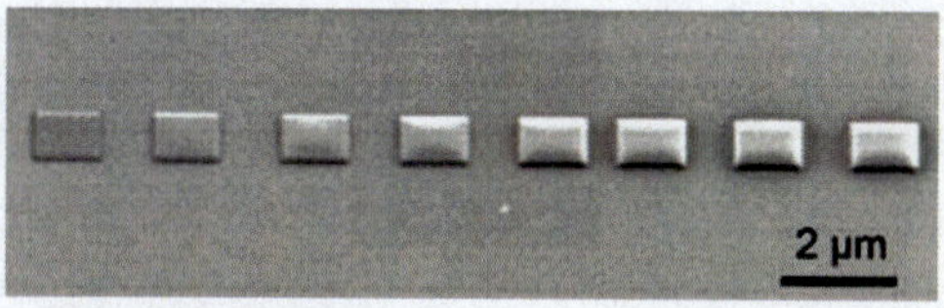

Fig. 11.10 Boxes grown at various current by He-IBID, using $C_9H_{16}Pt$; numbers indicate current in pA (HIM image, with 45° sample tilt.) (Reproduced from S. Boden [9])

30 keV. Beam dwell time was 1 μs and beam step size 1 nm. The exposure time for all boxes was 100 s. The height of the boxes was measured by atomic force microscopy (AFM).

Not surprisingly, higher doses imply higher boxes. In this example, they range from 2 nm at 0.5 pA to 65 nm at 4.0 pA. Sanford et al. have grown 500–1,000-nm high boxes, using the same precursor and similar conditions. The substrate, however, was chromium-coated glass [4]. These authors studied the dependence of the growth rate and the Pt content on ion current, box size, and beam step size. These issues will be discussed below.

11.3.2 Deposition with a Stationary Beam

The small sub-nanometer size of the primary beam would, in principle, allow the fabrication of structures with features below 1 nm. Indeed, Van Dorp et al. [25] have made the smallest EBID structures ever: 1.0 nm in diameter using a 0.3-nm STEM beam, a thin Si_3N_4 membrane as substrate, and $W(CO)_6$ as precursor gas. However, these deposits were also low. During continued exposure, small EBID structures grow in height and become tiny pillars, but they grow also in width up to a saturation value of approximately 25 nm [22]. For Ga-IBID, the best pillars are much broader, typically at least 100 nm. The width of Ga-IBID pillars cannot be explained by the lateral spreading of the ions in the deposited material, which is typically only several tens of nanometers – see Fig. 11.3 – nor by the range of the secondary particles, which is even less. The poor spatial resolution of Ga-IBID is still not understood, but could be related to the inevitable occurrence of sputtering during growth.

There are only a few studies of He-IBID with a stationary beam. A remarkable example is from Hill et al. [5], shown in Fig. 11.11: a tungsten whisker with an aspect ratio of 130. The thinnest He-IBID pillar made so far is 36 nm wide [7] – see Fig. 11.12 at 0.8 pA. So far, no group has studied the nucleation and initial outgrowth of a He-IBID deposit. It is not clear yet whether He-IBID in the nucleation phase can reach the same limit of 1 nm as EBID [25].

The top of the 36-nm wide PtC pillar of Fig. 11.12 has a radius of curvature of 9 ± 2 nm. A TEM image of the top of a comparable pillar with the same radius of curvature is shown in Fig. 11.13. For this pillar, the diameter reaches a constant width of 52 nm at a distance of 200 nm from the top.

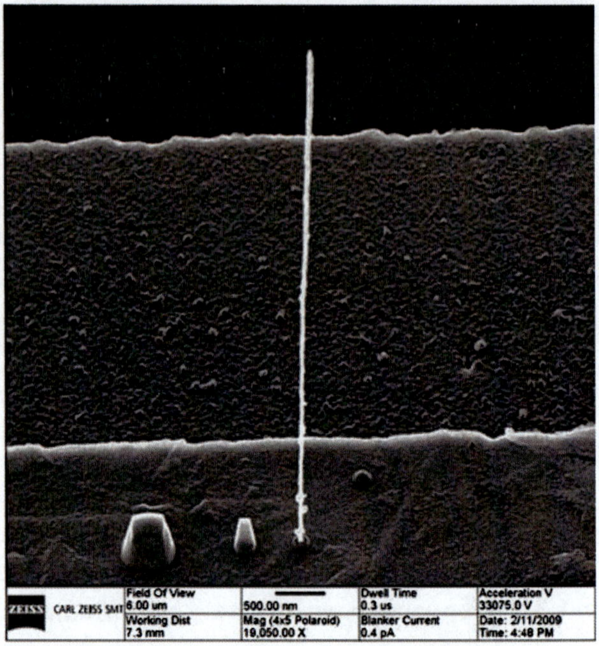

Fig. 11.11 Tungsten pillar, 6.5 μm high and 50 nm wide, grown with a stationary 30 keV He^+ beam (Reproduced from R. Hill [5])

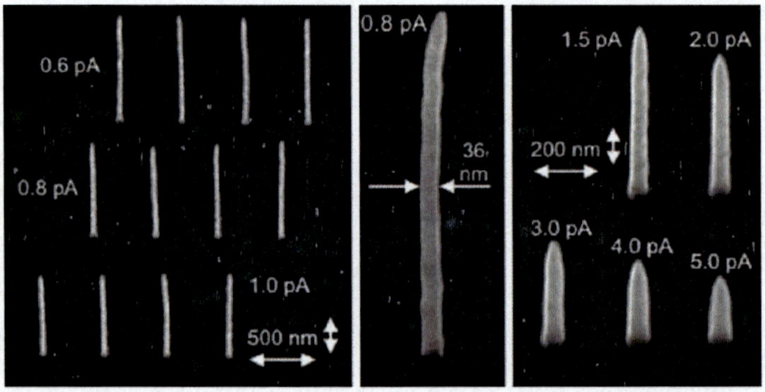

Fig. 11.12 PtC pillars grown with a stationary beam of 25 keV He^+ and a fixed dose of 6.0 pC. The current was varied between 0.6 and 5.0 pA (Reproduced from P. Chen [7])

Ions scattered in the growing pillar might escape from the pillar's shaft, where they and their associated secondary electrons can induce lateral growth. The typical penetration depth and lateral deflection in bulk PtC material is 200 and 80 nm, respectively [7]. Alkemade et al. [6] and Chen et al. [7] concluded that the penetration depth and the mean lateral deflection determine the final, saturated pillar width.

Fig. 11.13 TEM image of a He-IBID PtC pillar. The radius of curvature of the apex is 9 nm (Maas, Van Veldhoven, and Tichelaar, unpublished)

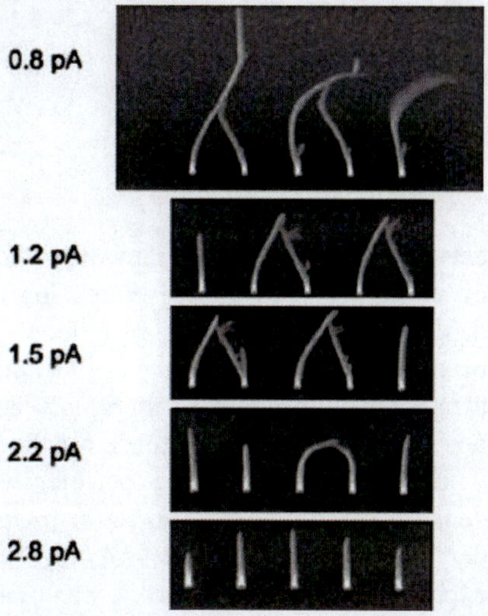

Fig. 11.14 Silicon oxide whiskers grown by He-IBID with TEOS. The irregular shapes are caused by charging during growth (Maas and Van Veldhoven, unpublished)

With TEOS as precursor and with a stationary He^+ beam, Maas and Van Veldhoven grew whiskers of SiO_2 – see Fig. 11.14. Once again, sharp structures were formed, but their shapes were irregular: branches developed and the growing whiskers were apparently attracted to nearby previously grown whiskers. Furthermore, the whiskers were moving during imaging, likely also a consequence of charging.

11.3.3 Growth Rates

A key process parameter of any fabrication technology is the processing time. Long processing times imply extra costs. Therefore, a key performance indicator is the growth or removal rate, expressed in deposited or removed volume per incident ion or per unit of charge, see Eq. (11.1). Figure 11.15 shows a compilation of all published data on the current dependence of the He-IBID growth rate of Pt. Apart from one data point, the growth rate in the experiment by Boden et al. [9] –indicated by the red squares – is almost constant: 0.18 μm^3/nC. The low rate at 0.5 pA might indicate an initial phase of slower growth. Sanford et al. measured higher rates:

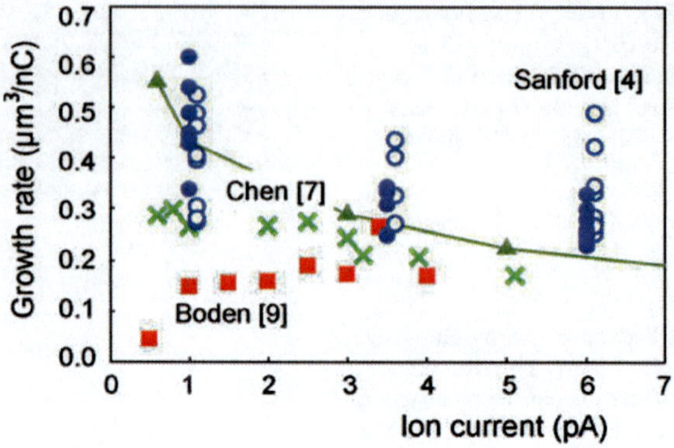

Fig. 11.15 Compilation of three published studies [4, 7, 9] of the current dependence of the He-IBID growth rate for Pt. The *open circles* are derived from the *filled circles* by correcting for variations in Pt content. (For clarity, the *open circles* are shifted horizontally by 0.1 pA.) The *triangles* are simulation results (Reproduced from P. Chen [7])

between 0.20 and 0.60 μm^3/nC – see the filled circles [4]. Furthermore, they observed a decrease with increasing current, which they tentatively attributed to depletion of the precursor molecules. The authors varied the beam step size between 1 and 3 nm, but that did not affect the growth rate. Figure 11.15 shows also the growth rate for a stationary beam – see the crosses. Surprisingly, the rates do not differ much from those for the boxes.

In the scanning-beam experiments by Sanford et al., the time-averaged current density was between 1 and 100 pA/μm^2 [4], whereas Boden et al. used current densities between 0.5 and 4 pA/μm^2 [9]. The current density in the stationary-beam experiment by Chen et al. [7] was many orders of magnitude higher: between 600 and 5,000 nA/μm^2, assuming a beam diameter of 1 nm. Despite this large variation in incident current densities, the growth rates in these studies are very similar. The current dependence of the growth rate in the work by Sanford, *i.e.* the filled circles, might be related to the decrease in C content at higher currents, see the next paragraph. The open circles in Fig. 11.15 are Sanford's data, but corrected for the variations in Pt content.[1] The lack of any current-density dependence suggests either a very fast precursor supply to the growing area or a much larger growth area than solely the primary beam spot.

11.3.4 Composition

Structures grown by electron or ion beam induced deposition are rarely pure. Various elements of the original precursor molecules are often incorporated in the deposits. In addition, oxygen and hydrogen from the ambient can be incorporated. Structures grown by EBID have a composition that resembles the original precursor more than the structures grown by Ga-IBID [15]. Sanford et al. [4] have measured by EDX the

[1] The data points were divided by the average Pt content; hence, the corrected data points scatter around the original ones.

composition of the boxes deposited by $C_9H_{16}Pt$ He-IBID. The measured platinum contents ranged between 8.5 and 20 at.%, much lower than for Ga-IBID (typically 35–40 at.%), but similar to EBID. Furthermore, there was a negative correlation between Pt content and growth rate. For increasing current, the deposition rate decreased but the Pt content increased. Possibly, the lower growth rate at the higher currents in Fig. 11.15 is due to removal of carbon by ion beam heating.

EDX analysis in TEM by Maas, Van Veldhoven and Tichelaar (unpublished) of the pillar of Fig. 11.13 revealed a Pt composition between 6 and 15 at.%, the remainder being C. The measured Pt-C ratio for He-IBID is thus close to that of the original precursor molecule.

11.3.5 Proximity Effects

The growth of SiO_2 whiskers close to each other, Fig. 11.14, showed that proximity effects can influence beam induced deposition. Here, proximity effects are very likely mediated by charging or discharging via neighboring structures. In work by Chen et al. [26] and by Maas and Van Veldhoven (unpublished), Pt pillars were grown at variable separation or pitch. Figure 11.16 shows sets of nine sequentially grown pillars at separations of between 80 and 500 nm and Fig. 11.17 shows the corresponding pillar height and diameter as a function of separation. The gain in height and width is about 25% at a pitch of 80 nm, corresponding to a near-doubling in volume. Furthermore, closely packed pillars tend to grow toward each other.

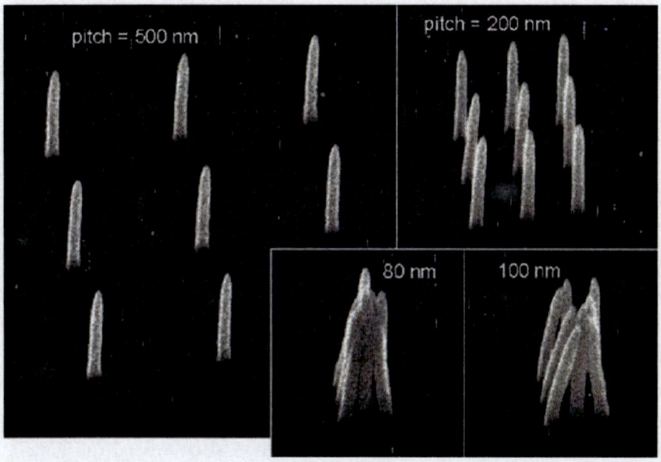

Fig. 11.16 Proximity effects: the pillar at the lower left is grown first and the one at the upper right last. If pillars are grown at small separations, their size is affected. Furthermore, they tend to bend toward each other (Maas and Van Veldhoven, unpublished)

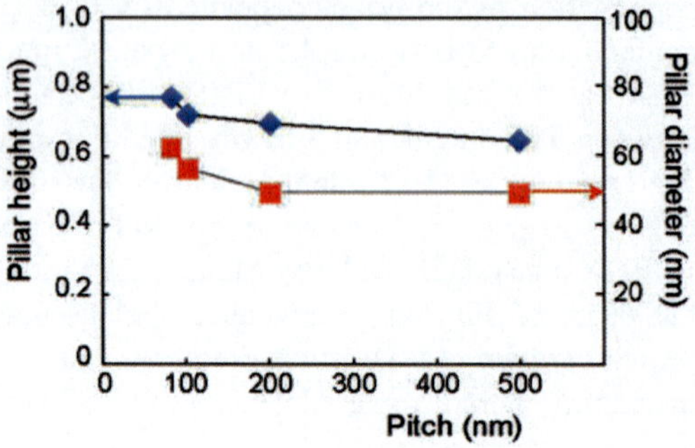

Fig. 11.17 Proximity effects: pillars become higher and wider when they are grown at smaller separation (Maas and Van Veldhoven, unpublished)

Similar proximity effects are known for Ga-IBID [27]. In Ref. [27], growth in the vicinity of the beam spot is attributed to secondary electrons and atoms. Ga-IBID pillars are wider than He-IBID and larger separations (>3 μm) are needed to avoid observable proximity effects. The cause of this difference is not well known, but could be related to sputtering during Ga-IBID. Sputtering slows down deposition at the beam spot, but not in the beam spot's vicinity. On the contrary, sputtered atoms might cause decomposition at their landing sites, contributing to the proximity effect.

11.3.6 Complex Structures

Nanofabrication is being utilized to fabricate functional structures or devices with one or more spatial dimensions in the sub-100-nm range. Before one can use a new fabrication technology, processing conditions need to be investigated and, at least partially or heuristically, understood. Helium ion beam induced deposition is still in exploration and so far only some basic devices have been made. Systems equipped with a pattern generator allow deposition of more complex structures. Figure 11.18 shows an oblique view of five parallel lines, 50 nm high, 15 nm wide, and with a pitch of 50 nm. Note that the two outermost lines are virtually identical to the three central ones. Apparently, proximity effects did not result in geometric irregularities. The line-edge roughness (3σ) was 3 nm.

In a follow-up experiment by Scipioni et al. [28], pairs of lines were grown. The separation was reduced in steps down to 12 nm, see Fig. 11.19. The smallest resolved line width was 13 nm and the smallest gap 6 nm. When the pitch was reduced to 8 nm, the pair of lines merged. This experiment seems to be the current record of smallest feature size for He-IBID.

Another example with the $C_9H_{16}Pt$ precursor is shown in Fig. 11.20. The word 'Best' is written in 60-nm wide letters; the line width is 20 nm. Although the letters are densely spaced, they do not merge, leaving for example a small gap of again 6 nm between the letters 'B' and 'e'. The fact that the smallest gaps appear to be

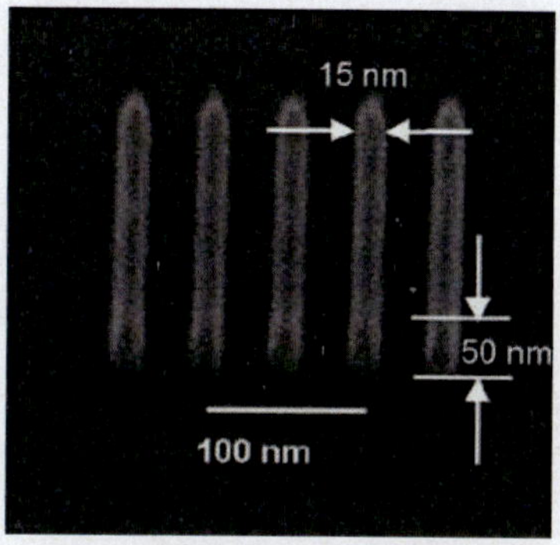

Fig. 11.18 HIM image of Pt lines made with He-IBIP imaged at 30°-sample tilt. The lines were made in parallel with a 0.5 pA and 25 keV He^+ beam. The line dose was 0.50 pC/μm (Reproduced from D. Maas [8])

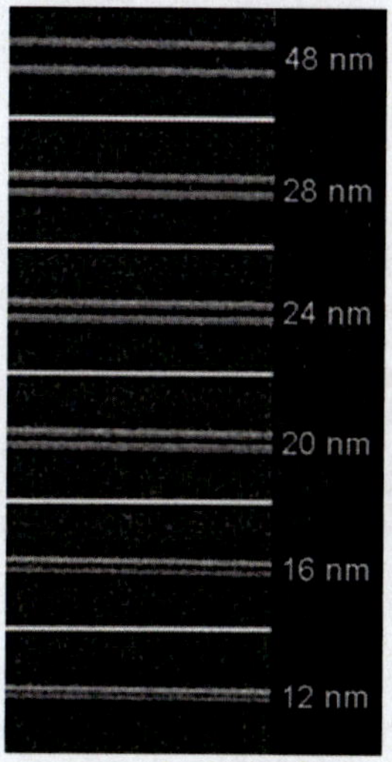

Fig. 11.19 Top-view HIM images of pairs of He-IBID Pt lines with various separations. A 0.5 pA and 25 keV beam and a line dose of 0.48 pC/μm were used. At a separation of 16 nm, the lines are still resolved (Reproduced from L. Scipioni [28])

only about half as large as the smallest lines is probably caused by the obstructed delivery of the precursor molecules to the gap region.

Figure 11.21 shows a four-point probe device made of Pt. Preliminary measurements revealed resistivities of between 2 and 60 Ω-cm, comparable to that of EBID-grown Pt [29], but much higher than for Ga-IBID. Hall bar structures in Pt and in W have been made by Boden et al. [9] – see Fig. 11.22.

11.3.7 Helium Ion Beam Milling

Several authors have shown the feasibility of direct milling by helium ions for nanostructuring. As material removal by helium ion beam milling is inefficient and

Fig. 11.20 Extended He-EBID structure in Pt; the bar is 100 nm long. The gap between the letters 'B' and 'e' is only 6 nm wide (Maas and Van Veldhoven, unpublished)

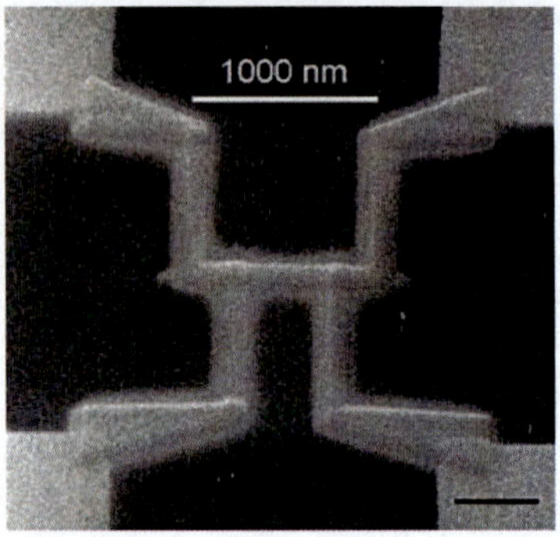

Fig. 11.21 He-IBID grown Pt lines in a four-point probe device. Resistivity is similar to EBID grown Pt (Maas and Van Veldhoven, unpublished)

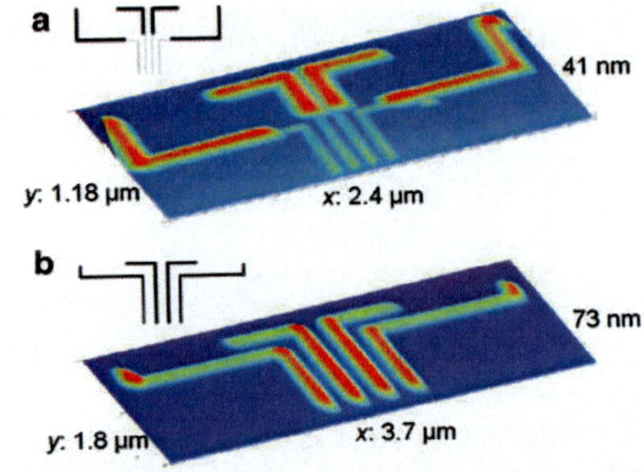

Fig. 11.22 AFM measurements of He-IBID grown Pt (**a**) and W (**b**) four-point Hall bars. The insets show the designs (Reproduced from S. Boden [9])

the maximum current of the available instruments is low, most studies so far have been limited to the shaping of materials in the sub-10 nm range. An example is shown in Fig. 11.23: the cutting of a supported thin gold wire [8]. The line cut has a width of only 8 nm. The high precision cut was achieved thanks to the narrow probe size, the limited beam spread, and the high sputtering yield at high angles of incidence. Scipioni et al. have made vias as narrow as 10 nm in a 100-nm-thick gold layer, see Fig. 11.24 [12].

Rudneva et al. [13] have milled rectangular holes with very steep walls in small particles of a number of materials – see for example Fig. 11.25. The milling of some of

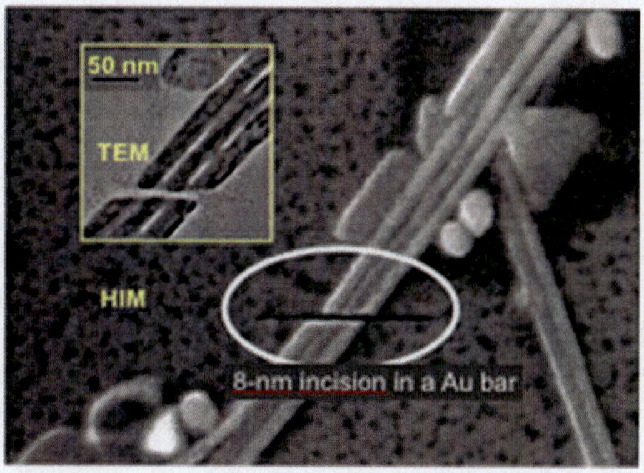

Fig. 11.23 HIM image of a gold wire, cut by the helium beam. The inset is a TEM image of the same area. The gap width of 8 nm is much larger than the beam diameter; nevertheless, this precision is difficult to achieve by means of Ga-milling (Reproduced from D. Maas [8])

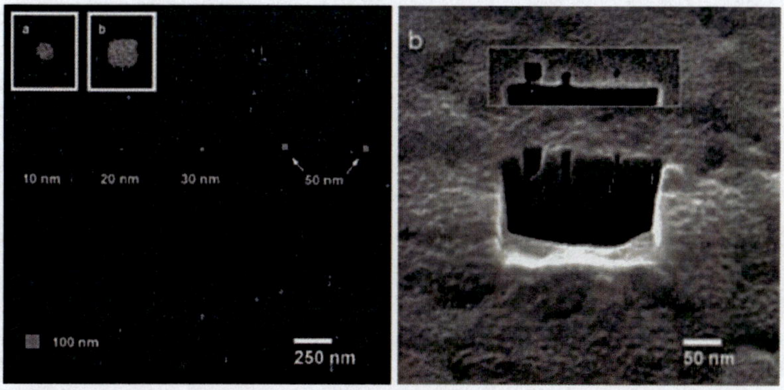

Fig. 11.24 Milling of vias in gold. Left: vias made in a 100-nm-thick Au foil, imaged in transmission mode; right: vias in a Au film, imaged in scanning mode, after removal of obstructing material; the inset is a top view (Reproduced from L. Scipioni [12])

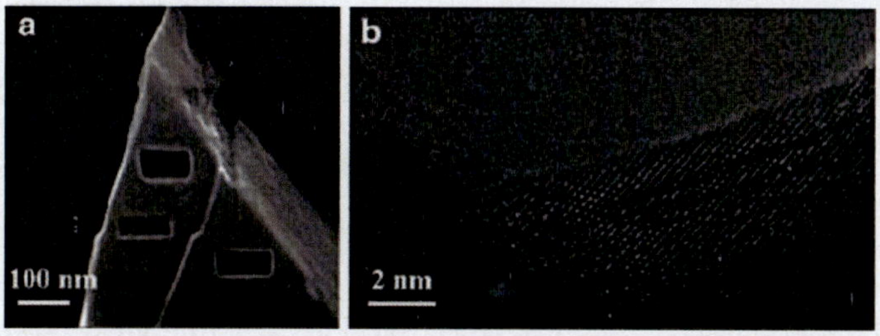

Fig. 11.25 (a) HIM images of rectangular holes milled with a 30 keV He^+ beam in a small $Cu_xBi_2Se_3$ particle. (b) High-resolution TEM image, suggesting that little damage was caused by the ion beam (Reproduced from M. Rudneva [13])

these holes was terminated just before the particle was completely punctured. High-resolution transmission electron microscopy revealed that the remaining lamellae were polycrystalline in the case of Au, Pt, and $Cu_xBi_2Se_3$ and amorphous in the case of Si. Apparently, either damage in the first three materials did not occur or the material annealed spontaneously or under ion beam bombardment. Note that, in conventional TEM lamella preparation with a Ga focused ion beam, the material is removed from the sides, not top-down. The high-aspect ratios of the vias or craters made by a helium beam are remarkable. In Ga milling, craters have a typical V shape, and aspect ratios are rarely above 5. Figure 11.4b shows two major differences between He- and Ga-milling: the latter is one to two orders of magnitude more efficient than the former. However, He-milling is relatively very efficient at grazing angles, implying that any protrusion in the sidewall of a crater is quickly removed by the ion beam. In Ga-milling, the steepness of the crater sidewall is largely determined by redeposition. The milling of high-aspect ratio holes might have important applications in nanotechnology, for instance as sieves or sensors for single-molecule studies.

Graphene is a single-atomic layer of graphite with potential applications in nanoelectronic devices, but graphene structuring is a delicate process. Several authors have reported success in structuring graphene with a focused helium ion beam [9–11]. So far, the smallest graphene structures are about 10-nm-wide ribbons [10], see Fig. 11.31.

11.3.8 He^+ Ion Beam Induced Etching

Removal and addition of material are the fundamental physical processes of nanofabrication. Removal by direct helium ion erosion (*milling*) is possible, but slow, about one to two orders of magnitude slower than with Ga^+ ions. The Ga^+ ions stop in the top layer of the material – see Fig. 11.3. As the top layer is constantly being removed, damage by the Ga^+ ions does not accumulate. In contrast, the He^+ ions are stopped significantly deeper in the material and damage accumulation does occur [18]. One can reduce the damage by lowering the beam energy and thereby reducing the implant depth. However, ion beam milling at lower energy requires higher doses. Fortunately, addition of an etchant gas during ion beam bombardment can enhance erosion rates dramatically [30]. One can also remove material by using an etchant gas combined with an electron beam [14].

Helium ion beam induced etching (He-IBIE) is one of the potent materials processing capabilities of the new helium ion microscope. Although the implementation of chemical etching increases the possibilities for materials processing substantially, every combination of material and shape might require its own recipe. However, finding useful recipes is time consuming and the added chemistries make the instrument both more complex as well as susceptible to degradation. Recipe development and impact on the instrument preclude rapid progress of helium ion beam materials processing on the nanometer scale. Fortunately, there is ample experience with conventional chemistry-enhanced plasma etching of materials [31]. The main difference between chemistry-enhanced plasma etching and helium ion

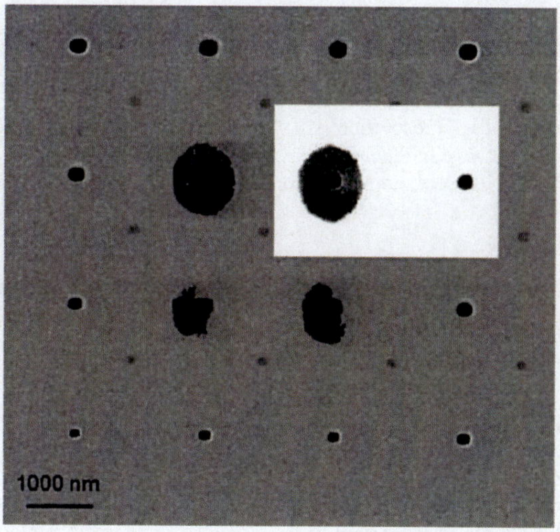

Fig. 11.26 He^+ ion beam induced etching of SiO_2. The image contrast of one segment has been enhanced to make the Si cavity below the hole in the oxide visible (Maas and Van Veldhoven, unpublished)

beam induced etching is the method of pattern definition. Plasma-etching is based on the use of masks, whereas He-IBIE is based on controllable focused ion beams.

In the experiment shown in Fig. 11.26, XeF_2 was used as an etchant for the surface oxide layer of Si. The ion beam dose was increased in steps from the lower left to the upper right. The opening in the oxide caused spontaneous underetching of Si by the influx of XeF_2, see the gray cavity in the enhanced-contrast segment. Note that spontaneous chemical etching is usually isotropic, whereas beam-induced etching is anisotropic. For good processing results, the flux of etchant molecules and the flux of ions must be well tuned. For instance, one can in some cases adjust the anisotropy of the etch and thus influence the slope of the sidewalls in the pattern by tweaking the balance between the two fluxes.

Figure 11.27 shows the surface of a TaN film, exposed to a 0.7 pA beam of 25 keV He^+ ions with and without the presence of XeF_2. Clearly, beam induced etching took place. This processed material can be used as a mask layer with a high spatial resolution. For instance, the good resolution and the steepness of the crater walls make He-IBIE ideally suitable for manufacturing photonic crystals with short periodicity that can operate at the important range of blue or ultraviolet wavelengths.

11.3.9 Modeling of He-IBID

Equations (11.1) and (11.2) provide a theoretical basis to describe the growth by beam induced deposition. Similar equations apply for beam induced etching. However, for a non-flat and non-constant geometry, such as a growing pillar-like deposit or an etch pit, the solutions of the differential equations might become very complex. Moreover, many processes are not quantitatively known, for instance spontaneous desorption or

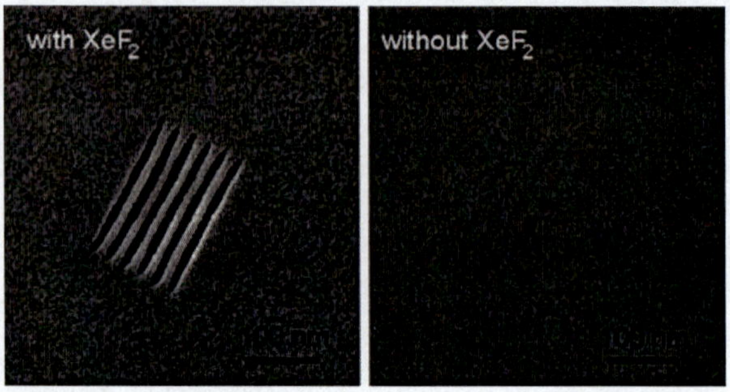

Fig. 11.27 He^+ ion beam induced etching of an 80-nm thick TaN film. The line dose was 3.9 pC/μm. The demonstrated resolution is 12 nm half-pitch (Maas and Van Veldhoven, unpublished)

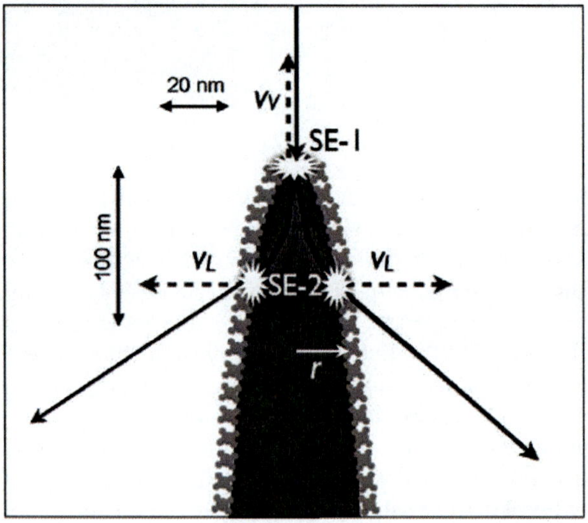

Fig. 11.28 Vertical growth occurs when ions penetrate a deposit, whereas lateral growth occurs when they escape. The final width W of the deposit depends on the average spreading of the ions and the penetration range L (Modified from P. Alkemade [6])

surface diffusion. Therefore, simplified or qualitative models [15, 32] or Monte Carlo simulations [14–16] are used. The current-dependence of the He-IBID pillar height and width has been explained qualitatively by a simplified model by Alkemade et al. [6] and by Monte Carlo simulation by Chen et al. [7]. Both models assume that precursor decomposition takes place wherever a particle enters or leaves the growing structure – see Fig. 11.28. The primary ions enter the apex of the growing deposit within an area of about 1 nm in diameter. However, their associated secondary electrons –denoted as SE1's– may escape from a region around the entrance point that is as wide as their escape depth: up to ~10 nm – see Fig. 7 in Chap. 4. The value of 10 nm corresponds well with the observed radius of curvature of the pillar's apex – see Fig. 11.13.

Hence, the growth area is not the beam impact region of ~1 nm^2, nor the region from where SE1's escape (~100 nm^2), but the entire pillar top, up to a distance L of ~100 nm below the apex, where L is the penetration range. Hence, the growing area is roughly πWL, ~10,000 nm^2 – see Fig. 11.28. At 1 pA, the average current density

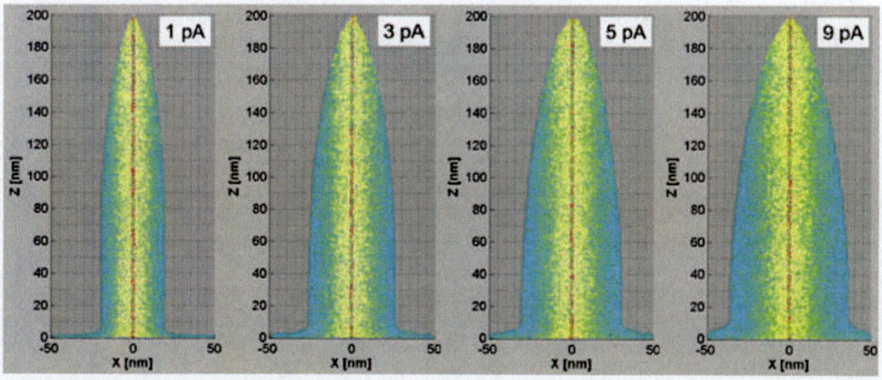

Fig. 11.29 Simulated He-IBID pillars of PtC_4 grown at different He^+ currents. Various colors denote various decomposition mechanisms: *red*: primary ions; *yellow*: SE1's; *green*: forward scattered ions; *cyan*: SE2's (Reproduced from P. Chen [7])

over this area is ~0.1 nA/μm^2, comparable to the highest density in the box growth experiment of Sanford et al. [4]. A dominating contribution of secondary particles can explain the similar growth rates for box and for pillar deposition in Fig. 11.15, despite the orders of magnitude difference in primary current densities.

Figures 11.29 and 11.30 show the simulated growth of PtC_4 pillars by He-IBID [7] obtained with the EnvisION Monte Carlo code of Smith et al. [24]. In this code, decomposition is assumed to occur either by ions or by secondary electrons. The ions can cause decomposition when they enter and when they leave the pillar after one or more scattering events. Chen et al. concluded that the simulation explains the measured trends in the current dependence of the pillar volume – see the triangles in Fig. 11.15 [7]. In particular, the decreasing height with increasing current is a direct consequence of precursor depletion at the pillar apex. Furthermore, the simulated pillar shape nicely resembles the actual one – see Fig. 11.13.

11.4 Conclusion and Outlook

Experiments conducted in various laboratories with different precursor gases and substrate materials have shown that helium ion beam induced processing (He-IBIP) is an attractive technique for structuring materials on the scale between a few nanometers to a few micrometers. It has several features in common with electron beam induced processing (EBIP) and heavy-ion induced processing, in particular Ga-IBIP. With EBIP it shares high spatial resolution and with Ga-IBIP high processing rates. The weak concurrent ion-beam erosion is an additional advantage of He-IBIP over Ga-IBIP. However, the composition of helium-ion beam deposited materials is not pure, but resembles the original composition of the precursor gas. This situation is similar to that of EBID. In contrast, Ga-IBID produces purer deposits. Despite the low sputtering yield – usually well below 1 atom per ion – direct

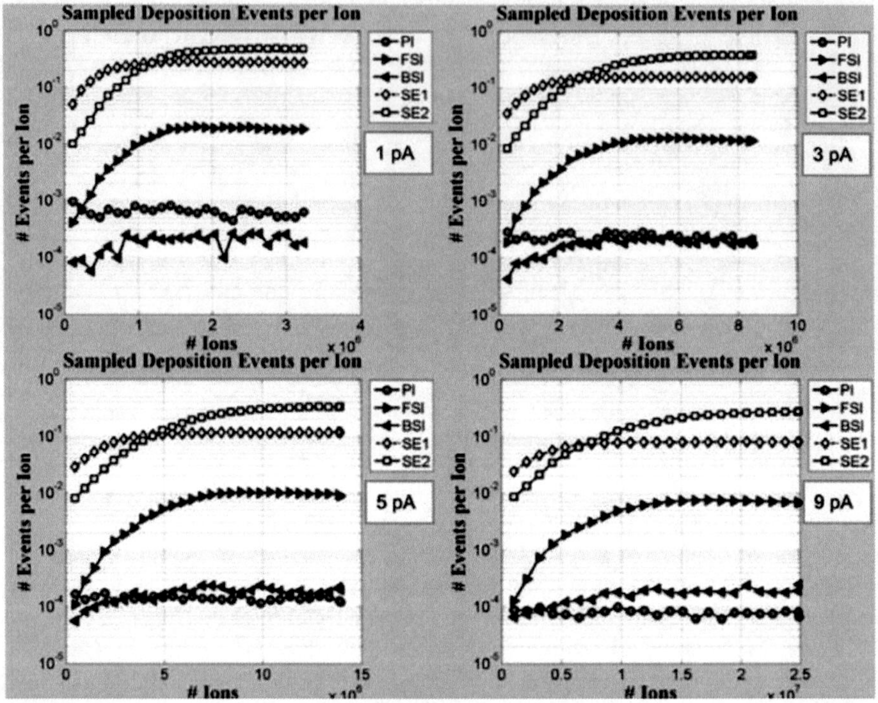

Fig. 11.30 Simulated dose-dependence of He-IBID growth of PtC_4 pillars. Various decomposition mechanisms are plotted (Reproduced from P. Chen [7])

structuring of materials by helium ion beam milling is possible. Acceptable processing times and very sharp features appear to be the consequence of a relatively high sputtering yield (up to 1 atom/ion) at grazing incidence. Especially when very thin layers such as graphene are to be structured, focused helium ion beam milling is probably the best technique, see Fig. 11.31.

The mechanism of Ga-IBID is thought to be precursor decomposition by collisions of excited substrate atoms [1, 15], although contributions from secondary electrons cannot be excluded [21]. It is evident that this collisional mechanism is inherently weak in helium ion beam processing, and therefore we attribute the high processing rates of He-IBID to a relatively high contribution from a large number of low-energy secondary electrons. The smallest structures made so far with He-IBID are a 13-nm wide line and a 6-nm gap. Although the best demonstrated resolution in EBID is 1 nm [25], we expect the same limit because the mechanisms of EBID and He-IBID are similar. Furthermore, the relatively high deposition rate and the weak sputtering effects imply favorable conditions for studying the processes by simulation. Indeed, recent Monte Carlo simulations explain measured trends in He-IBID growth of simple structures.

We foresee good opportunities for He-IBIP in areas where Ga-IBIP reaches its limits, mainly concerning mediocre resolution, unacceptable contamination by implanted Ga, and damage formation by sputtering. In particular, He-IBIP might

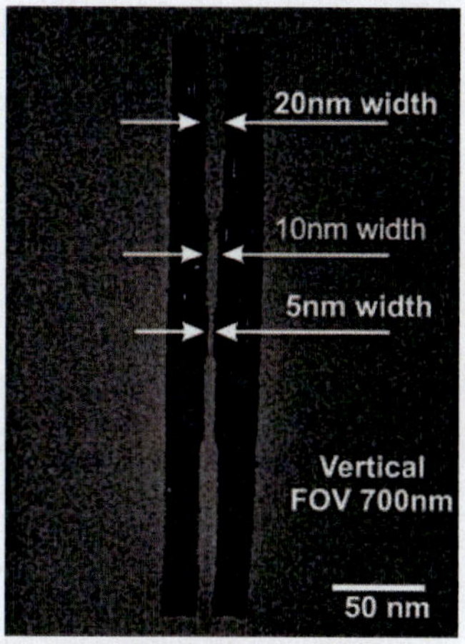

Fig. 11.31 A nano-ribbon with variable width, cut with He ions in graphene (Reproduced from D. Pickard [10])

provide an alternative to standard Ga-IBIP for photomask repair and semiconductor circuit edit. Here, He-IBIP will have to compete with EBIP, which is also being considered as an alternative to Ga-IBIP. A shortcoming is, however, the deep penetration of the ions. Implanted helium can cause large damage to the integrity of the substrate hundreds of nanometers deep. On the other hand, damage near the surface is less or might even disappear by self-annealing [13]. The rapid growth of tips –such as the extremely long tungsten whisker in Fig. 11.11 – might find many applications in nanosensing; for example as tips in atomic force microscopy or in functionalized sensors, nano-antennas, field-emission tips, etc. Moreover, He-IBIP materials might be used as mechanical or electrical connecting elements in nanodevices. Finally, He-IBIP structured materials can be used as patterned seed layers for other growth techniques such as chemical vapor deposition or atomic layer deposition.

Helium ion beam induced etching with a good spatial resolution has been demonstrated as well. High-precision photomask repair and fabrication of short-wavelength photonic crystals are examples for which He-IBIE might be the technique of choice.

Probably the most tedious part of research and development in this field is the time-consuming search for proper chemistries and post-processing steps that meet the requirements for dedicated applications. The situation for He-IBIP, however, is no different from that for EBIP and Ga-IBIP. Nevertheless, the availability of a third technique that combines the advantages of EBIP and Ga-IBIP could accelerate progress in this field greatly. Finally, other noble gas ion sources such as neon with its higher sputtering yields and shallower penetration depths [5, 33] or other light ion sources such as hydrogen with very low damage formation might provide the means to structure materials on the nanoscale with the best and finest tool: an atomically sharp and easily controllable beam of inert but powerful particles.

Acknowledgements This research is part of NanoNed, a national research program on nanotechnology, funded by the Dutch ministry of Economic Affairs in the Netherlands. Our colleagues P. Chen, E. van der Drift, H. Salemink from Delft University of Technology and D. Maas from

TNO are gratefully acknowledged for their contributions and discussions. The authors acknowledge L. Scipioni, S. Boden, R. Hill, R. Livengood, S. Tan, I. Utke, C. Sanford, M. Rudneva, and F. Tichelaar for giving permission to use their figures.

References

1. Dubner AD, Wagner A, Melngailis J, Thompson CV. J Appl Phys. 1991;70:665–73.
2. Ward BW, Notte JA, Economou NP. J Vac Sci Technol B. 2006;24:2871–4.
3. Morgan J, Notte J, Hill R, Ward B. Microsc Today. 2006;14(4):24–31.
4. Sanford CA, Stern L, Barriss L, Farkas L, DiManna M, Mello R, Maas DJ, Alkemade PFA. J Vac Sci Technol B. 2009;27:2660–7.
5. Hill R, Faridur Rahman FHM. Nucl Instr Meth A. Nucl Instr Meth A. 2011;645:96–101.
6. Alkemade PFA, Chen P, van Veldhoven E, Maas D. J Vac Sci Technol B. 2010;28:C6F22–5.
7. Chen P, van Veldhoven E, Sanford CA, Salemink HWM, Maas DJ, Smith DA, Rack PD, Alkemade PFA. Nanotechnology. 2010;21:455302. 7 pp.
8. Maas D, van Veldhoven E, Chen P, Sidorkin V, Salemink H, van der Drift E, Alkemade P. Proc SPIE. 2010;7638:763814. 10 pp.
9. Boden SA, Moktadir Z, Bagnall DM, Mizuta H, Rutt HN. Microelectron Eng. Microelectron Eng. 2011;88:2452–5.
10. Pickard D, Scipioni L. Graphene nano-ribbon patterning in the orion plus (Zeiss Application Note, Oct 2009).
11. Bell DC, Lemme MC, Stern LA, Marcus CM. J Vac Sci Technol B. 2009;27:2755–8.
12. Scipioni L, Ferranti DC, Smentkowski VS, Potyrailo RA. J Vac Sci Technol B. 2010;28: C6P18–23.
13. Rudneva MI, van Veldhoven E, Shu MS, Maas D, Zandbergen HW. Abstract 17th international microscopy conference, Rio de Janeiro; 2010.
14. Randolph SJ, Fowlkes JD, Rack PD. Crit Rev Solid State Mater Sci. 2006;31:55–89.
15. Utke I, Hoffmann P, Melngailis J. J Vac Sci Technol B. 2008;26:1197–276.
16. van Dorp WF, Hagen CW. J Appl Phys. 2008;104:081301. 42 pp.
17. Rabalais JW. Principles and applications of ion scattering spectrometry. New York: Wiley-Interscience; 2003.
18. Livengood R, Tan S, Greenzweig Y, Notte J, McVey S. J Vac Sci Technol B. 2009;27:3244–9.
19. Castaldo V, Hagen CW, Kruit P, van Veldhoven E, Maas D. J Vac Sci Technol B. 2009;27:3196–202.
20. Eckstein W, Behrisch R, editors. 'Sputtering yields' in sputtering by particle bombardment. Berlin: Springer; 2007.
21. Chen P, Salemink HWM, Alkemade PFA. J Vac Sci Technol B. 2009;27:2718–21.
22. Silvis-Cividjian N, Hagen CW, Teunissen LH, Kruit P. Microelectron Eng. 2002;61–62:693–9.
23. Fowlkes JD, Randolph SJ, Rack PD. J Vac Sci Technol B. 2005;23:2825–32.
24. Smith DA, Joy DC, Rack PD. Nanotechnology. 2010;21:175302. 7 pp.
25. van Dorp WF, van Someren B, Hagen CW, Kruit P, Crozier PA. Nano Lett. 2005;5:1303–7.
26. Chen P. PhD thesis, Delft University of Technology; 2010.
27. Chen P, Salemink HWM, Alkemade PFA. J Vac Sci Technol B. 2009;27:1838–43.
28. Scipioni L, Sanford C, van Veldhoven E, Maas D. Microsc Today. 2011;19(3):22–6.
29. Botman A, Mulders JJL, Weemaes R, Mentink S. Nanotechnology. 2006;17:3779–85.
30. Winters HF, Coburn JW. Appl Phys Lett. 1979;34:70–3.
31. Flamm DL, Donnelly VM. Plasma Chem Plasma Process. 1981;1:317–63.
32. Lobo CJ, Toth M, Wagner R, Thiel BL, Lysaght M. Nanotechnology. 2008;19:025303. 6 pp.
33. Livengood RH, Tan S, Hallstein R, Notte J, McVey S, Faridur Rahman FHM. Nucl Instr Meth A. 2011;645:136–40.

Laser Nanopatterning 12

Robert Fedosejevs, Ying Tsui, Zhijiang Chen,
and Shyama Banerjee

Abstract

Over the past decade, a variety of techniques have been developed to allow flexible writing of nanopatterns and structures using visible, infrared and ultraviolet laser radiation on a size scale well below the wavelength of light employed. These include the use of subwavelength near field optical elements, nonlinear interactions such as two photon absorption, nonlinear response of the medium via contrast enhancement agents and coupling to plasmon modes which have shorter wavelengths than the incident radiation. These can be used for writing of surface features, internal features or complete 3D structures via photopolymerization. Also, nanoablation can be employed both for precision nanomilling of surfaces and direct production of nanoparticles. Laser induced forward transfer of micro- and nano-dots of material is under development for the direct deposition of materials onto surfaces with feature sizes down to 100 nm. Finally, a whole new generation of VUV, XUV and x-ray lasers is emerging, promising even smaller feature sizes in the near future.

12.1 Introduction

Lasers are currently available throughout the ultraviolet to infrared wavelength range and with pulse lengths from femtoseconds to continuous wave. This leads to a large parameter space of potential interaction conditions which can be exploited in order to support a variety of patterning techniques that can be used to produce a wide range of nanostructures. In particular, the short interaction times obtainable using femtosecond laser pulses can be exploited to give minimal lateral or

R. Fedosejevs (✉) • Y. Tsui • Z. Chen • S. Banerjee
Department of Electrical and Computer Engineering, University of Alberta, Edmonton, Alberta, Canada
e-mail: rfed@ece.ualberta.ca

M. Stepanova and S. Dew (eds.), *Nanofabrication*,
DOI 10.1007/978-3-7091-0424-8_12, © Springer-Verlag/Wien 2012

volumetric energy spread and generate some of the smallest features possible. In order to obtain nanometer scale features, a number of approaches have been developed to go beyond the wavelength limit of the incident light. These include the use of nonlinear processes which will reduce the interaction zone to a fraction of the wavelength, near field focusing optics, interference effects between two beams or coupling to shorter wavelength plasma waves both on the surface and inside the volume using femtosecond pulses. By using ultraviolet wavelength lasers, feature sizes can be made even smaller by these same approaches. New techniques for efficiently generating coherent vacuum ultraviolet (VUV) and extreme ultraviolet (XUV) pulses are advancing rapidly femtosecond and attosecond pulses are advancing rapidly by means of high harmonic generation (HHG) [1–5], capillary discharge lasers [6, 7] and short pulse laser-plasma pumped lasers at 13.9 nm [8]. New free electron laser (FEL) sources operating in the x-ray regime [9–11] are opening up the possibility of direct patterning at the nanometer wavelength scale. This requires the development of new optical components which can be used to focus such short wavelength radiation and in many cases necessitates working in vacuum. In addition to techniques of ablative or chemically patterning of materials, it is also possible to use laser ablation for additive deposition of materials on a nanometer scale in a process called Laser Induced Forward Transfer (LIFT). All of these approaches will be touched on in the following sections.

The limiting factor for any laser based approach is the resolution limit, of the order of a wavelength, which can be obtained for an imaging lens. For an imaging system, the standard Rayleigh criterion for resolving two spots for uniform plane wave input beams is based on the peak intensity of one spot coinciding with the first minimum of the Airy intensity distribution function of the second spot [12] leading to a limiting resolution of:

$$\mathrm{Res} = 0.61\ \lambda/\mathrm{n}\ \sin\theta \tag{12.1}$$

where λ is the vacuum wavelength, n is the refractive index of the ambient medium, ranging from 1.000 for air to 1.515 for immersion oil, and θ is the angle subtended by the edge of the lens aperture to the image point. The product of n sin θ is defined as the numerical aperture (NA) of the imaging system. The full width at half maximum (FWHM) intensity for the Airy spot distribution function of a single focused flat top laser spot is slightly smaller than this resolution criterion and is given by a coefficient of 0.515 in Eq. (12.1). Also, real laser beams will be more Gaussian in profile and thus the actual focal spot will be the Fourier transform of a truncated Gaussian beam resulting in a different shape function than the ideal Airy function. High power microscope objectives with numerical apertures up to 1.4 have been used for laser nanopatterning applications [13] which for visible and ultraviolet wavelengths of 500 and 250 nm would give Rayleigh resolutions of 218 and 109 nm.

Further improvement in resolution is obtained in many applications by using nonlinear processes to transfer the pattern. In these cases, a process which responds as a function of the second or higher power of intensity is employed, leading to a much smaller region of response and modification than the original focal spot diameter. One common such process is two photon absorption for writing 3D structures inside of transparent media. Also, nonlinear material response can be used to further enhance the feature sizes obtained.

12.2 Laser Material Interactions

The initial coupling of the laser radiation to the material surface comes through either normal linear absorption or, for ultrashort pulses, through nonlinear absorption processes such as two-photon absorption. The latter are particularly useful for transparent dielectric materials [14–16]. In many cases, the surface heats rapidly and a thermal heat wave is launched into the material governed by the heat diffusion equation. This sets an important length scale for the interaction which is governed by the heat diffusivity coefficient:

$$D = \kappa/(\rho C) \tag{12.2}$$

where κ is the thermal conductivity, ρ is the density of the target material and C is its heat capacity. The heat diffusion process leads to spreading of the heat during an interaction time t over a distance given by the approximate relation:

$$X = a_0(D\, t)^{1/2} \tag{12.3}$$

The coefficient a_0, of the order of unity, depends on the exact geometry of the system and exact time dependent profiles can easily be calculated numerically. For example, in silicon which has a diffusivity of 0.8 $cm^2\, s^{-1}$, the heat will spread over a distance of ~280 nm in a nanosecond and 9 nm in a picosecond, while for fused silica with a diffusivity of 0.008 $cm^2\, s^{-1}$, the heat only spreads over a distance of 28 nm in a nanosecond and 0.9 nm in a picosecond. Typically, a time scale of the order of the laser pulse width is chosen to estimate the heat diffusion length which then defines the volume affected by this heat. In the ablation regime, the depth of the interaction feature size is related to either the heat diffusion length or optical skin depth, whichever dominates [17]. This is why pulses of picosecond or femtosecond duration are typically used for nanoprocessing.

The high transient surface temperature can induce phase transitions in the material as well as chemical reactions with ambient gases. Such phase transitions can include melting and vaporization. These, in turn, can lead to physical or crystallographic changes upon resolidification [18, 19]. All of these processes can modify the surface chemical composition and crystal structure of the interaction region and surrounding region giving desired results in the interaction zone and undesirable results outside of the interaction zone. The overall affected region is called the heat affected zone (HAZ) [20–24]. In addition, shock waves also will be launched by the rapid temperature rise which leads to a large local pressure jump in the interaction region since the inertial time response of the material to expand and reach pressure equilibrium lags behind the energy absorption time scale. Long pulse (multi-nanosecond) and lower intensity interactions will not lead to much shock pressure. However, higher intensity short pulse interactions will lead to such pressure jumps and also can lead to avalanche ionization and plasma formation. If the latter occurs within the laser pulse duration, the absorption also rises rapidly and the pressure jump becomes much greater. In these cases, the pressure jump can become large enough to launch a shock wave which can lead to a larger damage affected zone (DAZ) than the interaction region itself. For sensitive device applications, such a damage affected zone may lead to device degradation or failure. Thus the intensity range employed should be chosen to minimize these deleterious effects.

At high intensities, free electrons initially created by multiphoton ionization [25, 26] will be rapidly heated due to collisional or inverse Bremsstrahlung absorption which in turn causes further ionization leading to an avalanche in the electron and ion density and the creation of a plasma [17, 27, 28]. The plasma causes ablation of a significant amount of material due to an inward propagating strong shock wave and rapidly expanding plasma plume. Typically, this leads to ablation regions of several microns in size and only in the case of femtosecond pulses can the characteristic dimensions be maintained in the nanometer range. Typical shock velocities are on the order of 10^6 cm s^{-1} and for the plasma plume expansion of the order of 10^7 cm s^{-1}. The ablated species will have energies per particle ranging from 0.1 eV for vaporized species for interactions at around 10^9 W cm^{-2}, up to 100 eV for high intensity interactions at around 10^{14} W cm^{-2}. These ablated species themselves can be used for redeposition of material to form laser deposited thin films and in direct writing of materials using laser induced forward transfer (LIFT).

Advances in techniques to generate coherent VUV radiation are now at the threshold of power levels which can be used for nanomachining applications. Such sources include the XUV and x-ray free electron lasers at the Deutsches Elektronen-Synchrotron (DESY) and Stanford Linear accelerator laboratories (SLAC) [10, 11]. They also include new laser based sources using high harmonic generation from femtosecond pulses both in gas jet targets and from solid surface interactions [1, 2, 4, 5, 29–31]. The wavelengths accessible to such sources now extend from 100 to 0.1 nm with pulse energies on the order of microjoules per pulse [11, 32]. In the XUV wavelength range, absorption skin depths of many materials are in the range of 10's of nanometers allowing for the direct writing of patterns on a nanometer length scale both through linear interaction with chemical resist materials and through ablative interactions resulting in material removal.

12.3 2D Nanowriting Techniques

Nanometer surface feature sizes can be obtained using near field optical techniques, microlenses and ultrashort wavelengths. In addition, shorter wavelength sources are becoming available for direct writing in the sub 100 nm range.

A number of techniques have been demonstrated to allow reduction of the interaction region well below a wavelength. Nearfield optical microscopes have been employed to reduce the optical source size down to 100 nm scale size dimensions and lower. Nolte demonstrated this [33] as shown in Fig. 12.1, where 200 nm wide lines are written in a metal chrome layer on a quartz substrate by writing at 1 kHz repetition rate with 100 fs pulses at 266 nm using a fluence of 1.5 times the observed ablation threshold of 70 mJ cm^{-2}. Taylor et al. were able to demonstrate 100 nm holes on a glass surface by irradiating and then chemically etching the resulting damaged region [34].

Further reduction in size has been demonstrated in more recent publications of the group of Russo et al. [35, 36] where nanoablation spots of down to 27 nm using near field processing were obtained as shown in Fig. 12.2. By focusing on the rear surface of glass cover slips with 600 fs pulses, Joglekar et al. demonstrated direct creation of holes with exit crater diameters of down to 15–30 nm feature size [37]. In this case, self focusing or interaction with the glass material appears to play a role in reducing the interaction spot size.

An alternative technique to concentrate the radiation to a small working point is through the use of microlenses. These can be fabricated by a variety of standard lithographic techniques and then used to concentrate an illuminating beam to a high intensity working spot. An example is shown in Fig. 12.3 where a microlens array has been used to ablate holes in a chrome metal layer with diameters of 100 nm and depths of 20 nm at an incident fluence of 4.76 mJ cm^{-2} [38]. The lens arrays were fabricated by fluidic assisted processing and illuminated with nanosecond 532 nm laser pulses. In the same publication they also used continuous wave argon ion laser radiation and sintered small spots below the lenses. By further heating and sintering

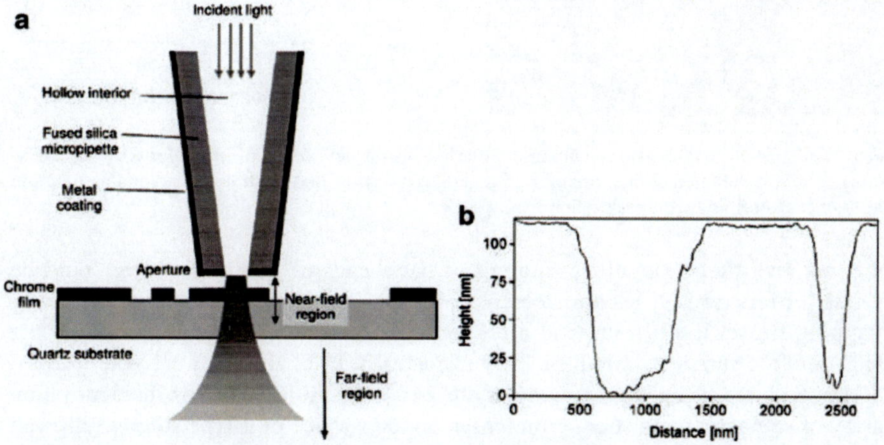

Fig. 12.1 Use of Nearfield Scanning Optical Microscope (NSOM) to produce subwavelength features: (**a**) schematic layout and (**b**) grooves written in a chrome layer on fused silica at scan rates of 2 and 10 mm/s at a fluence of ~100 mJ/cm^2, at 266 nm wavelength and 1 kHz repetition rate. Reprinted with permission from [33]

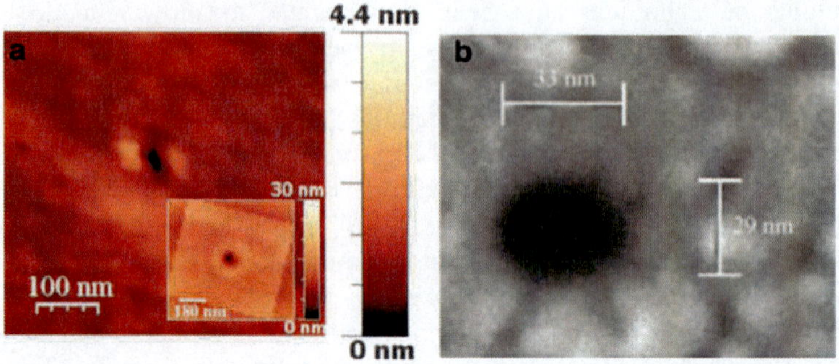

Fig. 12.2 (**a**) Use of NSOM to produce 27 nm FWHM holes with single 0.18 nJ laser pulses, reprinted with permission from [35], and (**b**) exit hole of 30 nm diameter produced on Corning 0211 cover slips by focusing 4 nJ 600 fs pulses using a 1.3 NA oil immersion objective. Reprinted with permission from [37]

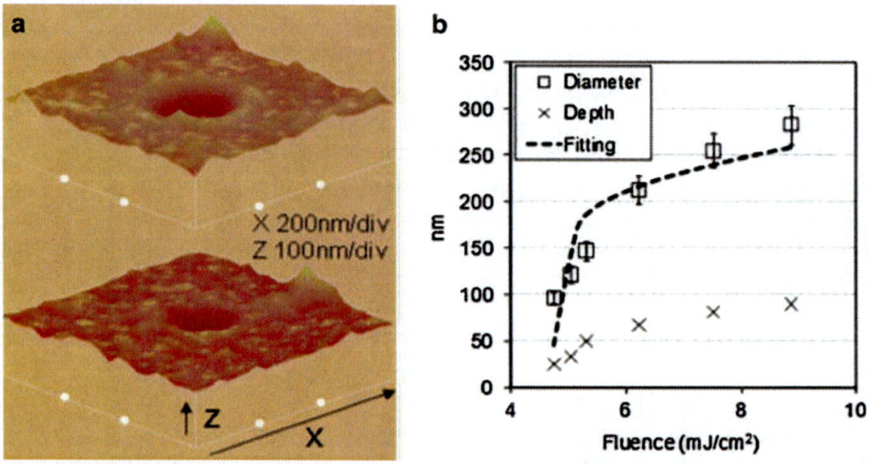

Fig. 12.3 Microlens-assisted nanohole ablation using an array of microlenses: (**a**) AFM images of holes ablated at fluences of 5.05 and 4.76 mJ cm^{-2} and (**b**) hole diameter as a function of fluence. Reprinted with permission from [38]

the material after irradiation, an array of nanobeads of 50 nm diameter could be formed. McLeod [39] used a steerable optical microlens, steered by an optically trapping Bessel laser beam, and a femtosecond pulse then irradiated through the lens onto the sample to produce 110 nm features. Li et al. [40] were able to form 30 nm features using shorter wavelength 248 nm KrF laser pulses incident on an array of self assembled submicron lenses on the surface of n-type silicon followed by etching. Further examples of such microlens and microlens array ablation and etching can be found in Refs. [41, 42].

Another technique to give very small interaction regions involves the use of an AFM probe tip to concentrate the laser radiation in the gap between the tip and

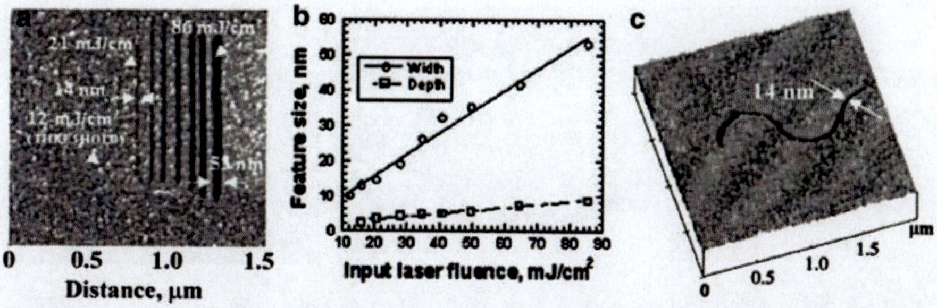

Fig. 12.4 Ablative features written into gold film using 1 kHz, 83 fs, 800 nm laser pulses concentrated with an AMF tip: (**a**) SEM image of 14 nm to 86 nm line features, (**b**) dependence of feature size on laser fluence, (**c**) 14 nm continuous line. Reprinted with permission from [43]

surface. This has been shown to give feature sizes in a gold film down to 10 nm as shown in Fig. 12.4. In this case, a kHz, femtosecond laser at 800 nm wavelength was used to irradiate the probe tip at a fluence of around 70 mJ cm^{-2} while the probe tip was kept a few nm from the surface. Further examples of AFM tip writing can be found in Refs. [43, 44].

The direct approach to achieving nanometer feature sizes is to use very short wavelength sources. Argon fluoride lasers operating at 193 nm have been used to machine nanopores with exit diameters of 90 nm [45]. Fluorine lasers operating at a wavelength of 157 nm in the vacuum ultraviolet region have been used to write structures with submicron feature sizes [46, 47]. An F_2 laser has also been used to carry out interference lithography using high contrast resists and high index immersion fluid yielding a half pitch linewidth of 22 nm resolution [48]. UV interference lithography has been used to write arrays of silicon nanowires of the order of 65 to 100 nm diameter using laser wavelengths of 244 and 325 nm combined with metal assisted etching [49–51]. Three beam interferometry was also used with 800 nm femtosecond laser pulses to create periodic patterns with feature sizes of 100 nm [52]. Applications of such interference lithography of textured structures include the control of cell behavior [53].

Many self organized surface nanostructures can be formed by carefully controlling the fluence and polarization of a short pulse irradiating laser. Typically, rippled or grating like structures are formed. Usually, many pulses are required in order to allow the structures to propagate and grow. It is thought that these structures are formed through the interaction of the incident radiation and surface plasma waves excited on the surface, particularly in the case of metal surfaces but also for semiconductors and dielectrics once the surface is ionized and becomes conducting with each laser shot. Such rippled structures have been observed on various semiconductor materials including GaP, GaAs, GaN and InP [54–56] as shown in Fig. 12.5a,b with feature sizes down to 170 nm. More complex textured and random structures with feature sizes of the order of 40 to 200 nm can be produced by the rapid melting and cooling of the surface [57], such as shown in

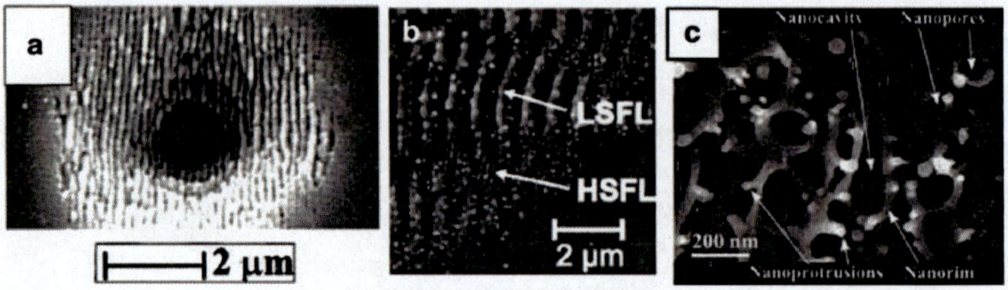

Fig. 12.5 Subwavelength Low spatial frequency (LSFL) and high spatial frequency (HSFL) rippled structures on (**a**) GaP with 170 nm HSFL structures, reprinted with permission from [54] and (**b**) InP with both LSFL and HSFL structures, reprinted with permission from [55] and (**c**) complex structures in the ablation spot on copper at 0.35 J cm^{-2} for 65 fs, 800 nm pulses. Reprinted with permission from [57]

Fig. 12.5c. While ripple structures with periods down to 110 nm have also been observed on graphite surfaces [58] and down to 170 nm on silica glass [59].

The next further step in wavelength for laser sources are the newly emerging sources in the EUV and x-ray regime. Direct output of pulses on the order of a millijoule at 46.9 nm generated in argon capillary discharge driven lasers has been demonstrated and applied to fabricate 55 nm grating structures and 60 nm pillars using interference lithography [60–62]. This source has also been employed for nanomachining with ablation spots sizes of 82 nm full width on PMMA surfaces using zone plate imaging [63]. High harmonic generation using gas jet, capillary interaction or ionized plasma media [5, 29–31] and solid surface targets [64] can generate nanojoule to microjoule level pulses. It is expected that micromachining with such short wavelength sources will be demonstrated in the near future.

Much more intense free electron laser sources now exist at two accelerator laboratories in the world, FLASH at DESY and LCLS at SLAC, which can produce microjoules of pulse energy at wavelengths of 30 to 0.1 nm. Initial micromachining experiments have begun at FLASH producing ablation holes with diameters of a few microns using submicron focused beams [65]. It is expected that much smaller ablation features will be produced in the near future and there will be significant growth in the application of such systems over the next decade as the technology becomes available.

12.4 3D Nonlinear Nanowriting Techniques

In order to write inside a volume, it is generally necessary to use a nonlinear enhancement technique usually through an intensity dependent process such as multiphoton absorption. By focusing below the surface, the average reaction rates in the focal spot are much higher than in the intermediate planes between the surface and the focal plane because the average intensity is so much higher in the focal spot. Thus, a response can be obtained in the focal region without affecting the rest of the material. This was first demonstrated by writing patterns in chemical resists and photopolymers which are not sensitive to the fundamental laser

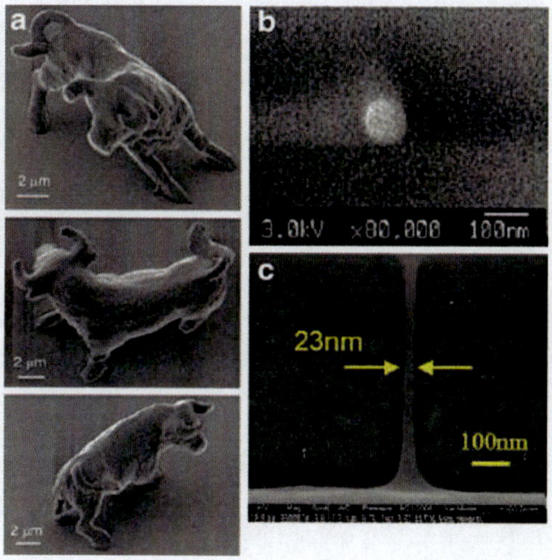

Fig. 12.6 (a) 10 by 7 micron bull written in SCR500 photopolymer with 120 nm resolution and (b) 100 nm dot written in SCR500 photopolymer with the addition of polymerization quenchers to enhance the contrast response. Reprinted with permission from [66]; (c) 23 nm post written in SCR500 Photopolymer. Reprinted with permission from [67]

wavelength but which will respond to the second harmonic wavelength. Kawata et al. [14, 66] set the initial standard for 2-photon writing in solids with their demonstration of 3D writing of a micro statue of a bull whose total size was 7 by 10 μm as shown in Fig. 12.6a. They obtained a transverse resolution of 120 nm when writing with 120 fs 800 nm pulses. Subsequently, the same group was able to demonstrate writing of voxels with dimensions of 100 nm with the same technique but by enhancing the contrast of the resist using quenching agents [15, 66] as shown in Fig. 12.6b. Subsequently using more precise control, 23 nm features have been demonstrated in the same resist as shown in Fig. 12.6c [67].

Other groups also followed with similar demonstrations with SU-8 resist including the writing of 100 nm resolution features with 800 nm 100 fs pulses by the group of Chichkov [68] and writing of 30 nm diameter nanorods in SU-8 resist [69]. A comparison of the various resists is summarized in Refs. [70, 71].

3D interference lithography using a nearfield diffractive optical element has also been used to write photonic crystal structures into SU-8 resist which was subsequently infiltrated with amorphous silicon to produce complex 3D photonic bandgap structures with feature sizes of the order of 150 nm [72].

Nanograting structures could also be produced inside transparent dielectric media by focusing light below the entrance surface of the material. Such structures require many repeated laser shots while scanning the sample and again are thought to arise from an interaction between the incident laser radiation and plasma waves produced inside the medium by ionization of the material. These interactions produce densification and nanovoids within the material which act as grating structures [73, 74]. In order to visualize them, the glass can be sectioned and etched revealing the damage grating structure as shown in Fig. 12.7.

Fig. 12.7 Nanograting structures in glass after sectioning and etching for different polarizations and scan directions. Reprinted with permission from [73]

12.5 Nanomilling and Nanoparticle Synthesis

By operating below the single shot damage threshold and using repetitive pulses, an incubation phenomenon is observed in most materials where eventually macroscopic damage becomes visible on the surface. The number of shots required increases as a power law of the ratio of the operating fluence versus the single shot damage fluence, and the strength of the dependence is given by an incubation parameter, ξ [75, 76]. Such behavior is similar to fatigue failure in metals where continuous flexing below the mechanical yield limit of a metal will eventually lead to failure. In the case of laser excitation, the thermal pressure stress and subsequent expansion and contraction with each laser pulse eventually leads to crystal dislocations and micro-damage on the atomic level which grows with every laser pulse. Operating with a single pulse right at threshold or with multiple pulses slightly below threshold fluences can lead to the ablation of a very thin layer of material which allows for the fine nanomilling of a surface to achieve a precise thickness. The ability to nanomill down to an average rate of 2 nm was demonstrated by Kirkwood et al. on copper surfaces [77]. Nanomilling can be employed for fine tuning surface acoustic wave devices or optical ring resonator devices. Recent work is starting to demonstrate this capability [78]. Other applications of nanomilling can be found in surface patterning and processing [79–81].

Another major application of nanomilling is the production of nanoparticles from surfaces. Typically, ablation is carried out in liquid in order to collect the nanoparticles produced. A variety of techniques and material systems have been demonstrated. These include the generation of nanoparticles of gold ranging in size from 2 to 30 nm [82–84], nanoparticles of silicon [85, 86], core–shell gold–silver, platinum, silver, copper and organic samples [87–89]. Images of some of the nanoparticles which can be produced are shown in Fig. 12.8.

12.6 Laser Induced Forward Transfer

Instead of nanomilling surfaces, it is also possible to add material to surfaces using laser induced forward transfer (LIFT). In this case, a portion of a thin film is transferred in the form of dots from a transparent donor substrate onto a nearby acceptor substrate by laser pulses. This technique was first demonstrated by

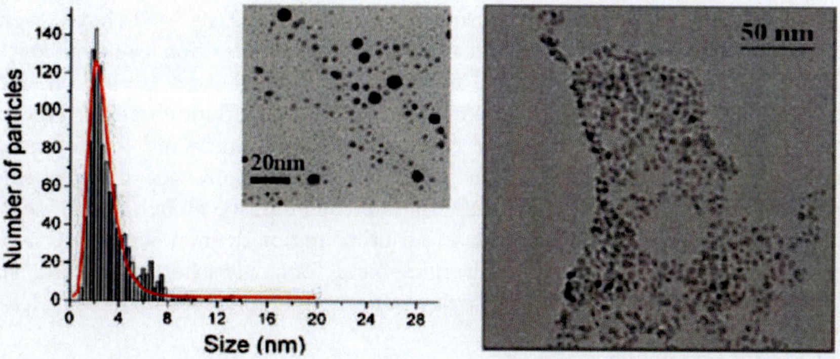

Fig. 12.8 Nanoparticles produced by laser ablation in liquids. Left: dextran nanoparticles, reprinted with permission from [89] and right: gold nanoparticles. Reprinted with permission from [88]

Fig. 12.9 A schematic diagram showing the LIFT process

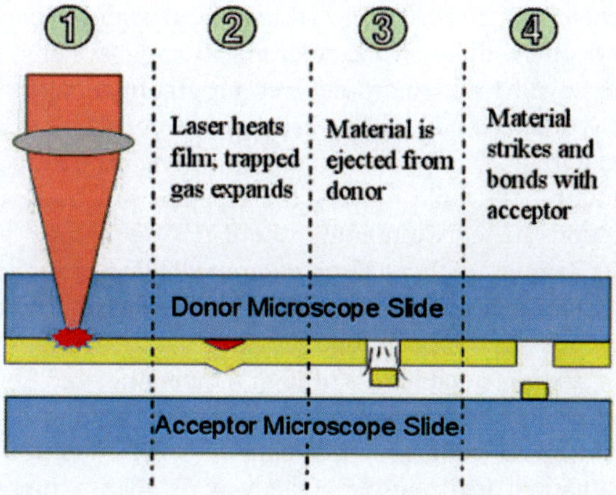

Bohandy et al. [90] to produce direct writing of 50 μm wide Cu lines on silicon and fused silica substrates by using single nanosecond (ns) excimer laser pulses (193 nm). Recently, feature sizes down to 200 nm have been demonstrated [91].

Figure 12.9 shows the typical LIFT process. A continuous thin film is first deposited onto a substrate which is transparent to the laser. The film is usually between a few tens of nm up to a micron thick; and this coated substrate is called the donor substrate. An acceptor substrate is placed underneath, in close proximity to the film to the donor substrate. A laser pulse, typically a hundred femtoseconds to tens of nanoseconds in duration, is focused onto the film through the donor substrate, causing formation of vapor at the interface. The force tears out a portion of the film causing it to eject from the donor substrate at high speed. The ejected film travels through the gap between the donor and acceptor substrates, striking and binding to the acceptor substrate.

Metallic materials transferred by using nanosecond pulses (ns-LIFT) have several shortcomings, such as poor film quality due to the energy deposition leading to cracks and debris around the transferred spot, and oxidation of metal and delamination of the transferred layers because of the melting and solidification during transfer process. Typically these shortcomings can be overcome by using ultrashort picosecond or femtosecond laser pulses (ps-LIFT or fs-LIFT). For a ns pulse, the absorbed laser energy will diffuse during the laser pulse duration and thus the molten region will be over a much larger volume than the laser absorbed region defined by the laser focal spot and laser penetration depth. On the other hand, for an ultrashort laser pulse, heat diffusion is very small during the pulse duration which is conducive to the transfer of submicron patterns.

Since the LIFT process was first proposed in 1986, a large variety of metals, inorganic and organic materials have been transferred by LIFT. Even living cells were reported to be transferred by LIFT. Laser Induced Forward Transfer process is ideal for applications in prototyping and custom device fabrication, as well as in the modification and repair of existing devices or surfaces whose topography or chemistry make traditional micro-patterning techniques impossible. Using a LIFT system with an XYZ micropositioning system, material can be added onto or removed from micro-devices programmatically as needed. Various applications have been realized by directly transferring metals, other inorganic and organic materials. Complicated patterns can be printed to a substrate by building up LIFT micro-dots while moving the acceptor substrate with a computer-controlled translation stage. Combining the LIFT technique with other laser direct writing techniques such as laser micromachining leads to a complete capability of both adding and removing materials, allowing the fabrication of embedded electronic devices as has been reported.

Extreme conditions of high temperature and high pressure exist during the LIFT process. Many delicate materials such as functional nanostructured materials and biological cells may not survive such condition. Thus, ways of mitigating the situation are required. One way is to use more sophisticated design of LIFT donor substrate. Employing multilayered films for LIFT was first applied by Tolbert, originally called laser ablation transfer [92]. In this method, a thin intermediate layer consisting of laser absorbing materials is first deposited on the transparent donor substrate. The target material which is going to be transferred is then deposited on top of this laser absorption material. During the LIFT process, the incident laser pulse interacts with the absorption layer, resulting in vaporization, which will then force the target material that is in contact with the absorption layer to be removed and transferred towards the acceptor substrate. This approach will reduce the damage of the target material, since laser energy mainly reacts with the intermediate layer and make the transfer of materials with weak absorption of the laser radiation possible. By using this approach, laser dye [93], phosphor powders (Y_2O_3:Eu and Zn_2SiO_4:Mn) [94], polymers [95], organic conducting polymer [96], biomolecule microarrays [97–99], peptides [100], proteins [101] and living cells [102] were successfully transferred with the use of metals (Au, Cr or Ti) as the absorption layer (several tens of nanometers thick). Organic light emitting pixels were successfully transferred by LIFT using 30–40 nm Ag nanoparticles as an absorption material on the donor substrate. Polymers can also be used as absorption materials on donor substrates. Stem cells were successfully transferred using a commercial polymer as the absorption material [103]. Cells [104], organic light-emitting diode (OLED) pixels [105], and semiconductor nanocrystal quantum dots (NQD's) [106] were successfully transferred using LIFT with photosensitive triazene polymer (TP) as a sacrificial layer. For the transfer of OLED pixels and semiconductor NQD's, the donor substrate used consists of three layers: a layer of TP, a layer of metal, and a layer of LIFT material (OLED pixels or CdSe NQDs). The TP sacrificial layer is vaporized by the UV laser pulse and the rapidly expanding organic vapor pushes the metal-LIFT material bi-layer towards the acceptor substrate. The metal layer serves as an electrode and it also prevents the exposure of the OLED pixels or CdSe NQD's to the UV radiation. Pique's group from Naval Research Laboratory (NRL) have developed the MAPLE DW (matrix-assisted pulsed laser evaporation direct write) technique which combined the laser-induced forward transfer (LIFT) and matrix-assisted pulsed laser evaporation (MAPLE) [107]. This approach employs the mixture of soluble materials to be deposited and a solvent phase which is usually pre-cooled to low temperature as the target material on a donor substrate. When the target is irradiated with a laser pulse, the solvent is rapidly vaporized and pumped away, propelling the solute toward to the acceptor substrate to form a uniform thin film with minimal decomposition. This approach has been applied to transfer metals, ceramics and polymeric materials such as Ag, Au, $BaTiO_3$ and BTO [108, 109]. Various types of physical and chemical sensor devices, microbatteries and biosensors have been also prototyped by this approach [110]. More recently, an approach named Laser Decal Transfer was proposed by the same NRL research group [111, 112]. This approach uses viscous nanoparticle suspensions (Ag in their case) as inks for the target material. It was found that the viscosity of the suspension plays an important role in the ability to perform the decal transfers. Another technique called blister-based laser induced forward transfer was used to transfer diamond nanopowder [113]. In this study, the diamond nanopowder is spread over the metal surface on a Ti-coated donor substrate and a 50 ps visible laser pulse is used to heat the metal film on the donor substrate. The laser heated metal film works like a flexible membrane as it moves outward it pushes the nanopowder towards the donor substrate. Not all functional materials require complex donor substrates. Carbon nanotube (CNT) field emission cathodes have been successfully transferred by ns-LIFT using a simple CNT-coated glass substrate [114]. The field emission properties of the transferred CNT are not degraded.

The spatial resolution of the LIFT process depends on the size of the LIFT spot. Typically a laser with Gaussian spatial profile is used in LIFT. Since laser ablation typically has a sharp energy threshold, the region of film that is exposed to the laser above the ablation threshold will be removed. For the region of the film below this threshold, little permanent change occurs in the film. Heat diffusion blurs the energy threshold by allowing the transfer of heat through the film. Thus, ultrashort pulses permit ablation spots smaller than the laser beam waist. If the ablation threshold is close to the peak intensity of the laser spot, the deposition spot may be significantly smaller than the beam waist of the laser profile.

The size of the LIFT spot depends on the fluence, as shown in Fig. 12.10. In these experiments a donor substrate coated with 80 nm thick Cr is used. The 800 nm, 130 fs laser has a focused beam waist diameter of 3.7 μm. The diameter of the LIFT spots increases rapidly with fluence from 0.2 to 2 J/cm^2 and more slowly for fluences from 2 to 20 J/cm^2. The smallest Cr LIFT spots have a diameter around 700 nm, which is five times smaller than the laser beam waist diameter, when the laser fluence is kept around the transfer threshold at near 0.2 J/cm^2, as shown in Fig. 12.11.

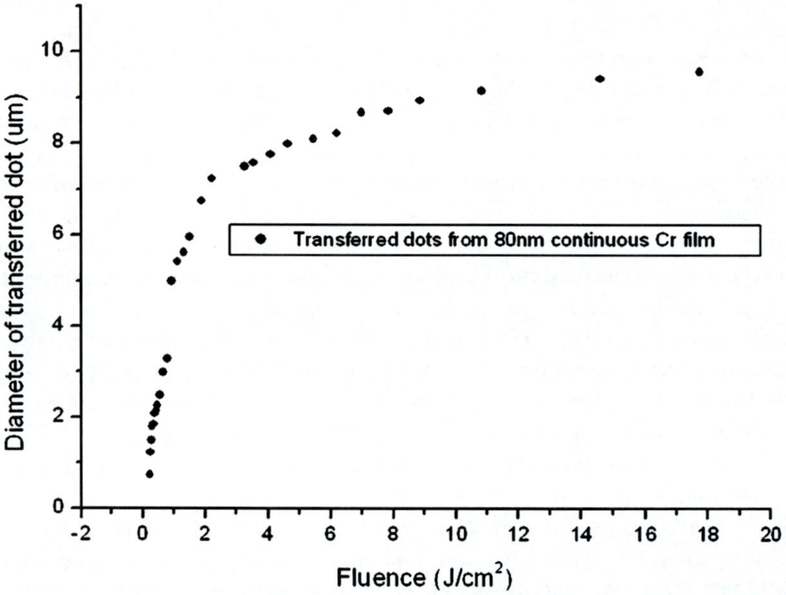

Fig. 12.10 Diameter of transferred Cr dots versus fluence. Reprinted with permission from [115]

Fig. 12.11 SEM images of transferred spots from 80 nm continuous Cr film at F = 0.24 J/cm^2 (the scale bar is 1 μm). Reprinted with permission from [115]

The thickness of the film on the donor substrate is an important parameter. It has been reported that the minimum achievable spatial resolution of transferred Cr material by fs-LIFT is around 330 nm [116]. In that study, the film on the donor substrate is very thin (30 nm) so that the laser is expected to melt through the entire thickness of the film and would cause a molten Cr droplet to deposit on the acceptor substrate. Numerical simulations confirm [115] the melting depth is approximately 40 nm for experimental conditions as shown by Banks et al. [116]. The technique using a film less than the melting depth is called nanodroplet LIFT. In such a technique, the size of the LIFT droplet is expected to be dictated by the thickness of the film and the material properties (in particular its liquid surface tension).

The size of a nanodroplet depends on the molten volume which is defined by the laser focal spot size and the film thickness. As shown in Fig. 12.12, the diameter of

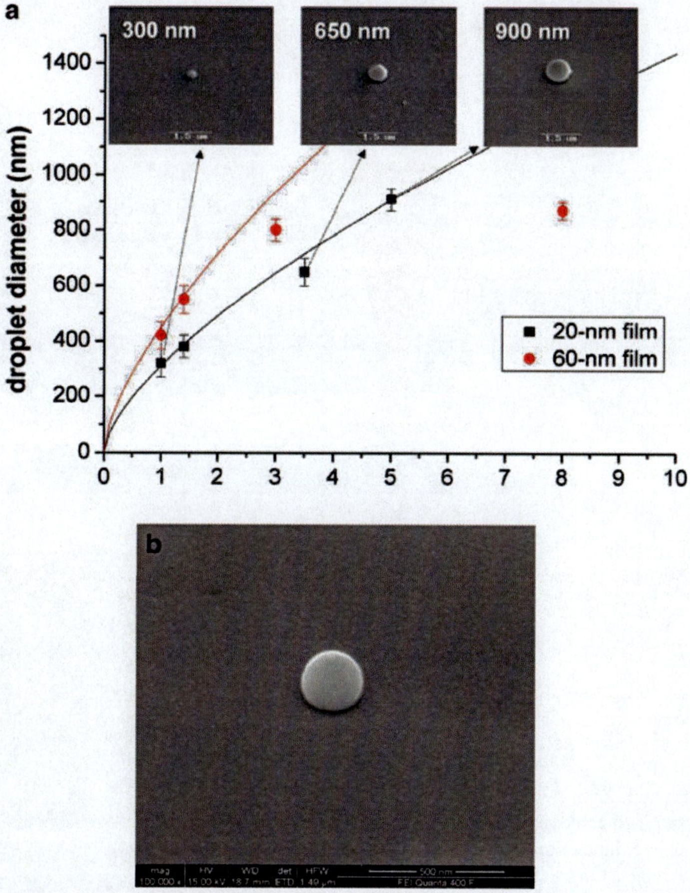

Fig. 12.12 LIFT of gold droplets: (**a**) dependence of the transferred droplet size on the focusing conditions and gold film thickness and (**b**) a 220 nm Au droplet transferred by using a 10 nm thick film. Reprinted with permission from [91]

the Au droplets decreases when smaller focal spots or thinner films are used. A 220 nm Au nanodroplet was obtained when an even thinner 10 nm thick Au film was used. Even smaller nanodroplets are expected if one would use the near field optical techniques discussed in Sect. 12.3.

Another technique [115] to transfer even smaller sub-micron dots is illustrated schematically in Fig. 12.13a. An array of sub-micron size material dots are fabricated on the donor substrate by using e-beam lithography, EUV lithography or laser nanopatterning techniques. The laser pulses are then focused on the top of these pre-patterned material spots in order to transfer them onto the acceptor substrate. This technique has recently allowed transfer of prepatterned spots with diameters of the order of 100 nm as shown in Fig. 12.13b.

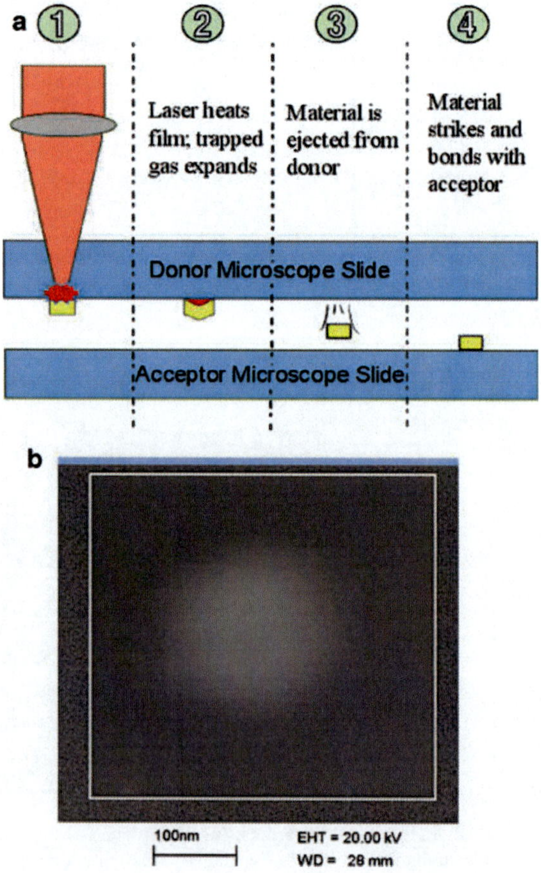

Fig. 12.13 LIFT process with pre-patterned donor substrate in order to achieve 100 nm scale size deposition spots: (**a**) schematic diagram showing the process; (**b**) a 100 nm Cr dot transferred using the technique

12.7 Summary

It is clear that optical techniques are pushing towards smaller feature sizes in all the different application areas. Feature sizes down to the order of 100 nm are available by ablative patterning, nonlinear writing and laser induced forward transfer of materials, particularly using femtosecond lasers and ultraviolet wavelengths. Features down to 10 nm scale size are achievable using NSOM and AFM field concentration techniques and other nonlinear enhancement techniques. Working close to the ablation threshold, nanomilling thin surface layers is possible and low intensity ablation also leads to the production of nanoparticles from a range of materials. In the future, new coherent EUV and x-ray radiation sources, currently under development, should allow direct writing of nanostructures approaching a 10 nm size limit. It is expected that the use of laser nanopatterning techniques will continue to grow in the coming years.

References

1. Gaudiosi DM, Reagan B, Popmintchev T, et al. Phys Rev Lett. 2006;96:203001.
2. Reagan BA, Popmintchev T, Grisham ME, et al. Phys Rev A. 2007;76:013816.
3. Popmintchev T, Chen MC, Arpin P, et al. Nat Photonics. 2010;4:822–32.
4. Chen MC, Arpin P, Popmintchev T, et al. Phys Rev Lett. 2010;105:173901-1-4.
5. Pertot Y, Elouga Bom LB, Bhardwaj VR, et al. Appl Phys Lett. 2011;98:101104-1-3.
6. Benware BR, Moreno CH, Burd CJ, et al. Opt Lett. 1997;22:796–8.
7. Macchietto CD, Benware BR, Rocca JJ. Opt Lett. 1999;24(16):1115–7.
8. Martz DH, Alessi D, Luther BM, et al. Opt Lett. 2010;35:1632–4.
9. Ayvazyan V, Baboi N, Bohnet I, et al. Phys Rev Lett. 2002;88:104802.
10. Ayvazyan V, Baboi N, Bahr J, et al. Eur Phys J D. 2006;37:297–303.
11. Emma P, Akre R, Arthur J, et al. Nat Photonics. 2010;4:641–7.
12. Born M, Wolf E. Principles of optics. Oxford: Pergamon Press; 1975.
13. Iversen L, Metzler OY, Martinez KL, et al. Langmuir. 2009;25:12819–24.
14. Kawata S, Sun HB, Tanaka T, et al. Nature. 2001;412:697–8.
15. Takada K, Sun HB, Kawata S. Appl Phys Lett. 2005;86:071122.
16. Luo H, Li Y, Cui HB, et al. Appl Phys A. 2009;97:709–12.
17. Liu X, Du D, Mourou G. IEEE J Quantum Elec. 1997;33:1706–16.
18. Bonse J, Brzezinka KW, Meixner AJ. Appl Surf Sci. 2004;221:215–30.
19. Dassow R, Kohler JR, Helen Y, et al. Semicond Sci Technol. 2000;15:L31–4.
20. Le Harzig R, Huot N, Audouard E, et al. Appl Phys Lett. 2002;80:3886–8.
21. Beresna M, Gertus T, Tomasiunas R et al. Laser Chem. 2008 doi: 10.1155/2008/976205.
22. Singh R, Alberts MJ, Melkote SN. Int J Mach Tool Manu. 2008;48:994–1004.
23. Tan B, Dalili A, Venkatakrishnan K. Appl Phys A. 2009;95:537–45.
24. Bonse J, Krüger J. J Appl Phys. 2010;107:054902.
25. Mainfray G, Manus C. Rep Prog Phys. 1991;54:1333–72.
26. Schaffer CB, Brodeur A, Mazur E. Meas Sci Technol. 2001;12:1784–94.
27. Stuart BC, Feit MD, Herman S, et al. Phys Rev B. 1996;53:1749–61.
28. von der Linde D, Schuler HJ. Opt Soc Am B. 1996;13:216–22.
29. Ganeev RA, Bom LBE, Kieffer JC, et al. Phys Rev A. 2007;76:023831.
30. Shiner AD, Trallero-Herrero C, Kajumba N, et al. Phys Rev B. 2009;103:073902.
31. McFarland BK, Farrell JP, Bucksbaum PH, et al. Phys Rev A. 2009;80:033412.
32. Hergott JF, Kovacev M, Merdji H, et al. Phys Rev A. 2002;66:021801.

33. Nolte S, Chichkov BN, Welling H, et al. Opt Lett. 1999;24:914–6.
34. Taylor RS, Hnatovsky C, Simova E, et al. Opt Lett. 2003;28:1043–5.
35. Zorba V, Mao X, Russo RE. Appl Phys Lett. 2009;95:041110.
36. Zorba V, Mao X, Russo RE. Anal Bioanal Chem. 2010;396:173–80.
37. Joglekar AP, Liu H, Spooner GJ, et al. Appl Phys B. 2003;77:25–30.
38. Pan H, Hwang DJ, Ko SH, et al. Small. 2010;6:1812–21.
39. McLeod E, Arnold CB. Nat Nanotechnol. 2008;3:413–7.
40. Li L, Guo W, Wang ZB, et al. J Micromech Microeng. 2009;19:054002.
41. Brodoceanu D, Landstrom L, Bauerle DB. Appl Phys A. 2007;86:313–4.
42. Tan LS, Hong M. Int J Optomechatronics. 2008;2:382–289.
43. Chimmalgi A, Grigoropoulos CP, Komvopoulos K. J Appl Phys. 2005;97:104319.
44. Milner AA, Zhang K, Prior Y. Nano Lett. 2008;8:2017–22.
45. Yu M, Kim HS, Blick RH. Opt Express. 2009;17:10044–9.
46. Herman PR, Chen KP, Wei M, et al. Proc SPIE. 2001;4274:149–57.
47. Ihlemann J, Uller SM, Puschmann S, et al. Appl Phys A. 2003;76:751–3.
48. Bloomstein TM, Marchant MF, Deneault S, et al. Opt Express. 2006;14(14):6434–43.
49. Choi WK, Liew TH, Dawood MK. Nano Lett. 2008;8:3799–802.
50. Choi WK, Liew TH, Chew HG, et al. Small. 2008;4:330–3.
51. de Boor J, Geyer N, Wittemann JV, et al. Nanotechnology. 2010;21:095302.
52. Li X, Feng D, Jia T, et al. Micro Nano Lett. 2011;6:177–80.
53. Zhu M, Zhou L, Li B et al. Nanoscale (2011) doi: 10.1039/C1NR00015B.
54. Borowiec A, Haugen HK. Appl Phys Lett. 2003;82:4462–4.
55. Bonse J, Munz M, Sturm H. J Appl Phys. 2005;97:013538-1-9.
56. Wang XC, Lim GC, Ng FL, et al. Appl Surf Sci. 2005;252:1492–7.
57. Vorobyev AY, Guo C. Opt Express. 2006;14:2164–9.
58. Golosov EV, Ionin AA, Kolobov YR, et al. Phys Rev B. 2011;83:115426.
59. Sun Q, Liang F, Vallée R, et al. Opt Lett. 2008;33:2713–5.
60. Capeluto MG, Vaschenko G, Grisham M, et al. IEEE Trans Nanotechnol. 2006;5:3–7.
61. Capeluto MG, Wachulak P, Marconi MC, et al. Microelectron Eng. 2007;84:721–4.
62. Wachulak PW, Capeluto MG, Menoni CS, et al. Opto-Electron Rev. 2008;16:444–50.
63. Vaschenko G, Garcia EA, Menoni CS, et al. Opt Lett. 2006;31:3615–7.
64. Dromey B, Kar S, Bellei C, et al. Phys Rev Lett. 2007;99:085001.
65. Andreasson J, Iwan B, Andrejczuk A, et al. Phys Rev E. 2011;83:016403.
66. Sun HB, Kawata S. Adv Polym Sci. 2004;170:169–273.
67. Tan D, Li Y, Qi F, et al. Appl Phys Lett. 2007;90:071106 -1-3.
68. Ovsianikov A, Ostendorf A, Chichkov BN. Appl Surf Sci. 2007;253:6599–602.
69. Juodkazis S, Mizeikis V, Seet KK, et al. Nanotechnology. 2005;16:846–9.
70. Lee KS, Kim RH, Yang DY, et al. Prog Polym Sci. 2008;33:631–81.
71. Zhang YL, Chen QD, Xia H, et al. Nano Today. 2010;5:435–48.
72. Chanda D, Zachari N, Haque M, et al. Opt Lett. 2009;34:3920–2.
73. Hnatovsky C, Taylor RS, Simova E, et al. Appl Phys A. 2006;84:47–61.
74. Bhardwaj VR, Simova E, Rajeev PP, et al. Phys Rev Lett. 2006;96:057404-1-4.
75. Jee Y, Beckera MF, Walser RM. J Opt Soc Am B. 1988;5:648–59.
76. Kirkwood SE, van Popta AC, Tsui YY et al. Appl Phys A 2004. doi: 10.1007/s00339-004-3135-7.
77. Kirkwood SE, Taschuk MT, Tsui YY, et al. J Phys Conf Ser. 2007;59:591–4.
78. Bachman D, Chen Z, Prabhu A et al. Integrated photonics research, silicon and nano photonics (IPR) topical meeting (2011), June 12–16, 2011, Toronto.
79. Kirkwood SE, Shadnam MR, Amirfazli A, et al. J Phys Conf Ser. 2007;59:428–31.
80. Hartmann N, Franzka S, Koch J, et al. Appl Phys Lett. 2008;92:223111.
81. Klingebiel B, Scheres L, Franzka S, et al. Langmuir. 2010;26(9):6826–31.
82. Compagnini G, Scalisi AA, Puglisi O. Phys Chem Chem Phys. 2002;4:2787–91.
83. Kabashin AV, Meunier M. J Appl Phys. 2003;94(12):7941–3.

84. Muto H, Miyajima K, Mafuné F. J Phys Chem C. 2008;112:5810–5.
85. Khang Y, Lee J. J Nanopart Res. 2010;12:1349–54.
86. Semaltianos NG, Logothetidis S, Perrie W, et al. J Nanopart Res. 2010;12:573–80.
87. Abid JP, Girault HH, Brevet PF. Chem Commun. 2001:829–30.
88. Amendola V, Meneghetti M. Phys Chem Chem Phys. 2009;11:3805–21.
89. Besner S, Kabashin AV, Winnik FM, et al. Appl Phys A. 2008;93:955–9.
90. Bohandy J, Kim BF, Adrian J, et al. J Appl Phys. 1986;60:1538–9.
91. Kuznetsov AI, Koch J, Chichkov BN, et al. Opt Express. 2009;17:18820–5.
92. Toth Z, Szörényia T, Tóthb AL. App Surf Sci. 1993;69:317–20.
93. Nakate Y, Okada T, Maeda M. Jpn J Appl Phys. 2002;41:L839–41.
94. Fitz-Gerald JM, Pique A, Chrisey DB, et al. Appl Phys Lett. 2000;76:1386–8.
95. Boutopoules C, Tsouti V, Goustouridis D, et al. Appl Phys Lett. 2008;93:191109.
96. Rapp L, Cibert C, Alloncle PA, et al. Appl Surf Sci. 2009;255:5439–43.
97. Fernández-Pradas JM, Colina M, Serra P, et al. Thin Solid Films. 2004;27:453–4.
98. Serra P, Fernández-Pradas JM, Berthet FX, et al. Appl Phys A. 2004;79:949–52.
99. Duocastella M, Fernández-Pradas JM, Serra P, et al. J Laser Micro/Nanoeng. 2008;3:1–4.
100. Dinca V, Kasotakis E, Catherine J, et al. Appl Surf Sci. 2007;254:1160–3.
101. Barron JA, Young HD, Dlott DD, et al. Proteomics. 2005;5:4138–44.
102. Barron JA, Rosen R, Jones-Meehan J, et al. Biosens Bioelectron. 2004;20:246–52.
103. Doraiswamy A, Narayan RJ, Lippert T, et al. Appl Sufi Sci. 2006;252:4743–7.
104. Kattamis NT, Purnic PE, Weiss R, et al. Appl Phys Lett. 2007;91:171120.
105. Fardel R, Nagel M, Nuesch F, et al. Appl Phys Lett. 2007;91:061103.
106. Up J, Liu J, Cui D, et al. Nanotechnology. 2007;18:025403.
107. Chrisey DB, Piqué A, Fitz-Gerald JM, et al. Appl Surf Sci. 2000;154–155:593–600.
108. Chrisey DB, Piqué A, Mode R, et al. Appl Surf Sci. 2000;168:345–52.
109. Piqué A, Chrisey DB, Fitz-Gerald JM, et al. J Mater Res. 2000;15:1872–5.
110. Piqué A, Arnold CB, Warden RC, et al. RIKEN Rev. 2003;50:57–62.
111. Piqué A, Eyeing RCY, Kim H, et al. J Laser Micro/Nanoeng. 2008;3:163–9.
112. Piqué A, Eyeing RC, Markus KM, et al. Proc SPIE. 2008;6879:687911.
113. Kononenko TV, Alloncle P,Konov VI, et al. Appl Phys A. 2009;94:531–6.
114. Chang-Jian SK, Ho JR, Cheng JWJ, et al. Nanotechnology. 2006;17:1184–7.
115. Wang Q. MSc thesis, University of Alberta, 2009.
116. Banks DP, Grivas C, Mills JD, et al. Appl Phys Lett. 2006;89:192107.

13 Templating and Pattern Transfer Using Anodized Nanoporous Alumina/Titania

Karthik Shankar

Abstract

This chapter provides an overview of the non-lithographic nanofabrication process known as "hard templating". Nanoporous alumina fabricated by electrochemical anodization continues to be the most widely used hard template although anodically formed nanotubular titania is increasing in importance for templating applications. Hard templates sustain almost no deformation to minor mechanical loads and are unchanged under the action of organic solvents and neutral salt solutions. Their mechanical robustness and relative chemical inertness allows hard templates to be compatible with a variety of chemical, electrochemical and mechanical processes typically used in nanofabrication, several of which are covered in other chapters of this book. Hard templates typically consist of a self-organized array of nanochannels of similar or identical size oriented orthogonally to a substrate. The last 15 years have seen immense progress in the construction of thinner, more versatile hard templates of greater pattern order prepared on ever more diverse substrates. Hard templating has been at the forefront of nanotechnology research as a method to direct the creation of a wide variety of metallic, semiconducting and organic nanostructures. This chapter is organized as follows: The processes used to form hard templates and to improve their pattern order are presented in Sect. 13.2, the use of the template nanochannels to grow ordered functional one-dimensional nanomaterials in Sects. 13.3 and 13.4, and the use of hard templates to affect nanoscale pattern transfer is presented in Sect. 13.5.

K. Shankar (✉)
Department of Electrical & Computer Engineering, University of Alberta, Edmonton, AB, Canada
e-mail: kshankar@ualberta.ca

M. Stepanova and S. Dew (eds.), *Nanofabrication*,
DOI 10.1007/978-3-7091-0424-8_13,

13.1 Introduction

Self-organized nanopore arrays of valve metal oxides can be formed non-lithographically by the electrochemical anodization of valve metals such as Al, Ti, Ta, Hf, Zr, etc. The anodization process is simple and economical and the resulting structures are mechanically robust and chemically resistant even at elevated temperatures. Therefore, anodically formed nanoporous valve metal oxides are excellent architectures for templating and pattern transfer and a wide variety of functional nanostructures have been formed using nanoporous alumina and titania. The anodic formation of porous alumina has been known since 1956 [1, 2] but has been extended to the other valve metals only in the last decade [3–7]. We shall restrict our discussion to anodic aluminum oxide (AAO) and nanotubular TiO_2. In AAO, the thickness of the nanoporous film, the size of the nanopores and their spacing are the morphological parameters of interest and these can be controlled by tuning the anodization potential, the duration of the anodization and by choosing the appropriate electrolyte to perform the anodization. In TiO_2 nanotube (TNT) arrays, the tubular architecture results in an additional morphological parameter, namely the wall-thickness. Both AAO and TNT's have been fabricated on a variety of different substrates such as glass [8–12], Si wafers [13–15], flexible polymeric substrates [16] and even curved surfaces such as metallic pipes [17, 18]. Also, both AAO and TNT's can be transformed into free-standing membranes several hundreds of μm in thickness by detaching the nanoporous film from the underlying substrate. It has been difficult to form high quality nanoporous Al_2O_3 on the technologically important transparent conducting oxide (TCO) coated glass substrates due to issues of adhesion, though a relatively recent technique that uses a very thin Ti adhesion promoter appears to be promising [19]. On the other hand, the formation of TNT's on TCO coated glass substrates, while challenging, has been successfully achieved and utilized for devices [11]. Another important difference between AAO and TNT's lies in their differing conductivities; Al_2O_3 is an insulator whereas crystalline TiO_2 is an n-type semiconductor.

13.2 Formation of Nanoporous Alumina and Nanotubular Titania

13.2.1 Formation of Nanoporous Anodic Aluminum Oxide

Nanoporous alumina is formed by the electrochemical anodization of aluminum in acidic aqueous electrolytes. Figure 13.1a shows a commercially available electrochemical cell frequently employed to perform the anodic synthesis of nanoporous alumina while Fig. 13.1b is a schematic of an electrochemical cell used to perform the anodic synthesis of titania nanotubes. The cathode of the electrochemical cell is formed from an inert material that does not corrode in the anodization electrolyte. Platinum is typically used as the cathodic material. The valve metal undergoing anodization constitutes the anode of the electrochemical cell. A reference electrode

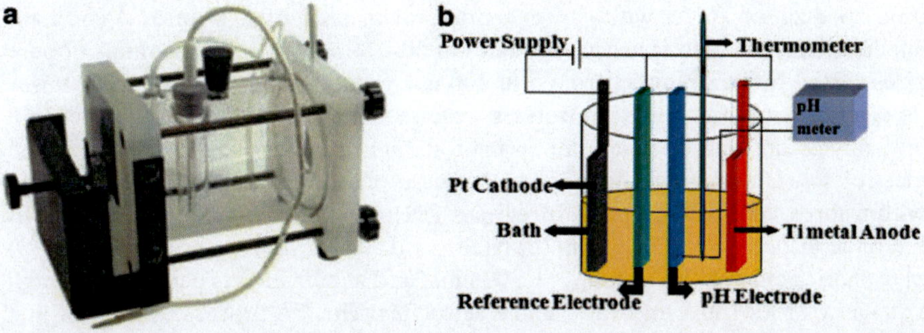

Fig. 13.1 (**a**) Electrochemical cell model K0235 from Princeton Applied Research, commonly used to anodically form nanoporous alumina. (**b**) Schematic of an anodization cell used to form TiO_2 nanotube arrays (Reprinted with permission from Ref. [25])

such as Ag/AgCl or saturated calomel is sometimes used as a third electrode. For temperature sensitive applications where thermal runaway during anodization presents a problem, cooling water lines fed to a glass sheath surrounding the electrochemical cell are used and a thermometer is inserted into the electrolyte to monitor the temperature (see Fig. 13.1b). The bubbling of inert gas through the electrolyte to control the concentrations of oxygen and moisture (for organic electrolytes) in the anodization bath is commonly applied (see Fig. 13.1a). During anodization, oxide growth and oxide etching occur simultaneously as field assisted oxidation of Al at the aluminum/oxide interface competes with field assisted dissolution of the oxide at the oxide/electrolyte interface. The porous structure occurs as a result of the balance between the competing reactions, a delicate equilibrium which is only achieved in certain electrolytes of oxalic, sulfuric and phosphoric acids. More extensive studies of the mechanism of pore initiation and growth may be found elsewhere [20–24].

Hexagonally ordered AAO membrane with monodispersed nanopores in which the range of order extends to areas of the order of μm^2, can be formed by a two-step anodization process [26–28]. In the two-step process, the first step consists of anodizing a polished Al foil in the requisite electrolyte for several hours and then etching away the nanoporous alumina thus formed using wet etchants. Chromic acid, phosphoric acid and mixtures thereof are commonly used as the wet etchants. In the second step, the Al foil subjected to the first step above is re-anodized in the same electrolyte as before and at the same potential. The concave dimples remaining on the surface of the Al foil act as nucleation sites to guide the formation of highly ordered self-organized pores in the second anodization step. The pore size and pore spacing of these alumina membranes are proportional to the anodization voltage. Thus by controlling the anodization voltage, the pore size and distribution of the pores can be easily tailored. Pore densities as high as 10^{11} pores per cm^2 can be achieved. AAO membranes are sold commercially, but only a limited number of pore diameters are available. Depending on the speed of formation of the porous layer, the processes to form self-ordered alumina nanopore arrays are classified into mild anodization (MA) with slower oxide growth rates of 2–6 μm h^{-1} and hard anodization (HA) with faster growth rates of 50–100 μm h^{-1}. A wide range of pore sizes can be formed ranging from 5 to 200 nm. Smaller pores are formed at lower voltages in sulfuric acid electrolytes, intermediate-sized pores in oxalic acid electrolytes and larger pores are formed at higher voltages in phosphoric acid electrolytes. However, a high degree of spatial ordering for MA is only obtained within three well-defined growth regimes [29]: (1) sulfuric acid at 25 V for an interpore distance (D_{int}) of 63 nm (2) oxalic acid at 40 V for $D_{int} = 100$ nm and (3) phosphoric acid at 195 V for $D_{int} = 500$ nm. Hard anodization is performed at high voltages of 100–150 V in oxalic acid electrolytes. The HA process consists of first forming a thick protective oxide layer of thickness >400 nm by mild anodization at lower voltages (e.g. 40 V) followed by gradually ramping the anodization voltage to the final target voltage (e.g. 150 V), where it is then held constant for the reminder of the HA process. The purpose of the thick oxide layer is to suppress breakdown effects and to enable uniform oxide film growth even at the high voltages used in the HA process. When a free-standing AAO membrane is required, the aluminum substrate is removed after the final anodization by selective dissolution in a saturated solution of mercuric chloride.

13.2.2 Large-Scale Highly Ordered Nanoporous Alumina by Pretexturing the Aluminum

Nanoporous alumina is perfectly ordered when the pores form a hexagonal honeycomb structure consisting of close-packed nanochannels of high-aspect ratio (shown in Fig. 13.2). As mentioned in Sect. 13.2.1, the largest sizes of domains with defect-free ordering is restricted to a few μm^2. To demonstrate scalability and increase throughput for templating applications, nanoporous alumina with much larger defect-free domains is demanded. The size of self-ordered domains have been increased by prepatterning the aluminum (Al) surface prior to the anodization process and using the patterns as the nucleation sites to guide the growth of the nanochannels [31]. In 1997, Masuda pioneered the use of hard-stamping to form the necessary nucleation sites on the surface of polished aluminum. In Masuda's process, conventional electron beam lithography was used to pattern a hexagonally arranged array of convexes on a master mold [32]. The mold was made from a mechanically hard material such as SiC. The master mold was then pressed onto the aluminum surface using an oil press at room temperature to generate the required array of concaves on the surface of aluminum. The soft-imprinting technique [16] introduced by the Gao group uses Ar^+ plasma etching through a free standing nanoporous alumina membrane etch mask to create ordered nanoindentations on the Al surface. The AAO etch mask was itself formed by a regular two-step anodization such as described in Sect. 13.2.1. Using this technique, highly ordered porous anodic alumina templates were fabricated on different substrates (such as Si, glass slides, and flexible polyimide films) over large areas (>1.5 cm^2). Another interesting technique employing self-assembly is based on the spontaneous

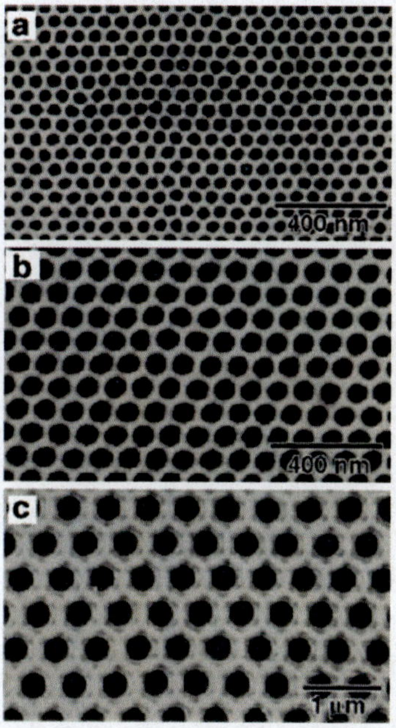

Fig. 13.2 SEM micrographs of naturally occurring long-range ordered anodic porous alumina formed in three types of acid electrolytes: (**a**) sulfuric acid, (**b**) oxalic acid and (**c**) phosphoric acid (Reprinted with permission from Ref. [30])

organization of monodisperse polystyrene nanoparticles into a 2D array. In this process, shallow concaves were formed on Al by replicating the ordered structure of the 2D array of polystyrene particles on the submicrometer scale and initiating hole development during anodization [33]. Other approaches in the service of the same objective include direct focused ion beam (FIB) lithography [34], interference lithography [35], holographic lithography [36], colloidal lithography [37] and block copolymer self-assembly [38]. Tables 13.1 and 13.2 summarize the approaches discussed above and provide details about the techniques themselves, the areas over which order is preserved and substrates used. A more recent technique namely, step and flash imprint lithography (SFIL) [31] was used to demonstrate near-perfect ordered AAO with square and hexagonal lattice configuration on silicon substrate over 4 in. wafer areas (see Fig. 13.3).

13.2.3 Formation of Titania Nanotube Arrays

The key processes responsible for the formation of a nanoporous structure in alumina and titania are similar with one exception. In the anodic formation of TiO_2 nanotube arrays, in addition to the field assisted oxidation and field-enhanced oxide dissolution reactions, a third reaction is present, namely chemical etching by HF [25]. Chemical dissolution of titania in the HF electrolyte plays a key role in the formation of nanotubes rather than a nanoporous structure. Similar to AAO, the

Table 13.1 Summary of methods to pretexture bulk Al

Al pre-patterning methods	Ordered pattern area	Remarks
Two-steps anodization[13]	Not available	This method requires the use of a thick al film (>40 μm).
Hard stamping using SiC stamp[14]	3 × 3 mm	SiC stamp was fabricated using electron beam lithography. Stamping pressure: 28 kN/cm^2.
Hard stamping using Ni stamp[15]	Wafer scale	Ni stamp can be mass-replicated from master pattern, fabricated using laser interference lithography, Stamping pressure: 25 kN/cm^2.
Hard stamping using optical diffraction grating[16]	5 × 5 mm	Close packed arrays of rhomb-shaped ridges were formed on Al surface by using a two-step press-in procedure of a right angle friangular master grating.
Hard stamping using Si_3N_4 stamp[17]	4-in. wafer	Silicon nitride stamp was replicated from a silicon master, fabricated using deep-UV lithography and KOH wet etching. Stamping pressure: 5 kN/cm^2.
Focused-ion-beam[18,20]	Not available	Resist-assisted or direct patteming onto Al surface.
Colloidal lithography[21]	>cm^2 areas	Colloidal crystals were deposited onto mica surface using an accelerated evaporation induced self-assembly. Stamping pressure: 100 kN/cm^2.
Block-copolymer self-assembly[22]	Not available	Pore size of 12 nm with interval of 45 nm was fabrica ted using this method.

Reprinted with permission from Ref. [31]

Table 13.2 Summary of methods to pretexture Al on substrates

Holographic lithography[19]	1 × 1 cm	Nanoscale surface corrugations were created by evaporating Al onto the photoresist-grating-patterns on a substrate, developed by holographic lithography.
Soft imprinting[23]	>1.5 cm^2	This method utilizes argon milling technique to transfer the pattern from a master template, fabricated using a two-steps anodization process, onto the Al surface.
Nanoimprint lithography[24]	2.5 in. wafer	A 2.5″ nickel mold was replicated from a master pattern, fabricated using electron beam lithography, and imprinted into a thermoplastic resist. The resist patterns were then transferred to the Al surface *via* argon ion beam milling.

Reprinted with permission from Ref. [31]

pore diameter in anodically formed TNT's is mainly determined by the anodization voltage. The wall-thickness of the nanotubes is known to be influenced by the temperature of the anodization electrolyte, the strength and duration of chemical etching and the nature of the solvent medium [39]. The exact dependence of the wall-thickness on the anodization parameters is not yet well-understood. TNT's form in fluoride ion bearing aqueous electrolytes over the pH range 0–5. In addition, TNT's also form in various organic electrolytes containing F^- ions. Formamide, ethylene glycol (EG), glycerol and dimethyl sulfoxide (DMSO) are solvents commonly used to prepare organic electrolytes. The nanotube length is determined by the strength of chemical etching and by the thickness of the barrier

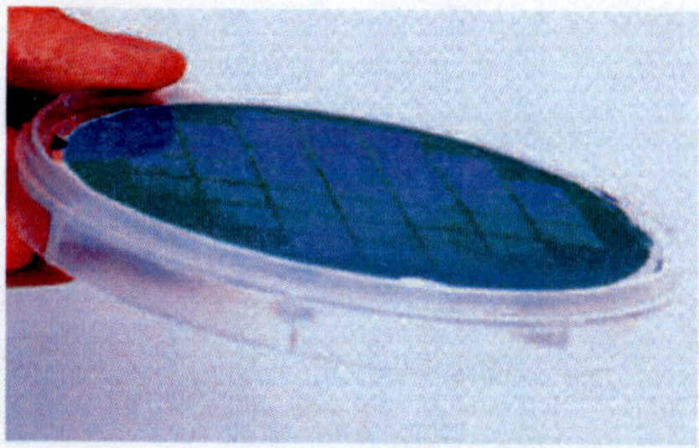

Fig. 13.3 Photograph of a near-perfect AAO template on a 4 in. silicon wafer. The 10 mm × 10 mm square areas with bright light diffraction indicate the anodized sample that was prepatterned using SFIL (Reprinted with permission from Ref. [31])

layer present at the interface between the Ti metal and the electrolyte. Strong chemical etching reduces the length of the nanotubes. In electrolytes of low-to-moderate viscosity, solid state transport of reactant ions through the barrier layer is the rate limiting step capping the length of the nanotubes [40, 41]. In organic electrolytes, the barrier layer formed is thinner and consequently the nanotubes grow faster and much higher tube-lengths are achieved. While the maximum length of nanotubes obtained in aqueous electrolytes is 6.6 μm (at a pH of ~5), nanotubes as long as 100 μm are obtained in formamide-based electrolytes and tube-lengths up to 1 mm have been obtained in EG-based electrolytes. The range of pore sizes obtained extend from 15 nm (inner diameter), obtained in aqueous electrolytes at anodization voltages of 8 V, to nanotubes of 900 nm inner diameter obtained at 150 V in diethylene glycol (DEG)-based electrolytes [42]. Cylindrical nanotubes are obtained in all electrolytes excepting those based on EG. In EG based electrolytes, irregular hexagonal pores are obtained. Efforts towards achieving ideal hexagonal ordering in anodic nanotubular titania are ongoing [43]. Table 13.3 provides a summary of commonly used recipes for the synthesis of TNT's (Fig. 13.4).

13.2.4 Formation of Nanoporous Alumina and Nanotubular Titania Templates on Non-Native Substrates

The formation of nanopore arrays on a non-native substrate requires first the deposition of the metal to be anodized on the desired substrate. Typically, sputtering or evaporation is used for this step. The next step consists of anodizing the deposited Al/Ti film. In the case of TNT's, a third step, namely the induction of crystallinity by a thermal annealing process, may also be performed.

Dresselhaus et al. [8] obtained high aspect ratio nanoporous alumina with a good degree of pore ordering and high optical transparency using a two-step anodization method on glass substrates, Si wafers and Platinum-coated Si wafers. Thick aluminum films (~6-12 μm) were first deposited on substrates by thermal evaporation. Subsequently, the samples were electrochemically polished and then subjected to the first anodization for 10 minutes, corresponding to the formation of 1 μm of alumina. Next, this layer of Al_2O_3 was etched away and the sample was reanodized

Table 13.3 Summary of commonly used anodization recipes to fabricate TiO_2 nanotube arrays

	Electrolyte				Anodic parameters		Morphology of NT's		
References	Solvent	F^- carrier	F^- conc.	H_2O conc.	V (V)	t (h)	ID (nm)	L (μm)	WT (nm)
[44]	Water (pH < 1)	HF	0.25 wt%	–	10	1	22	0.2	13
[44]	Water (pH < 1)	HF	0.25 wt%	–	20	1	76	0.4	17
[45]	Water (pH = 4.5)	KF	0.1 M	–	10	90	30	2.3	–
[46]	Water (pH = 4.5)	KF	0.1 M	–	25	17	110	4.4	20
[46]	Water (pH = 5)	KF	0.1 M	–	25	17	110	6	20
[41]	Formamide	NH_4F	0.27 M	5% (vol)	20	24	56	19.6	17
[41]	Formamide	NH_4F	0.27 M	5% (vol)	35	30	129	37.4	15
[41]	Formamide	$N(n\text{-}Bu)_4F$	0.27 M	5% (vol)	35	48	146	68.9	22
[41]	Formamide	$N(n\text{-}Bu)_4F$	0.27 M	5% (vol)	15	46	50	20	15
[47]	Dimethyl sulfoxide (DMSO)	HF	1% (vol)	1% (vol)	20	70	50	10	–
[47]	Dimethyl sulfoxide (DMSO)	HF	1% (vol)	1% (vol)	60	70	150	93	–
[48]	Glycerol	NH_4F	0.5 wt%	0	20	13	40	2.5	12
[49]	Glycerol	NH_4F	0.27 M	50% (vol)	35	6	230	2	–
[40]	Ethylene glycol	NH_4F	0.3 wt%	2% (vol)	20	17	45	5	10
[40]	Ethylene glycol	NH_4F	0.3 wt%	2% (vol)	65	17	135	105	25

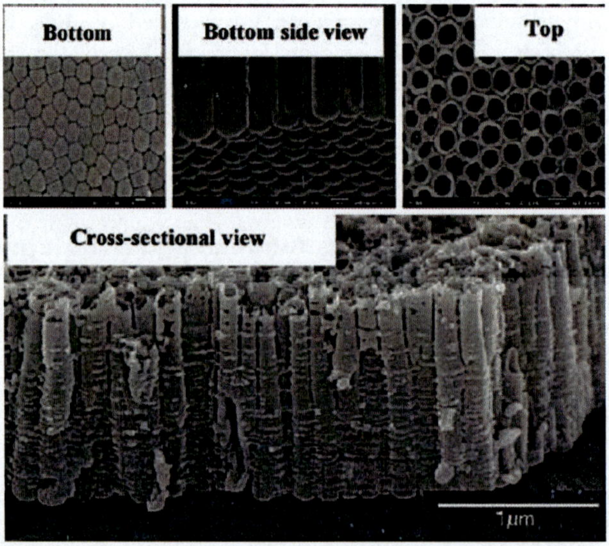

Fig. 13.4 SEM micrographs of TiO_2 nanotube array morphologies achieved by Ti anodization. The upper images are from Ti anodization in an ethylene glycol bath and show an irregular hexagonal architecture similar to nanoporous alumina; the lower image is the architecture resulting from Ti anodization in an aqueous KF bath and shows nanotubes with a circular cross-section (Reprinted with permission from Ref. [50])

under the same conditions until the entire Al film was converted into nanoporous alumina. However, only AAO templates smaller than 1 μm in thickness remained attached to the substrate at the end of the process. When the alumina layer was thicker than 1 μm, the build-up of compressive stress within the film due to the volume expansion during anodization, resulted in the AAO template detaching from the underlying substrate [8]. The detachment of thicker AAO templates was avoided by using silicon dioxide coated Si wafers at the expense of a thicker barrier layer. On the technologically important TCO-coated glass substrates (FTO and ITO), the growth of nanoporous alumina is problematic, due to which AAO-templated deposition has not been widely used in growing semiconductor nanowire- and nanotube arrays for use in optoelectronic devices. Even though Todoroki et al. [12] report the formation of nanoporous alumina on TCO-coated glass by a simple anodization of a sputtered Al film, others report the delamination and destruction of the template [19], also confirmed by our own experience. In this regard, the use of a thin Ti seed layer before the deposition of Al appears to be a promising route towards stabilizing the growth of nanoporous alumina on TCO-coated glass substrates and improving adhesion of the template to the substrate [19] (Fig. 13.5).

Mor et al. [9] formed highly transparent, well-ordered TiO_2 nanotube arrays on glass substrates by anodizing a vacuum deposited Ti film in an aqueous electrolyte containing 0.25 wt% HF and 12.5 vol% acetic acid. More recently, Varghese et al. [11] formed very long (1–20 μm) TNT's of high quality on conductive FTO-coated glass substrates by anodizing a vacuum deposited Ti film in a DMSO- based electrolyte containing 4% HF. In both cases, the method of vacuum deposition of the Ti film was found to be critical to the formation of high quality TNT's. TNT's were formed only when the Ti films were suitably dense and possessed coarse grains. Ti films deposited at room temperature using thermal evaporation or ordinary sputtering did not result in TNT's following anodization. Only Ti films deposited using magnetron sputtering at an elevated substrate temperature of 500°C or formed at room-temperature by ion-beam assisted magnetron sputtering were found to result in high-quality TNT's upon anodization (Fig. 13.6).

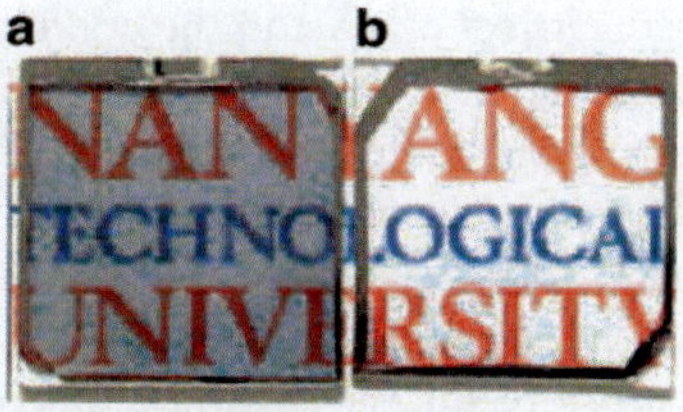

Fig. 13.5 Optical micrographs of AAO templates produced on ITO:glass substrates coated with a 0.3–10 nm Ti. The template without a barrier layer in (**b**) is more transparent than the one with a barrier layer in (**a**) (Reprinted with permission from Ref. [19])

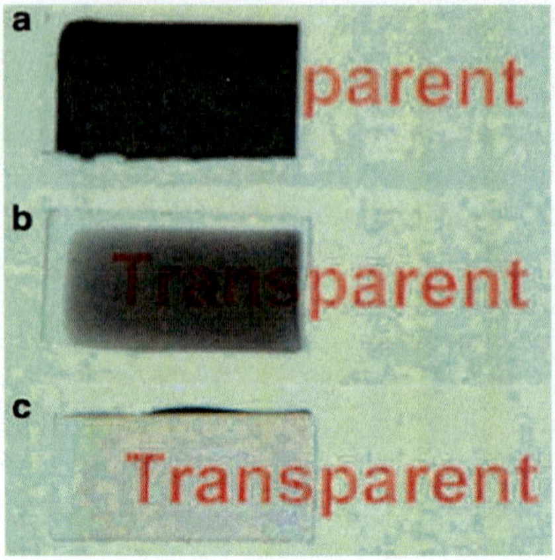

Fig. 13.6 Key stages in the fabrication of a transparent TiO_2 nanotube-array film: (**a**) sputter deposition of a high-quality Ti thin film, (**b**) anodization of resulting film, and (**c**) heat treatment to oxidize the remaining metallic islands (Reprinted with permission from Ref. [51])

13.3 Templated Growth of Functional Nanomaterials

Nanoporous alumina presents highly ordered templates of pre-determined channel length, diameter, width and density, due to which it is frequently used to form the templated growth of nanowires (NW's) and nanotubes (NT's) of various other materials. In this synthetic approach, nanocylinders of the desired material are formed by either filling the pores of AAO (for NW's) or coating the walls of AAO (for NT's). Whether NW's or NT's are formed depends on the wetting behavior of the deposited material and the chemistry of the pore wall. A subset of this technique is membrane-based nanomaterial synthesis, in which free-standing AAO membranes are used as the growth template. In comparison to track-etched polymeric membranes and mesoporous materials, AAO membranes have the advantages that they possess a larger pore density and the pores are organized in a non-random hexagonal array. The large thickness of AAO and TNT membranes (50–1,000 μm) in comparison with the pore size (15–200 nm) permits the formation of very high aspect ratio nanostructures. Typically, the AAO template is released by selective dissolution of the Al_2O_3 host solid subsequent to templated growth. However in some cases, the template is not released. When the underlying substrate is conducting (e.g. heavily doped Si or TCO-glass), the mechanical robustness of the Al_2O_3 host and built-in electrical contact to the conducting substrate are useful for the electrical transport measurements on template-grown nanowire arrays [8]. Conductivity of the underlying substrate also allows facile electrodeposition into the nanopores to form NW and NT arrays. However, the anodization process produces a dense and continuous alumina barrier layer along with the nanopores. This barrier layer is completely resistive, and therefore prevents direct physical and electrical contact between the pore and the substrate. There are three methods used to overcome this limitation. In one method, the nanoporous template is detached from the substrate by etching away the substrate, followed by subsequent etching of the exposed barrier layer to open up the blocked ends of the alumina channels.

A metal layer is vacuum deposited on one side of the template (which is now freestanding) and provides the desired electrical contact to the pore. Since this method produces an unsupported freestanding membrane which is brittle and therefore fragile, it works best for membranes several tens of micrometers thick. Thinner membranes are difficult to detach from the substrate intact [52]. A second method, which does not require the detachment of the template from the substrate, consists of thinning the barrier layer at the bottom of the pores by stepping down the anodization voltage to reduce its resistance. By exploiting the rectifying property of the barrier layer, alternating current (ac) electrodeposition is used to electrochemically deposit materials into the pores. The third method, which again does not require detachment, consists of selectively chemically etching the barrier layer to either provide direct access to the underlying substrate or at least create tiny conducting paths in the barrier layer. A 3 h etch in a 0.1 M phosphoric acid solution for 3 h at room temperature forms a through-pore membrane but also slightly widens the diameter of the nanopore. Plasma etching with CF_4 + O_2 gas system is also effective for removing the barrier layer without changing either the pore diameter or the period. The flow rates of CF4 and O2 used in one study were 18 and 8 sccm, respectively. The pressure was 100 mT with a power of 250 W. An etching time of 10 min opened the pores [53].

The nanomaterials whose templated growth is formed using AAO fall into the following broad categories:

1. Nanowires of ferromagnetic materials such as Ni, Fe, Co and their alloys for studies of magnetic properties and for use in high density memories and biosensors. Electrochemical deposition is the most commonly used technique to fill the pores of AAO with ferromagnetic metal [54–56]. Consequently, a conductive substrate is essential. Due to its simplicity, direct current electrodeposition is widely used. Alternating current electrodeposition provides more control and does not require the barrier layer at the bottom of the pores to be completely removed. Nanoporous alumina is the template of choice although recently, a TiO_2 nanotube array template was also used to form ferromagnetic Ni nanocylinder arrays [57].
2. The localized surface plasmon resonance (LSPR) of noble metal nanostructures occ3urs at the wavelengths of visible light. Consequently, nanorods of Au [58, 59], Ag [60, 61] and Cu [62] are being actively studied for use in plasmonics for applications including, but not limited to, high sensitivity label-free biosensing [63], metamaterials [64], enhancement of non-linear optical effects [65] and ultra-sensitive single-molecule sensing [66]. The higher surface area of noble metal nanotubes makes them more attractive for all the above mentioned applications but fewer reports exist of their fabrication and characterization [67]. Once again, electrodeposition into the pores of nanoporous alumina is a widely used technique to form noble metal nanotubes and nanowires, a technique pioneered in the nineties by the research group of C. R. Martin [68–72]. However, non-electrochemical methods have also been applied such as oblique angle evaporation of the noble metal into the AAO template [73], photochemical decomposition of Ag- and Au-containing

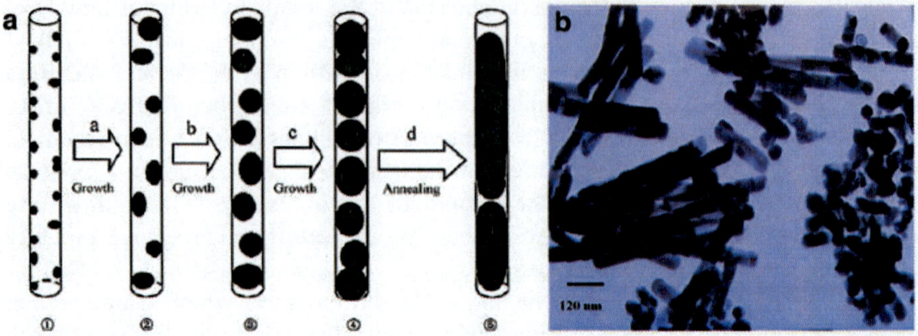

Fig. 13.7 (**a**) Schematic of the growth process for metal nanorods by fusion of metal nanoparticles and (**b**) TEM image of Ag nanorods released from the AAO template (Reprinted with permission from Ref. [75])

precursors in AAO pores [74] and fusion of Ag nanoparticle chains (see Fig. 13.7) formed in AAO templates by solvothermal synthesis [75].

3. Nanorods and nanotubes of inorganic semiconductors for applications in photovoltaics [76, 77], light emission [78] and lasing [79], single-electron devices [80] and thermoelectrics [81]. CdS nanorods has been grown by electrochemical deposition [82] and CdS nanotubes by chemical vapor deposition [83] into AAO templates. Recently, a number of papers have reported on the formation of single crystal CdS nanowires by techniques such as template-assisted vapor–liquid–solid (VLS) growth [84] and chemical precipitation in AAO templates [85]. Templated growth has been used to form 1-D nanostructures primarily in II-VI semiconductors such as CdS, CdSe [86], CdTe [87], ZnO [88, 89], ZnS [90–93], ZnTe [94], PbS [95], PbSe [96] and Bi_2Te_3 [97–99]. Reports on the templated formation of NR's/NT's from III-V semiconductors such as GaN [100, 101] exist but are far fewer. TiO_2 nanotube arrays are also emerging as hard-templates for II-VI nanostructure synthesis [102, 103].
4. Organic nanotubes and nanowires of π-conjugated organic small-molecules and polymers. The growth of carbon nanotubes (CNT's) by the catalytic pyrolysis of unsaturated hydrocarbons such as ethylene, propylene or acetylene on Co particles electrochemically embedded at the bottom of the pores of AAO templates was demonstrated in the late nineties [104–107]. Such CNT arrays are being considered for use as active materials in nanoscale transistors [108], as field-emitters [109–111] and as electrodes in fuel cells and Li-ion batteries [112]. Nanorods and nanotubes of π-conjugated organic small molecules such as phthalocyanines [113, 114], triphenylamines [115], linear acenes [116] and fullerenes [117] have also been formed by AAO templating. Similarly, semiconducting polymer NT's and NR's have been grown in AAO [118–123]. Such π-conjugated nanostructures find application in ordered bulk heterojunction solar cells, organic schottky diodes and devices employing polarized light emission.

13.4 Pore Differentiation: Highly Ordered Mosaic-Like Nanostructures

In 2003, Masuda et al. introduced a new method based on pretexturing the aluminum to engineer differentiation of the self-organized AAO pores into two types with a controllable period [124]. To appreciate how this is accomplished, consider the following process: A SiC master mold is patterned using conventional electron-beam lithography to create an ideally ordered arrangement of convex dots with a period of 200 nm but with the important difference that every sixth site had a defect (no dot). When this master mold is stamped onto a polished Al foil, the pattern on the mold is replicated on the Al surface as an ordered array of concave dimples but with a dimple missing at every sixth site. The imprinted Al foil is anodized at 80 V in 0.05 M oxalic acid at 16°C to form the nanoporous architecture [125]. When the aluminum is subsequently selectively removed to reveal the barrier layer, it is found that the barrier layer is thicker at the imprinted sites than at the non-imprinted sites. Likewise, the pores formed at the non-imprinted sites are smaller in diameter than the pores at the imprinted sites. The difference in thickness between the barrier layer at the imprinted and non-imprinted sites is exploited to selectively open the bottoms of the smaller non-imprinted pores. Figure 13.8L shows the SEM image taken from the barrier layer side of such an AAO membrane with selectively opened holes. It is clear from this image that pores of slightly smaller diameter, at every sixth site (corresponding to the non-imprinted sites) are selectively opened. In the next step, a contact layer is deposited on the top-side of the membrane to provide electrical contact during the subsequent electrodeposition of gold into the membrane. Since the pore bottoms have been selectively opened only at the non-imprinted sites, Au is selectively deposited into these pores as seen at the site labeled ‘b’.

Clearly, the selective opening of the pores is contingent upon performing the wet etch of the barrier layer for just enough time for the thinner barrier layer at the

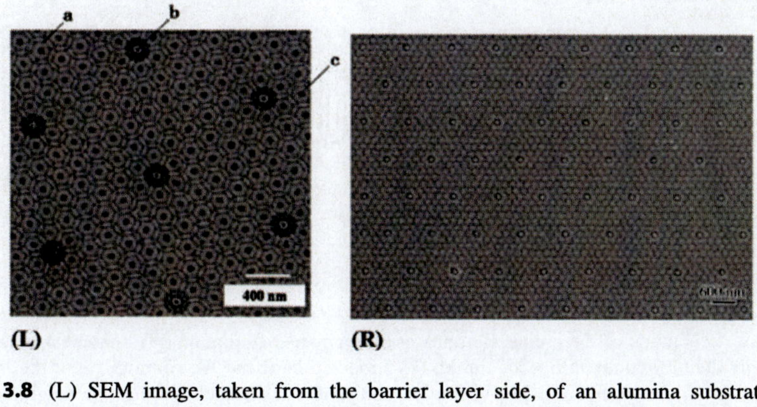

Fig. 13.8 (L) SEM image, taken from the barrier layer side, of an alumina substrate with selectively opened holes (**a**) porous alumina (**b**) selectively opened hole and (**c**) barrier layer (R) SEM image of the fabricated Au disk array (Reprinted with permission from Ref. [125])

bottom of the non-imprinted pores to be completely removed thus opening the pore but not so long as to also cause the remaining pores to open. In this particular report [125], a dipping time of 77 min in phosphoric acid was found to be optimal for selective opening of the pores. The entire process sequence is illustrated in the schematic diagram of Fig. 13.9.

The same pore differentiation process can be further continued to open all the remaining pores after the deposition of metal (say gold) into the selectively opened pores. In the subsequent step, a second material (say nickel) is deposited into the remaining pores (now open) thus resulting in a binary mosaic-like nanocomposite. It also follows that the pore differentiation process can be extended to vary the period of the differentiated pores. Also, the two materials used to form the mosaic composite need not be restricted to metals such as Ni and Au but could be any two materials that can be grown in the pores by electrochemical or electro phoretic deposition. The detailed fabrication sequence for such an elaborate pore differentiation process [124] is schematically depicted in Fig. 13.10. SEM images of the resulting binary mosaic-nanocomposite are shown in Fig. 13.11. In very recent work [126], this principle was extended to form an anodic nanoporous alumina mask with a checkerboard pattern.

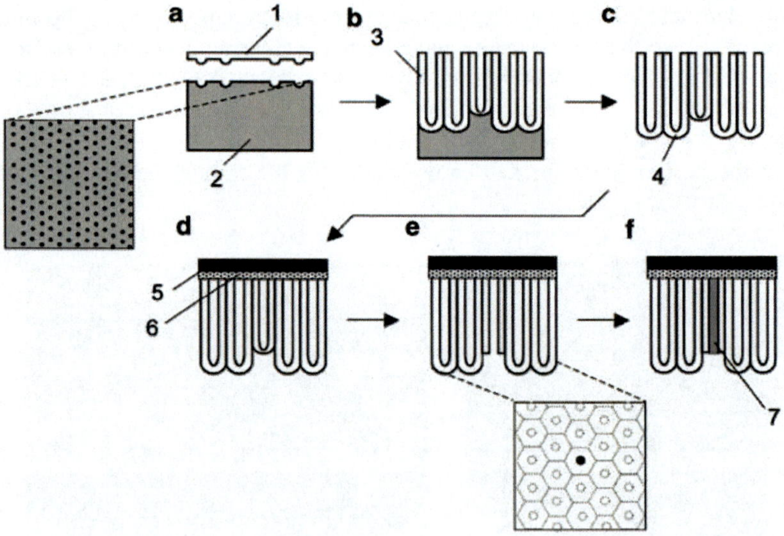

Fig. 13.9 Schematic of the preparation process for a pore-differentiated Au-nanodisk array: (**a**) imprinting the aluminum with a SiC mold; (**b**) anodization of the Al; (**c**) removal of the Al; (**d**) formation of an electrode on the top-surface of the AAO membrane; (**e**) selective etching of the barrier layer at shallow sites to form selectively opened holes; (**f**) deposition of Au (Reprinted with permission from Ref. [125])

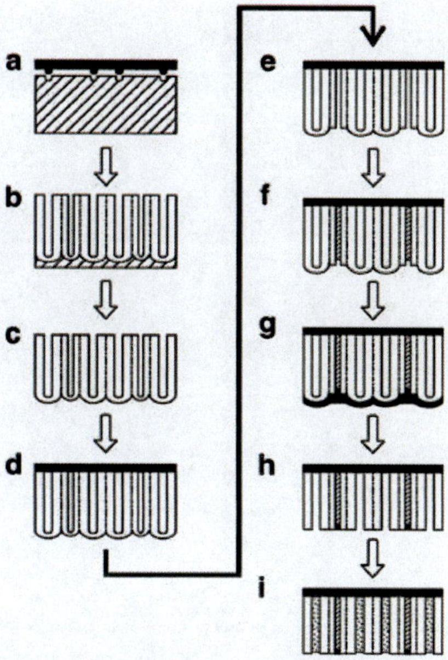

Fig. 13.10 Schematic diagram showing the preparation process of the Au-Ni mosaic nanostructure: (**a**) stamping the Al using a SiC mold; (**b**) Al anodization; (**c**) Al removal; (**d**) formation of a contact electrode on the surface side of the nanoporous alumina membrane; (**e**) selective penetration of the barrier layer at the non-imprinted sites; (**f**) selective deposition of Au; (**g**) formation of an insulator layer; (**h**) penetration of the barrier layer at the imprinted sites; (**i**) deposition of Ni resulting in a mosaic nanocomposite (Reprinted with permission from Ref. [124])

13.5 Pattern Transfer Using Anodically Formed Nanoporous Hard Templates

Self-organized anodically formed nanoporous templates offer an attractive high-yield low-cost route for large area sub-100 nm scale pattern production. The templated growth of nanomaterials considered in Sect. 13.3 is an additive process consisting of material deposition into the pores of AAO or TNT's. Pattern transfer, on the other hand can employ either additive or subtractive processes to replicate the nanoporous structure of a free-standing AAO membrane onto an underlying substrate. In subtractive processes, the nanopores of anodic alumina are used as a mask to etch the substrate. Thus, AAO has been used as a mask to perform pattern transfer by MeV ion irradiation [127]. It has also been used as a mask for wet etching [128] and for dry etching methods such as plasma ashing [129] and reactive ion etching (RIE) [53, 101, 130]. Subtractive processes are classified into positive transfer and negative transfer schemes. In positive transfer, the nanoporous pattern is directly transferred onto an underlying substrate (typically a metal or semiconductor) via etching. In negative transfer, the inverse structures of the nanoporous pattern are formed via evaporation of masking caps followed by subsequent etching.

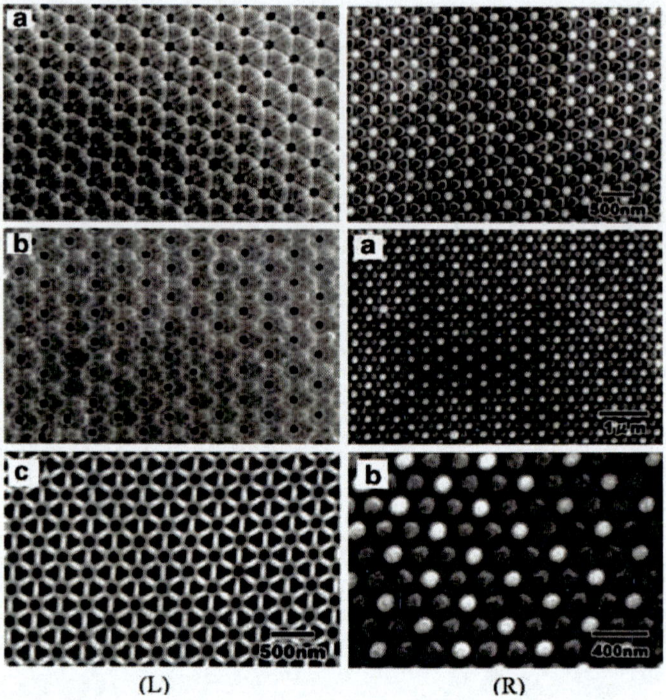

Fig. 13.11 (L) SEM images of the barrier layer of a pore-differentiated membrane highlighting the importance of the correct etching time (**a**) as-anodized membrane (**b**) after etching in 5 wt% phosphoric acid at 30°C for 60 min and (**c**) after etching in 5 wt% phosphoric acid at 30°C for 80 min. After etching for 60 min, only the barrier layer at the non-imprinted sites was penetrated as shown in (**b**). Additional etching resulted in penetration of the barrier layer for the imprinted sites as seen in (**c**). The etching time required to penetrate the barrier layer was thus dependent on whether or not the sites overlying the pores were imprinted or not, a unique property exploitable to selectively deposit two different materials inside the channels of anodic alumina templates. (R) (*top*) SEM image of the selectively embedded gold at the non-imprinted sites of the pore-differentiated AAIO membrane (*bottom*) SEM images of the mosaic composite composed of gold and nickel: (**a**) low magnification and (**b**) high magnification. Gold cylinders are the brighter dots. All images in Fig. 13.11 were reprinted with permission from Ref. [124]

13.5.1 AAO as Etch-Mask for Formation of Nanohole Arrays on Single Crystal Semiconductor Substrates

Hexagonally ordered AAO is attractive for use in 2D photonic crystal based micrometer optoelectronic devices. However, in order to utilize the highly ordered nano-channel structure of AAO as a photonic material, it must first be transformed into semiconductors, which have significantly higher dielectric constants and concomitantly larger photonic bandgap [131]. This has motivated the use of ultrathin alumina membranes as etch masks to form nanohole arrays on Si, GaAs, InP and ZnTe. Crouse et al. [13] formed nanoporous alumina with a pore spacing of 100 nm and a pore diameter of 50 nm on Si wafers by thermal evaporation of Al followed by single step anodization at 40 V in 0.3 M oxalic acid. After thinning the AAO template from 2 μm to 300 nm by ion-milling and removing the barrier layer by a 1 h long pore-widening etch in 5 wt% H_3PO_4, the hexagonal pattern was transferred to the Si substrate by using the AAO template as a mask for the reactive ion-etching of Si by Cl_2 + BCl_3. Very recently, InGaN-based light emitting diodes (LED's) were successfully grown and fabricated on Si (111) wafers patterned at the nanoscale using an AAO etch mask [79]. AAO templates with an average nanopore diameter of 150 nm and an interpore distance of 120 nm were formed by anodization at 6°C for 30 min in 0.3 M phosphoric acid under an applied voltage of 120 V. The AAO pattern was transferred to the Si substrate by inductively coupled plasma (ICP) etching through the template. The purpose of the nano-patterning was to reduce the high dislocation density that results when GaN is grown in Si substrates due to large mismatch in lattice constant and thermal conductivity between Si and GaN. In another report [132], antireflective sub-wavelength structured surfaces (for photovoltaic application) were fabricated on single-crystalline Si wafers patterned using an AAO etch-mask.

Figure 13.12a shows a schematic of the fabrication process used to achieve the sub-wavelength structured surfaces while Fig. 13.12b shows SEM images of the resulting SWS Si. The starting step, namely the thermal oxidation of Si, was performed in a standard quartz furnace to form a thin protective oxide barrier for anodic oxidation. The thickness of the deposited Al film was 500 nm. The anodization was performed in 0.3 M oxalic acid at 40 V to obtain a periodicity of ~100 nm and was followed by pore-widening in 5 wt% H_3PO_4 to remove the barrier layer. Fast atom beam etching (FAB) with SF_6 gas performed under an acceleration

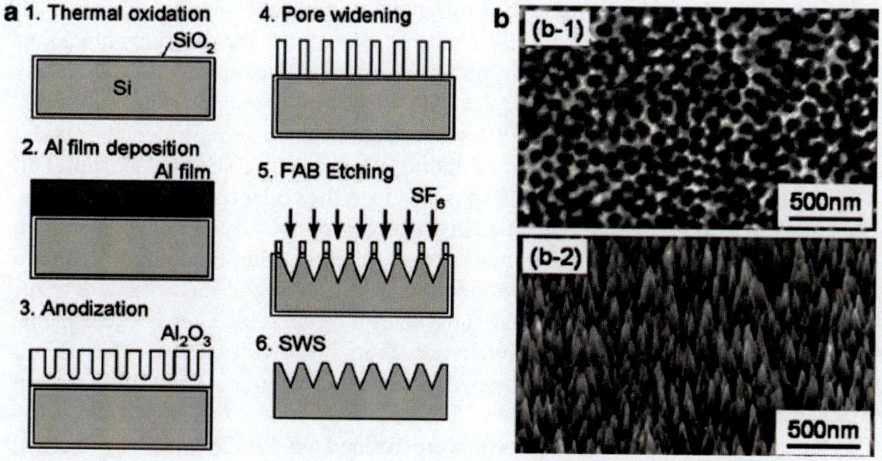

Fig. 13.12 Sub-wavelength structured (SWS) silicon wafers patterned using AAO: (**a**) fabrication sequence used to prepare SWS Si; (**b**) SEM images of the Si surface after fast atom beam etching through the AAO mask for 50 min showing (b-1) top-view and (b-2) cross-section (Reprinted with permission from Ref. [132])

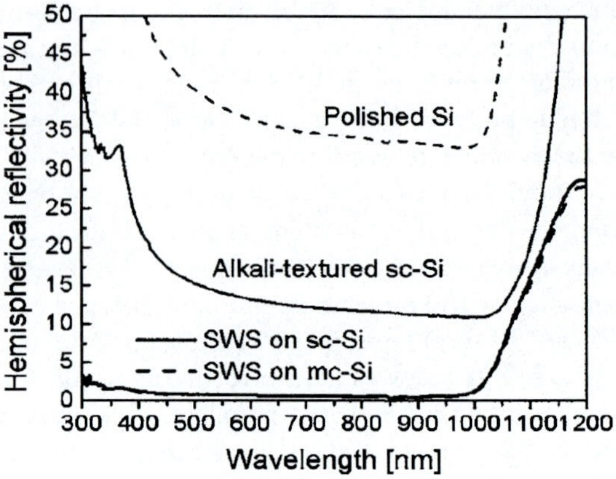

Fig. 13.13 Measured reflectivity spectra of AAO-patterned SWS Si samples fabricated by the process depicted in Fig. 13.12a. (Reprinted with permission from Ref. [132])

voltage was used to generate the SWS surface relief grating. Figure 13.13 demonstrates that the reflectivity of the resulting SWS Si is less than 1% across the entire wavelength range from 300 to 1,000 nm. Also shown in Fig. 13.13 is the high reflectance of a polished Si wafer and the reflectance of a conventionally alkali-textured Si wafer [132].

13.5.2 AAO as Stencil for Nano Dot Array Formation

The lateral ordering of semiconductor nanocrystals in 2D arrays is a subject of both fundamental and applied research in nanophotonics and nanoelectronics. Si nanocrystals embedded within thin dielectric layers form the building blocks of certain next generation non-volatile memory devices. Quantum dot (QD) lasers constructed from arrays of InP and InAs nano-islands on GaAs and other substrates are becoming increasingly important for applications such as high density optical data storage systems. The formation of lateral 2D arrays of QD's is also important in fundamental experiments designed to investigate the collective behavior of large assemblies of coupled nano-elements. In all these applications, high regularity in the arrangement of the QD arrays and control of their spacing is required to obtain precise control over the device properties. Furthermore, critical feature sizes smaller than 100 nm need to be fabricated over a large area. These requirements are beyond the reach of conventional nanofabrication techniques. The use of electrochemically formed self-organized nanoporous alumina stencils offers an attractive route towards the formation of 2D QD arrays.

In one report, CdS nanodot arrays were formed on Si substrates by thermal evaporation of CdS through an ultrathin AAO membrane. The resulting CdS nanodots (shown in Fig. 13.14) were polycrystalline with a (002) preferred orientation and possessed a mono-dispersed size distribution [133]. The photoluminescence (PL) characteristics of CdS NP's of two different sizes, 10 nm in (a)

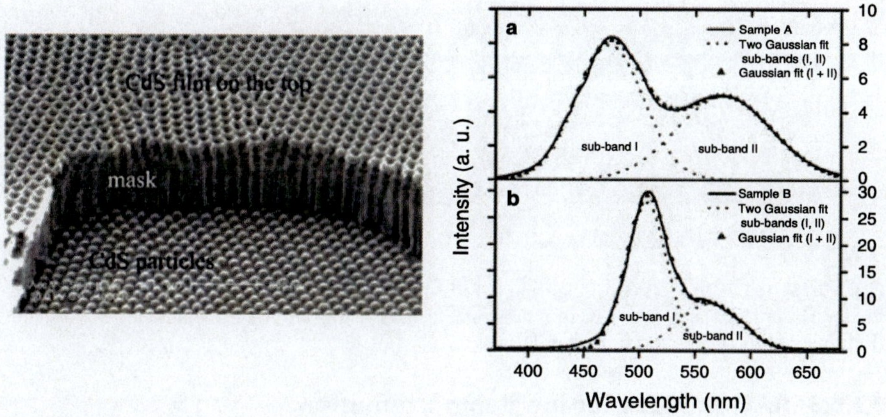

Fig. 13.14 Left, CdS nanodot arrays with an average diameter and spacing of 80 nm and 105 nm respectively, formed on Si substrates using an AAO mask; and right, photoluminescence spectra of CdS nanodot arrays of (**a**) 10 nm height and (**b**) 50 nm height and their two Gaussian fit subbands. The excitation wavelength is 350 nm. The peak positions of the subbands I and II are located at about 473 and 575 nm in sad and 506 and 563 nm in sbd, respectively (Reprinted with permission from Ref. [133])

and 50 nm in (b), are shown in Fig. 13.14R and each show two features, namely band-edge emission (sub-band I) and surface-defect emission (sub-band II). Band-edge emission is ascribed to the radiative recombination of excitons in the NP's and therefore the corresponding peak energy of such band-edge emission is usually slightly lower than the band-gap energy of the CdS NP's. The 10 nm nanodots had a particle crystallite size smaller than 4 times the Bohr exciton radius and therefore exhibited the effects of quantum confinement, which manifested itself in the strongly blue-shifted broad peak of sub-band I. On the other hand, the 50 nm nanodots, which were outside the excitonic confinement regime, exhibited a narrow size-independent band-edge emission at 506 nm [133].

By reactive ion etching (RIE) using BCl_3 through an AAO membrane etch-mask with a periodicity of 110 nm and a diameter of 55 nm, Liang et al. [134] transferred the nanopore array pattern onto a GaAs substrate. Subsequently, they grew a highly ordered InAs nanodot arrays by molecular-beam epitaxy on non-lithographically nanopatterned GaAs. The resulting dots (see Fig. 13.15) were organized in a dense hexagonal lateral superlattice.

For plasmonic device applications, it is desirable to create periodic arrays of size-controlled noble metal particles (Ag and Au) on arbitrary substrates. Here too, AAO templates have been used. Atwater et al. [135] demonstrated an 8% increase in short circuit density in an optically thin GaAs solar cell decorated with densely packed high-aspect ratio Ag nanoparticles (NP's) fabricated by masked deposition through an AAO template. Enhanced light absorption occurred in this device due to the longer optical path of incident light in the absorber layer, which was itself due to the strong scattering by interacting surface plasmons formed in the Ag NP's.

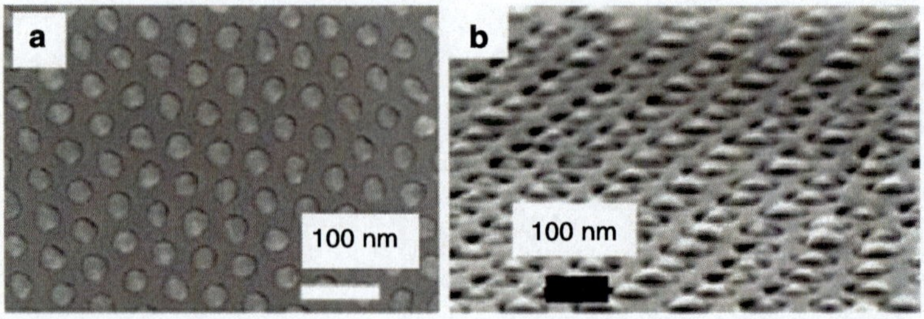

Fig. 13.15 (**a**) SEM top-view image and (**b**) SEM oblique view image of a hexagonally ordered InAs quantum dot array on a GaAs substrate formed using an AAO membrane as a stencil (Reprinted with permission from Ref. [134])

13.5.3 AAO as Mask for Ion Beam Irradiation

Matsuura et al. [136] performed ion beam irradiation at normal incidence to a polished single crystalline $SrTiO_3$ substrate through an AAO template bonded to the substrate with the beam aligned parallel to the pore axis to transfer the nanoporous pattern on to the substrate. The AAO template had 55 nm pores and a pitch of 100 nm. Different ion beam irradiation recipes were tried: Pt ions with energies of 200 keV, 500 keV and 1 MeV at fluences of 1×10^{14}, 2×10^{14} and 2×10^{14} cm^{-2} respectively were used to amorphize exposed areas on the $SrTiO_3$ substrate. During irradiation, the substrates were held at 77 K to precipitate the onset of amorphization. The amorphized $SrTiO_3$ also experienced a 20% reduction in density. Subsequently, the template was removed and the amorphized $SrTiO_3$ was selectively etched by immersing the substrate in concentrated aqueous nitric acid [136]. This patterning technique is powerful because ceramic materials such as $SrTiO_3$ are not easily etched or patterned by other nanofabrication methods.

References

1. Renshaw TA. J Appl Phys. 1958;29:1623–4.
2. Cosgrove LA. J Phys Chem. 1956;60:385–8.
3. Gong D, Grimes CA, Varghese OK, Hu WC, Singh RS, Chen Z, Dickey EC. J Mater Res. 2001;16:3331–4.
4. Mozalev A, Sakairi M, Saeki I, Takahashi H. Electrochim Acta. 2003;48:3155–70.
5. Tsuchiya H, Macak JM, Sieber I, Schmuki P. Small. 2005;1:722–5.
6. Tsuchiya H, Schmuki P. Electrochem Commun. 2005;7:49–52.
7. Allam NK, Feng XJ, Grimes CA. Chem Mater. 2008;20:6477–81.
8. Rabin O, Herz PR, Lin YM, Akinwande AI, Cronin SB, Dresselhaus MS. Adv Funct Mater. 2003;13:631–8.
9. Mor GK, Varghese OK, Paulose M, Grimes CA. Adv Funct Mater. 2005;15:1291–6.
10. Sadek AZ, Zheng HD, Latham K, Wlodarski W, Kalantar-Zadeh K. Langmuir. 2009;25:509–14.
11. Varghese OK, Paulose M, Grimes CA. Nat Nanotechnol. 2009;4:592–7.
12. Chu SZ, Wada K, Inoue S, Todoroki S. J Electrochem Soc. 2002;149:B321–7.

13. Crouse D, Lo YH, Miller AE, Crouse M. Appl Phys Lett. 2000;76:49–51.
14. Premchand YD, Djenizian T, Vacandio F, Knauth P. Electrochem Commun. 2006;8:1840–4.
15. Yu XF, Li YX, Ge WY, Yang QB, Zhu NF, Kalantar-Zadeh K. Nanotechnology. 2006;17:808–14.
16. Chong ASM, Tan LK, Deng J, Gao H. Adv Funct Mater. 2007;17:1629–35.
17. Yin AJ, Guico RS, Xu J. Nanotechnology. 2007;18:035304.
18. Shankar K, Basham JI, Allam NK, Varghese OK, Mor GK, Feng XJ, Paulose M, Seabold JA, Choi KS, Grimes CA. J Phys Chem C. 2009;113:6327–59.
19. Foong TRB, Sellinger A, Hu X. ACS Nano. 2008;2:2250–6.
20. Patermarakis G. J Electroanal Chem. 2009;635:39–50.
21. Houser JE, Hebert KR. Nat Mater. 2009;8:415–20.
22. Parkhutik VP, Shershulsky VI. J Phys D. 1992;25:1258–63.
23. Thompson GE, Furneaux RC, Wood GC, Richardson JA, Goode JS. Nature. 1978;272:433–5.
24. Osulliva J, Wood GC. Proc R Soc Lond Ser A Math Phys Sci. 1970;317:511.
25. Mor GK, Varghese OK, Paulose M, Shankar K, Grimes CA. Sol Energy Mater Sol C. 2006;90:2011–75.
26. Hwang SK, Jeong SH, Hwang HY, Lee OJ, Lee KH. Korean J Chem Eng. 2002;19:467–73.
27. Sui YC, Cui BZ, Martinez L, Perez R, Sellmyer DJ. Thin Solid Films. 2002;406:64–9.
28. Sulka GD, Stroobants S, Moshchalkov V, Borghs G, Celis JP. J Electrochem Soc. 2002;149: D97–103.
29. Lee W, Ji R, Gosele U, Nielsch K. Nat Mater. 2006;5:741–7.
30. Masuda H. In: Wehrspohn RB, editor. Ordered porous nanostructures and applications. New York: Springer; 2005. p. 37–56.
31. Kustandi TS, Loh WW, Gao H, Low HY. ACS Nano. 2010;4:2561–8.
32. Masuda H, Yamada H, Satoh M, Asoh H, Nakao M, Tamamura T. Appl Phys Lett. 1997;71:2770–2.
33. Matsui Y, Nishio K, Masuda H. Small. 2006;2:522–5.
34. Liu CY, Datta A, Wang YL. Appl Phys Lett. 2001;78:120–2.
35. Lee W, Ji R, Ross CA, Gosele U, Nielsch K. Small. 2006;2:978–82.
36. Sun ZJ, Kim HK. Appl Phys Lett. 2002;81:3458–60.
37. Fournier-Bidoz S, Kitaev V, Routkevitch D, Manners I, Ozin GA. Adv Mater. 2004;16:2193.
38. Kim BY, Park SJ, McCarthy TJ, Russell TP. Small. 2007;3(11):1869–72.
39. Mor GK, Shankar K, Paulose M, Varghese OK, Grimes CA. Nano Lett. 2005;5:191–5.
40. Prakasam HE, Shankar K, Paulose M, Varghese OK, Grimes CA. J Phys Chem C. 2007;111:7235–41.
41. Shankar K, Mor GK, Fitzgerald A, Grimes CA. J Phys Chem C. 2007;111:21–6.
42. Mohammadpour A, Shankar K. J Mater Chem. 2010;20:8474–7.
43. Macak JM, Albu SP, Schmuki P. Phys Status Solidi-Rapid Res Lett. 2007;1:181–3.
44. Mor GK, Shankar K, Varghese OK, Grimes CA. J Mater Res. 2004;19:2989–96.
45. Cai QY, Paulose M, Varghese OK, Grimes CA. J Mater Res. 2005;20:230–6.
46. Varghese CK, Paulose M, Shankar K, Mor GK, Grimes CA. J Nanosci Nanotechnol. 2005;5:1158–65.
47. Paulose M, Shankar K, Yoriya S, Prakasam HE, Varghese OK, Mor GK, Latempa TA, Fitzgerald A, Grimes CA. J Phys Chem B. 2006;110:16179–84.
48. Macak JM, Tsuchiya H, Taveira L, Aldabergerova S, Schmuki P. Angew Chem-Int Edit. 2005;44:7463–5.
49. Macak JM, Hildebrand H, Marten-Jahns U, Schmuki P. J Electroanal Chem. 2008;621:254–66.
50. Grimes CA, Mor GK. TiO_2 Nanotube Arrays. New York: Springer; 2009. ISBN 9781441900678.
51. Mor GK, Shankar K, Paulose M, Varghese OK, Grimes CA. Nano Lett. 2006;6:215–8.
52. Zhang JP, Kielbasa JE, Carroll DL. J Mater Res. 2009;24:1735–40.
53. Liang JY, Chik H, Yin AJ, Xu J. J Appl Phys. 2002;91:2544–6.

54. Cho SG, Yoo B, Kim KH, Kim J. IEEE Trans Magn. 2010;46:420–3.
55. Guo YG, Wan LJ, Zhu CF, Yang DL, Chen DM, Bai CL. Chem Mater. 2003;15:664–7.
56. Zeng H, Skomski R, Menon L, Liu Y, Bandyopadhyay S, Sellmyer DJ. Phys Rev B. 2002;65:134426.
57. Prida VM, Hernandez-Velez M, Pirota KR, Menendez A, Vazquez M. Nanotechnology. 2005;16:2696–702.
58. Shukla S, Kim KT, Baev A, Yoon YK, Litchinitser NM, Prasad PN. ACS Nano. 2010;4:2249–55.
59. Schmucker AL, Harris N, Banholzer MJ, Blaber MG, Osberg KD, Schatz GC, Mirkin CA. ACS Nano. 2010;4:5453–63.
60. Zong RL, Zhou J, Li Q, Du B, Li B, Fu M, Qi XW, Li LT, Buddhudu S. J Phys Chem B. 2004;108:16713–6.
61. Zhao C, Tang SL, Du YW. Chem Phys Lett. 2010;491:183–6.
62. Zong RL, Zhou J, Li B, Fu M, Shi SK, Li LT. J Chem Phys. 2005;123:094710.
63. McPhillips J, Murphy A, Jonsson MP, Hendren WR, Atkinson R, Hook F, Zayats AV, Pollard RJ. ACS Nano. 2010;4:2210–6.
64. Dickson W, Wurtz GA, Evans P, O'Connor D, Atkinson R, Pollard R, Zayats AV. Phys Rev B. 2007;76:115411.
65. Zong RL, Zhou J, Li Q, Li LT, Wang WT, Chen ZH. Chem Phys Lett. 2004;398:224–7.
66. Habouti S, Solterbeck CH, Es-Souni M. J Mater Chem. 2010;20:5215–9.
67. Hendren WR, Murphy A, Evans P, O'Connor D, Wurtz GA, Zayats AV, Atkinson R, Pollard RJ. J Phys Condens Matter. 2008;20:362203.
68. Foss CA, Hornyak GL, Stockert JA, Martin CR. J Phys Chem. 1992;96:7497–9.
69. Foss CA, Hornyak GL, Stockert JA, Martin CR. J Phys Chem. 1994;98:2963–71.
70. Hornyak GL, Patrissi CJ, Martin CR. J Phys Chem B. 1997;101:1548–55.
71. Cepak VM, Martin CR. J Phys Chem B. 1998;102:9985–90.
72. Wirtz M, Martin CR. Adv Mater. 2003;15:455–8.
73. Losic D, Shapter JG, Mitchell JG, Voelcker NH. Nanotechnology. 2005;16:2275–81.
74. Zhao WB, Zhu JJ, Chen HY. J Cryst Growth. 2003;258:176–80.
75. Li XM, Wang DS, Tang LB, Dong K, Wu YJ, Yang PZ, Zhang PX. Appl Surf Sci. 2009;255:7529–31.
76. Kang Y, Kim D. Sol Energy Mater Sol C. 2006;90:166–74.
77. Martinson ABF, Elam JW, Hupp JT, Pellin MJ. Nano Lett. 2007;7:2183–7.
78. Liu CH, Zapien JA, Yao Y, Meng XM, Lee CS, Fan SS, Lifshitz Y, Lee ST. Adv Mater. 2003;15:838.
79. Ding JX, Zapien JA, Chen WW, Lifshitz Y, Lee ST, Meng XM. Appl Phys Lett. 2004;85:2361–3.
80. Routkevitch D, Tager AA, Haruyama J, Almawlawi D, Moskovits M, Xu JM. IEEE Trans Electr Dev. 1996;43:1646–58.
81. Heremans JP, Thrush CM, Morelli DT, Phys MCWu. Rev Lett. 2002;88:216801.
82. Routkevitch D, Bigioni T, Moskovits M, Xu JM. J Phys Chem. 1996;100:14037–47.
83. Shen XP, Yuan AH, Wang F, Hong JM, Xu Z. Solid State Commun. 2005;133:19–22.
84. Ergen O, Ruebusch DJ, Fang H, Rathore AA, Kapadia R, Fan ZF, Takei K, Jamshidi A, Wu M, Javey A. J ACS. 2010;132:13972–4.
85. Mu C, He JH. J Nanosci Nanotechnol. 2010;10:8191–8.
86. Xu DS, Shi XS, Guo GL, Gui LL, Tang YQ. J Phys Chem B. 2000;104:5061–3.
87. Zhao AW, Meng GW, Zhang LD, Gao T, Sun SH, Pang YT. Appl Phys A Mater Sci Process. 2003;76:537–9.
88. Li Y, Meng GW, Zhang LD, Phillipp F. Appl Phys Lett. 2000;76:2011–3.
89. Zheng MJ, Zhang LD, Li GH, Shen WZ. Chem Phys Lett. 2002;363:123–8.
90. Shen XP, Han M, Hong JM, Xue ZL, Xu Z. Chem Vapor Depos. 2005;11:250–3.
91. Xu XJ, Fei GT, Yu WH, Wang XW, Chen L, Zhang LD. Nanotechnology. 2006;17:426–9.
92. Zhai TY, Gu ZJ, Ma Y, Yang WS, Zhao LY, Yao JN. Mater Chem Phys. 2006;100:281–4.

93. Farhangfar S, Yang RB, Pelletier M, Nielsch K. Nanotechnology. 2009;20:325602.
94. Li L, Yang YW, Huang XH, Li GH, Zhang LD. J Phys Chem B. 2005;109:12394–8.
95. Wu C, Shi JB, Chen CJ, Lin JY. Mater Lett. 2006;60:3618–21.
96. Peng XS, Meng GW, Zhang J, Wang XF, Wang CZ, Liu X, Zhang LD. J Mater Res. 2002;17:1283–6.
97. Sander MS, Gronsky R, Sands T, Stacy AM. Chem Mater. 2003;15:335–9.
98. Jin CG, Xiang XQ, Jia C, Liu WF, Cai WL, Yao LZ, Li XG. J Phys Chem B. 2004;108:1844–7.
99. Menke EJ, Li Q, Penner RM. Nano Lett. 2004;4:2009–14.
100. Zhang J, Zhang LD, Wang XF, Liang CH, Peng XS, Wang YW. J Chem Phys. 2001;115:5714–7.
101. Deb P, Kim H, Rawat V, Oliver M, Kim S, Marshall M, Stach E, Sands T. Nano Lett. 2005;5:1847–51.
102. Chen SG, Paulose M, Ruan C, Mor GK, Varghese OK, Kouzoudis D, Grimes CA. J Photochem Photobiol A Chem. 2006;177:177–84.
103. Seabold JA, Shankar K, Wilke RHT, Paulose M, Varghese OK, Grimes CA, Choi KS. Chem Mater. 2008;20:5266–73.
104. Hornyak GL, Dillon AC, Parilla PA, Schneider JJ, Czap N, Jones KM, Fasoon FS, Mason A, Heben MJ. Nanostruct Mater. 1999;12:83–8.
105. Iwasaki T, Motoi T, Den T. Appl Phys Lett. 1999;75:2044–6.
106. Jeong SH, Hwang HY, Lee KH, Jeong Y. Appl Phys Lett. 2001;78:2052–4.
107. Lee JS, Gu GH, Kim H, Jeong KS, Bae J, Suh JS. Chem Mater. 2001;13:2387–91.
108. Choi WB, Cheong BH, Kim JJ, Chu J, Bae E. Adv Funct Mater. 2003;13:80–4.
109. Gao H, Mu C, Wang F, Xu DS, Wu K, Xie YC, Liu S, Wang EG, Xu J, Yu DP. J Appl Phys. 2003;93:5602–5.
110. Jeong SH, Lee KH. Synth Met. 2003;139:385–90.
111. Hwang SK, Lee J, Jeong SH, Lee PS, Lee KH. Nanotechnology. 2005;16:850–8.
112. Reddy ALM, Shaijumon MM, Gowda SR, Ajayan PM. Nano Lett. 2009;9:1002–6.
113. Takami S, Shirai Y, Chikyow T, Wakayama Y. Thin Solid Films. 2009;518:692–4.
114. Chintakula G, Rajaputra S, Singh VP. Sol Energy Mater Sol C. 2010;94:34–9.
115. Haberkorn N, Gutmann JS, Theato P. ACS Nano. 2009;3:1415–22.
116. Al-Kaysi RO, Muller AM, Frisbee RJ, Bardeen CJ. Cryst Growth Des. 2009;9:1780–5.
117. Guo YG, Li CJ, Wan LJ, Chen DM, Wang CR, Bai CL, Wang YG. Adv Funct Mater. 2003;13:626–30.
118. Park DH, Kim BH, Jang MG, Bae KY, Joo J. Appl Phys Lett. 2005;86:113116.
119. Park DH, Kim BH, Jang MK, Bae KY, Lee SJ, Joo J. Synth Met. 2005;153:341–4.
120. Park DH, Lee YB, Kim BH, Hong YK, Lee SJ, Lee SH, Kim HS, Joo J. J Korean Phys Soc. 2006;48:1468–71.
121. Kim HS, Park DH, Lee YB, Kim DC, Kim HJ, Kim J, Joo J. Synth Met. 2007;157:910–3.
122. Wang HS, Lin LH, Chen SY, Wang YL, Wei KH. Nanotechnology. 2009;20:075201.
123. Haberkorn N, Weber SAL, Berger R, Theato P. ACS Appl Mater Interfaces. 2010;2:1573–80.
124. Masuda H, Abe A, Nakao M, Yokoo A, Tamamura T, Nishio K. Adv Mater. 2003;15:161.
125. Matsumoto F, Harada M, Nishio K, Masuda H. Adv Mater. 2005;17:1609.
126. Harada M, Kondo T, Yanagishita T, Nishio K, Masuda H. Appl Phys Express. 2010;3:015001.
127. Razpet A, Possnert G, Johansson A, Abid M, Hallen A. Radiation effects and ion-beam processing of materials. In: Wang LM, Fromknecht R, Snead LL, Downey DF, Takahashi H, editors. Materials research society symposium proceedings, 2004, vol 792, p. 575–580.
128. Zacharatos F, Gianneta V, Nassiopoulou AG. Nanotechnology. 2008;19:495306.
129. Menon L, Ram KB, Patibandla S, Aurongzeb D, Holtz M, Yun J, Kuryatkov V, Zhu K. J Electrochem Soc. 2004;151:C492–4.

130. Jung M, Lee S, Byun YT, Jhon YM, Kim SH, Mho SI, Woo DH. Advances in nanomaterials and processing, Pts 1 and 2.In: Ahn BT, Jeon H, Hur BY, Kim K, Park JW, editors. Solid state phenomena, 2007, vol 124–126, p. 1301–1304.
131. Nakao M, Oku S, Tamamura T, Yasui K, Masuda H. Jpn J Appl Phys Part 1. 1999;38:1052–5.
132. Sai H, Fujii H, Arafune K, Ohshita Y, Yamaguchi M, Kanamori Y, Yugami H. Appl Phys Lett. 2006;88:201116–201113.
133. Lei Y, Chim WK, Sun HP, Wilde G. Appl Phys Lett. 2005;86:103106.
134. Liang JY, Luo HL, Beresford R, Xu J. Appl Phys Lett. 2004;85:5974–6.
135. Nakayama K, Tanabe K, Atwater HA. Appl Phys Lett. 2008;93:121904.
136. Matsuura N, Simpson TW, Mitchell IV, Mei XY, Morales P, Ruda HE. Phys Lett. 2002;81:4826–8.